SYMPLECTIC METHODS FOR THE SYMPLECTIC EIGENPROBLEM

SYMPLECTIC METHODS FOR THE SYMPLECTIC EIGENPROBLEM

Heike Fassbender

Munich University of Technology
Munich, Germany

KLUWER ACADEMIC / PLENUM PUBLISHERS
New York, Boston, Dordrecht, London, Moscow

Library of Congress Cataloging-in-Publication Data

Fassbender, Heike.
    Symplectic methods for the symplectic eigenproblem/Heike Fassbender.
      p. cm.
    Includes bibliographical references and index.

      1. Eigenvalues. 2. Algorithms. 3. Title.

QA193 .F37 2000
512.9′434—dc21

ISBN 978-1-4419-3346-1   e-ISBN 978-0-306-46978-7 [00-058754]

© 2010 Kluwer Academic / Plenum Publishers, New York
233 Spring Street, New York, New York 10013

http://www.wkap.nl/

10 9 8 7 6 5 4 3 2 1

A C.I.P. record for this book is available from the Library of Congress

Printed in the United States of America

PREFACE

The solution of eigenvalue problems is an integral part of many scientific computations. For example, the numerical solution of problems in structural dynamics, electrical networks, macro-economics, quantum chemistry, and control theory often requires solving eigenvalue problems. The coefficient matrix of the eigenvalue problem may be small to medium sized and dense, or large and sparse (containing many zero elements). In the past tremendous advances have been achieved in the solution methods for symmetric eigenvalue problems. The state of the art for nonsymmetric problems is not so advanced; nonsymmetric eigenvalue problems can be hopelessly difficult to solve in some situations due, for example, to poor conditioning. Good numerical algorithms for nonsymmetric eigenvalue problems also tend to be far more complex than their symmetric counterparts.

This book deals with methods for solving a special nonsymmetric eigenvalue problem; the symplectic eigenvalue problem. The symplectic eigenvalue problem is helpful, e.g., in analyzing a number of different questions that arise in linear control theory for discrete-time systems. Certain quadratic eigenvalue problems arising, e.g., in finite element discretization in structural analysis, in acoustic simulation of poro-elastic materials, or in the elastic deformation of anisotropic materials can also lead to symplectic eigenvalue problems. The problem appears in other applications as well.

The solution of the symplectic eigenvalue problem has been the topic of numerous publications during the last decades. Even so, a numerically sound method is still not known. The numerical computation of an invariant subspace by a standard solver for nonsymmetric eigenvalue problems (e.g., the QR algorithm) is not satisfactory. Due to roundoff errors unavoidable in finite-precision arithmetic, the computed eigenvalues will in general not come in pairs $\{\lambda, \overline{\lambda^{-1}}\}$, although the exact eigenvalues have this property. Even worse, small perturbations may cause eigenvalues close to the unit circle to cross the unit circle such that the number of true and computed eigenvalues inside the open unit disk may differ. This problem is due to the fact that the standard solvers for nonsymmetric eigenvalue problems ignore the symplectic structure as the coefficient matrix is treated like any other nonsymmetric matrix.

In this book fast, efficient, reliable, and structure-preserving numerical methods for the symplectic eigenproblem are developed. For this, I make use of the rich mathematical structure of the problem as has been done successfully for symmetric/Hermitian and orthogonal/unitary eigenproblems. Such structure-preserving methods are desirable as important properties of the original problem are preserved during the actual computations and are not destroyed by rounding errors. Moreover, in general, such methods allow for faster computations than general-purpose methods.

This monograph describes up-to-date techniques for solving small to medium-sized as well as large and sparse symplectic eigenvalue problems. The algorithms preserve and exploit the symplectic structure of the coefficient matrix; they are reliable, faster, and more efficient than standard algorithms for non-symmetric eigenproblems. A detailed analysis of all algorithms presented here is given. All algorithms are presented in MATLAB-programming style. Numerical examples are given to demonstrate their abilities.

ACKNOWLEDGMENTS

This is an updated version of [50] which was written while I was with the Zentrum für Technomathematik of the Fachbereich 3 - Mathematik und Informatik at the Universität Bremen, Germany. A big thank you for the stimulating atmosphere and for the support everybody gave me.

I owe a debt of gratitude to Angelika Bunse-Gerstner who has been a source of support and encouragement. Thanks to my co-author David Watkins for many helpful discussions on Laurent polynomials and convergence properties of GR algorithms. A special thanks to him for being a great host and for arranging for a whole week of sunshine during my stay in Seattle in late October 97. Thanks to Alan Laub for making my visit to Santa Barbara possible. During my stay there, I started to work on the symplectic eigenproblem despite the fact that my office was located only 50 yards from the beach. My thanks go also to Volker Mehrmann and Hongguo Xu for their hospitality in Chemnitz. Sharing his office with me, Hongguo endured my grinding out the details of Section 4.2. Thanks also to Danny Sorensen, Greg Ammar, Ralph Byers, Nil and Steve Mackey, Lothar Reichel and Daniela Calvetti for their interest in my work and fruitful discussions at various meetings and visits. Special thanks go also to a number of anonymous referees for their knowledgeable and helpful suggestions which greatly helped to improve the presentation of the material. Moreover, thanks to George Anastassiou (book series editor) and Tom Cohn (former Senior Editor at Kluwer) for their support in publishing this book.

The person who selflessly stood behind me, bucked me up when I was glum, and endured my erratic ways, is Peter, my husband and fellow mathematician. My special thanks to him for his love and understanding.

CONTENTS

LIST OF FIGURES

LIST OF TABLES

SYMPLECTIC METHODS FOR THE SYMPLECTIC EIGENPROBLEM

Chapter 1

INTRODUCTION

The advances in computing power in recent years have had a significant impact in all areas of scientific computation. This fast development provides a challenge to those who wish to harness the now available resources. New fields of application for numerical techniques have been established.

'*In addition to its traditional role in basic research, computational simulation is playing an increasingly important role in the design and evaluation of a wide variety of manufactured products. Designing "virtual" products on a computer has a number of advantages:*

- *It is often faster and less expensive than building mock-ups or prototype versions of real products.*

- *It reduces the need to deal with potentially hazardous materials or processes.*

- *It permits a much more thorough exploration of the design space, as design parameters can be varied at will.*

This is not to say, however, that effective computational simulation is easier than more conventional design methods. If anything, an even wider array of capabilities is required:

- *Accurate mathematical models, usually expressed in the form of equations, must be developed to describe the components and processes involved.*

- *Accurate, efficient algorithms must be developed for solving these equations numerically.*

- *Software must be developed to implement these algorithms on computers.*

- *...*

- *Computed results must be validated by comparison with theory or experiment in known situations to ensure that the predictions can be trusted in new, previously untested situations.*'[1]

---

[1]Quote from [70].

1

Here we will concentrate on developing accurate, efficient algorithms for a certain special eigenvalue problem, which arises in a number of important applications. That is, in this book we will consider the symplectic eigenvalue problem, and we will develop structure-preserving, more efficient algorithms for solving this eigenvalue problem than the ones previously known.

To be precise, we consider the numerical solution of the real (generalized) symplectic eigenvalue problem. A matrix $M \in \mathbf{R}^{2n \times 2n}$ is called symplectic (or J–orthogonal) if

$$MJM^T = J$$

(or equivalently, $M^T JM = J$), where

$$J = \begin{bmatrix} 0 & I_n \\ -I_n & 0 \end{bmatrix}$$

and I_n is the $n \times n$ identity matrix. The symplectic matrices form a group under multiplication. The eigenvalues of symplectic matrices occur in reciprocal pairs: If λ is an eigenvalue of M with right eigenvector x, then λ^{-1} is an eigenvalue of M with left eigenvector $(Jx)^T$. A symplectic matrix pencil $K - \lambda N, K, N \in \mathbf{R}^{2n \times 2n}$ is defined by the property

$$KJK^T = NJN^T.$$

The symplectic eigenvalue problem is helpful, e.g., in analyzing a number of different questions that arise in linear control theory for discrete-time systems. Industrial production and technological processes may suffer from unwanted behavior, e.g., losses in the start-up and change-over phases of operation, pollution, emission of harmful elements and production of unwanted by-products. Control techniques offer the possibility to analyze such processes in order to detect the underlying causes of the unwanted behavior. Furthermore, these techniques are very well suited for finding controllers which enable processes to operate in such a way that:

- the resulting products have high quality and tighter tolerances;

- less energy and material are consumed during manufacturing;

- change-over times are drastically reduced so that smaller product series can be made;

- processes are operated to the limits of their physical, chemical, or biological possibilities.

In order to achieve these objectives, advanced control techniques, which are carefully tuned to the process to be controlled, are needed.

Progress in information technology has considerably facilitated the implementation of advanced control strategies and has been ever increasing in industrial applications during the last ten years. These applications range from home appliances (such as the CD player) to complex production processes (such as glass furnaces, assembly lines, or chemical plants). An essential part of a control strategy consists of numerical calculations. However, this aspect of the software frequently fails to satisfy the reliability criteria commonly used in numerical mathematics.

Linear discrete-time systems are usually described by state difference equations of the form

$$x(i + 1) = A(i)x(i) + B(i)u(i),$$

where $x(i)$ is the state and $u(i)$ is the input at time t_i. $A(i)$ and $B(i)$ are matrices of appropriate dimensions. The output at time t_i is given by the output equation

$$y(i) = C(i)x(i) + D(i)u(i).$$

If the matrices A, B, C, and D are independent of i, the system is time-invariant.

A very simple example for such a discrete-time system is a savings bank account [81]: Let the scalar quantity $x(n)$ be the balance of a savings bank account at the beginning of the nth month and let α be the monthly interest rate. Also, let the scalar quantity $u(n)$ be the total of deposits and withdrawals during the nth month. Assuming that the interest is computed monthly on the basis of the balance at the beginning of the month, the sequence $x(n)$, $n = 0, 1, 2, \ldots$, satisfies the linear difference equation

$$
\begin{aligned}
x(n + 1) &= (1 + \alpha)x(n) + u(n), \\
x(0) &= x_0,
\end{aligned}
$$

where x_0 is the initial balance. (Here, $y(n) = x(n)$, that is the output at time n is equal to the state at time n). These equations describe a linear time-invariant discrete-time system.

As mentioned above, there are a number of important problems associated with discrete-time systems. Some of these, which will lead us to symplectic eigenproblems, will be discussed here briefly.

EXAMPLE 1.1 *Consider a system described by the following linear difference equation*

$$x(i + 1) = Ax(i), \qquad i \in \mathbf{N}_0,$$

where A is an $n \times n$ matrix and x denotes the state vector in $\mathbf{R}^n$. Suppose that the spectrum of A lies in the open unit disk $C_1(0)$; in this case, one says A is stable. It follows from the continuity of the spectrum that this property is preserved under sufficiently small perturbations of the entries of A. An important problem *of robustness analysis is to determine to what extent stability is preserved when larger perturbations are applied to A. Assume that the perturbed system matrix has the form $A + BDC$ where $B \in \mathbf{R}^{n \times m}, C \in \mathbf{R}^{p \times n}$ are fixed matrices defining the structure of the perturbations and $D \in \mathbf{K}^{m \times p}$ is an unknown disturbance matrix. The perturbed system may be formally interpreted as a closed loop system with unknown static linear output feedback:*

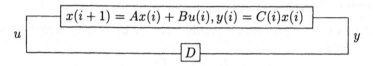

Figure 1.1.   Feedback interpretation of the perturbed system

The choice of B and C determines the perturbation structures of the matrix A to be studied. By different choices the effect of perturbations of certain entries, rows or columns of A can be investigated. In particular, if $B = C = I$, then all the elements of A are subject to independent perturbations.

In [73], the complex (discrete-time) structured stability radius r of A with respect to the perturbation structure (B, C), i.e.,

$$r = r(A, B, C) = \inf\{\|D\|; D \in \mathbf{C}^{m \times p}, \sigma(A + BDC) \not\subset C_1(0)\},$$

is studied, where $\|D\|$ is the operator norm of D with respect to an arbitrary pair of norms $\|\cdot\|_{\mathbf{C}^p}$ and $\|\cdot\|_{\mathbf{C}^m}$ on $\mathbf{C}^p$ resp. $\mathbf{C}^m$. The complex stability radius is a lower bound for the real stability radius $(D \in \mathbf{R}^{m \times p})$ and yields valuable information about the robustness of a system with respect to wider classes of perturbations (time-varying, non-linear and/or dynamic). The stability radius $r(A, B, C)$ is characterized in [73] with the help of the family of symplectic matrix pencils

$$W_\rho(\lambda) = K_\rho - \lambda N, \qquad \lambda \in \mathbf{C}, \; \rho \geq 0,$$

where

$$K_\rho = \begin{bmatrix} A & 0 \\ \rho^2 C^T C & I \end{bmatrix}, \qquad N = \begin{bmatrix} I & BB^T \\ 0 & A^T \end{bmatrix}.$$

It is shown that r is that value of ρ for which the spectrum of W_ρ hits the unit circle for the first time as ρ increases from 0 to ∞. Based on this fact, the authors propose a bisection algorithm to determine ρ: Suppose that in the kth step estimates ρ_k^- and ρ_k^+ are given such that

$$0 \leq \rho_k^- \leq r(A, B, C) \leq \rho_k^+ \leq \infty.$$

Consider $\rho = \frac{1}{2}(\rho_k^- + \rho_k^+)$. If W_ρ has eigenvalues on the unit circle, then set $\rho_{k+1}^- = \rho_k^-$ and $\rho_{k+1}^+ = \rho$. Otherwise set $\rho_{k+1}^- = \rho$, and $\rho_{k+1}^+ = \rho_k^+$. Hence in *every step of the algorithm the eigenvalues of the symplectic matrix pencil W_ρ have to be determined.*

EXAMPLE 1.2 *The standard (discrete-time) linear-quadratic optimization problem consists in finding a control trajectory $\{u(t), t \geq 0\}$, minimizing the cost functional*

$$J(x_0, u) = \sum_{t=0}^{\infty} [x(t)^T Q x(t) + u(t)^T R u(t)]$$

in terms of u subject to the dynamical constraint

$$x(t+1) = A x(t) + B u(t), \qquad x(0) := x_0.$$

Under certain conditions there is a unique control law,

$$u(t) = H x(t), \qquad H := -(R + B^T X B)^{-1} B^T X A,$$

minimizing J in terms of u subject to the dynamical constraint. The matrix X is the unique symmetric stabilizing solution of the algebraic matrix Riccati equation

$$X = A^T X A - A^T X B (R + B^T X B)^{-1} B^T X A + Q. \tag{1.0.1}$$

The last equation is usually referred to as discrete-time algebraic Riccati equation. It appears not only in the context presented, but also in numerous procedures for analysis, synthesis, and design of linear-quadratic Gaussian and H_∞ control and estimation systems, as well as in other branches of applied mathematics.

For convenience, let us assume that the matrix A is nonsingular. Consider the symplectic matrix M

$$M = \begin{bmatrix} A + B R^{-1} B^T A^{-T} Q & -B R^{-1} B^T A^{-T} \\ -A^{-T} Q & A^{-T} \end{bmatrix}.$$

Standard assumptions guarantee no eigenvalues on the unit circle and it is then easily seen that M has precisely n eigenvalues in the open unit disk and n outside. Moreover, the Riccati solution X can be computed from the invariant subspace of M corresponding to the n eigenvalues in the open unit disk. This can easily be seen from a Jordan form reduction of M: Compute a matrix of eigenvectors and principal vectors T to perform the following reduction

$$\begin{bmatrix} T_{11} & T_{12} \\ T_{21} & T_{22} \end{bmatrix}^{-1} \begin{bmatrix} A + F A^{-T} Q & -F A^{-T} \\ -A^{-T} Q & A^{-T} \end{bmatrix} \begin{bmatrix} T_{11} & T_{12} \\ T_{21} & T_{22} \end{bmatrix} = \begin{bmatrix} \Lambda & 0 \\ 0 & \Lambda^{-1} \end{bmatrix}$$

where $F = B R^{-1} B^T$, T_{ij} is $n \times n$, and Λ is composed of Jordan blocks corresponding to eigenvalues in the open unit disk. Finally, the Riccati solution X is found by solving a system of linear equations:

$$X = -T_{21} T_{11}^{-1}.$$

That this is true is easily seen by manipulating the invariant subspace equation

$$\begin{bmatrix} \overset{\star}{A} + FA^{-T}Q & -FA^{-T} \\ -A^{-T}Q & A^{-T} \end{bmatrix} \begin{bmatrix} T_{11} \\ T_{21} \end{bmatrix} = \begin{bmatrix} T_{11} \\ T_{21} \end{bmatrix} [\Lambda].$$

There are severe numerical difficulties with this approach when the symplectic matrix M has multiple or near-multiple eigenvalues. For a cogent discussion of the numerical difficulties associated with the numerical determination of Jordan forms, the reader is referred to the classic paper of Golub and Wilkinson [59].

To overcome these difficulties, Schur methods were proposed in 1978 by Laub [85, 86]. The numerical solution of Riccati equations by invariant subspace methods has been an extremely active area for a long time and much of the relevant literature is included in an extensive reference section of the tutorial paper [88]. Other solution techniques are also available for the numerical solution of Riccati equations. For a discussion of different aspects associated with Riccati equations (like, e.g., theory, solution techniques, applications in which Riccati equations appear) see, e.g., [7, 8, 9, 26, 66, 77, 80, 83, 104, 119, 124, 129, 146, 148].

The symplectic eigenproblem appears in applications other than in control theory as well. Two such examples are briefly discussed next.

EXAMPLE 1.3 *The matrix equation*

$$X = f(X), \qquad with \quad f(X) = Q + LX^{-1}L^{T}, \qquad (1.0.2)$$

where $Q = Q^{T} \in \mathbf{R}^{n \times n}$ is a positive definite matrix and $L \in \mathbf{R}^{n \times n}$ is nonsingular, arises in the analysis of stationary Gaussian reciprocal processes over a finite interval. The problem is to find a positive definite symmetric solution X. The steady-state distribution of the temperature along a heated ring or beam subjected to random loads along its length can be modeled in terms of such reciprocal processes. See [95] for references to this example as well as other examples of reciprocal processes.

In [1] it is noted that if X solves (1.0.2), then it also obeys the iterated equation

$$X = f(f(X)) = Q + F(R^{-1} + X^{-1})^{-1}F^{T} \qquad (1.0.3)$$

with

$$F = LL^{-T}, \qquad R = L^{T}Q^{-1}L.$$

Using the Sherman-Morrison-Woodbury formula to derive an expression for $(R^{-1} + X^{-1})^{-1}$, we obtain

$$X = Q + FXF^{T} - FX(X + R)^{-1}XF^{T}, \qquad (1.0.4)$$

a Riccati equation as in (1.0.1) with $F = A^{T}$ and $B = I$. Because (F, I) is observable and $(F, Q^{\frac{1}{2}})$ is controllable, (1.0.4) has a unique positive definite *solution $X^\star$. This unique solution coincides with that solution of (1.0.2) one is interested in. As L and hence F is nonsingular, one can follow the approach discussed in the last example in order to compute the desired solution $X^\star$. A slightly different approach is the following one which involves the solution of the eigenproblem for the symplectic matrix pencil*

$$K - \lambda N = \begin{bmatrix} F^T & 0 \\ Q & I \end{bmatrix} - \lambda \begin{bmatrix} I & R^{-1} \\ 0 & F \end{bmatrix}.$$

With an argument similar to the one used in the last example, on can show that the desired solution of the Riccati equation can be computed via a deflating subspace of $K - \lambda N$. As L is nonsingular, premultiplying $K - \lambda N$ with $\begin{bmatrix} L & 0 \\ 0 & L^{-1} \end{bmatrix}$ yields the equivalent symplectic matrix pencil

$$\widetilde{K} - \lambda \widetilde{N} = \begin{bmatrix} L^T & 0 \\ L^{-1}Q & L^{-1} \end{bmatrix} - \lambda \begin{bmatrix} L & QL^{-T} \\ 0 & L^{-T} \end{bmatrix},$$

where $\widetilde{K} = \widetilde{N}^T$ is symplectic.

EXAMPLE 1.4 *Quadratic eigenvalue problems of the form*

$$\lambda^2 Ax + \lambda Gx + Lx = 0 \tag{1.0.5}$$

where $A = A^T$ is positive definite, $L = L^T$ and $G = -G^T$ give rise to large, sparse symplectic eigenproblems. Such eigenvalue problems arise for example in finite element discretization in structural analysis [127], in acoustic simulation of poro-elastic materials [101, 125, 130], and in the elastic deformation of anisotropic materials [78, 90, 126]. In these applications, A is a mass matrix and $-L$ a stiffness matrix. Depending on the applications, different parts of the spectrum are of interest, typically one is interested in the eigenvalues with smallest real part or the eigenvalues smallest or largest in modulus.

It is well-known that the eigenvalues of (1.0.5) occur in quadruples λ, $\overline{\lambda}$, $-\lambda$, $-\overline{\lambda}$ or real or purely imaginary pairs λ, $-\lambda$ [82]; similar to eigenvalues of symplectic matrices which occur in quadruples λ, λ^{-1}, $\overline{\lambda}$, $\overline{\lambda}^{-1}$ or real or purely imaginary pairs λ, λ^{-1}.

If we make the substitution $y = \lambda x$ in (1.0.5), then we obtain (see [106])

$$H - \lambda Tz = \left(\begin{bmatrix} 0 & -L \\ A & 0 \end{bmatrix} - \lambda \begin{bmatrix} A & G \\ 0 & A \end{bmatrix} \right) \begin{bmatrix} y \\ x \end{bmatrix} = 0. \tag{1.0.6}$$

The matrix T can be written in factored form as

$$T = Z_1 Z_2 = \begin{bmatrix} I & \frac{1}{2}G \\ 0 & A \end{bmatrix} \begin{bmatrix} A & \frac{1}{2}G \\ 0 & I \end{bmatrix}.$$

If T is nonsingular, then the pencil (1.0.6) is equivalent to

$$Z_1^{-1}(H - \lambda T)Z_2^{-1} = \begin{bmatrix} -\frac{1}{2}GA^{-1} & \frac{1}{4}GA^{-1}G - L \\ A^{-1} & -\frac{1}{2}A^{-1}G \end{bmatrix} - \lambda I.$$

Hence the linearization (1.0.6) gives rise to a Hamiltonian operator $W = Z_1^{-1}HZ_2^{-1}$. Suppose one is interested in computing the eigenvalues of W (and hence of (1.0.5)) that lie nearest to some target value λ_0. The standard approach for this purpose is to consider the transformation matrix $(W - \lambda_0 I)^{-1}$, and to compute its eigenvalues. This new matrix is no longer Hamiltonian, its eigenvalue no longer appear in quadruples. Each eigenvalue of W near λ_0 is related to eigenvalues near $\overline{\lambda_0}$, $-\lambda_0$, and $-\overline{\lambda_0}$. Thus in order to preserve this structure-preserving structure, one should extract all four eigenvalues together. One possibility is to use a generalized Cayley transformation of the Hamiltonian matrix W which yields the symplectic operator

$$\begin{aligned} M &= (W - \lambda_0 I)^{-1}(W + \overline{\lambda_0}I)(W - \overline{\lambda_0}I)^{-1}(W + \lambda_0 I) \\ &= Z_2(H - \lambda_0 T)^{-1}(H + \overline{\lambda_0}T)(H + \overline{\lambda_0}T)(H + \lambda_0 T)Z_2^{-1}. \end{aligned}$$

The eigenvalues of M occur in quadruples λ, λ^{-1}, $\overline{\lambda}$, $\overline{\lambda^{-1}}$; the eigenvalues of W can easily be obtained from these via an inverse Cayley transformation.

Usually in the above mentioned applications, the matrices A, L and G are large and sparse, so that one should not build M explicitly, but use the structure of H, T and Z_2 to simplify the formulae for M when applying the symplectic operator M (see [106]).

For a discussion on other useful transformations of W and the pros and cons of the different transformations see [106].

The symplectic eigenproblem appears in applications other than the ones mentioned as well. For example, H_∞–norm computations (see, e.g., [148]) and discrete Sturm-Liouville equations (see, e.g., [27]) lead to the task of solving a symplectic eigenproblem.

The solution of the (generalized) symplectic eigenvalue problem has been the topic of numerous publications during the last decades. Even so, a numerically sound method, i.e., a strongly backward stable method in the sense of [30], is yet not known. The numerical computation of an invariant (deflating) subspace is usually carried out by an iterative procedure like the QR (QZ) algorithm; see, e.g., [104, 111]. The QR (QZ) algorithm is numerically backward stable but it ignores the symplectic structure. Applying the QR (QZ) algorithm to a symplectic matrix (symplectic matrix pencil) results in a general $2n \times 2n$ matrix (matrix pencil) in Schur form (generalized Schur form) from which the eigenvalues and invariant (deflating) subspaces can be read off. Due to round-off errors unavoidable in finite-precision arithmetic, the computed eigenvalues will in general not come in pairs $\{\lambda, \lambda^{-1}\}$, although the exact eigenvalues have this property. Even worse, small perturbations may cause eigenvalues close to the unit circle to cross the unit circle such that the number of true and computed eigenvalues inside the open unit disk may differ. Forcing the symplectic structure by a structure-preserving method, these problems are avoided.

In order to develop fast, efficient, reliable, and structure-preserving numerical methods for the symplectic eigenproblem one should make use of the rich mathematical structure of the problem in a similar way as it has been successfully done for symmetric/Hermitian and orthogonal/unitary eigenproblems. E.g., for the symmetric eigenproblem, one of the nowadays standard approaches involves first the reduction of the symmetric matrix to symmetric tridiagonal form followed by a sequence of implicit QR steps which preserve this symmetric tridiagonal form [58, 139]. Such structure preserving methods are desirable as important properties of the original problem are preserved during the actual computations and are not destroyed by rounding errors. Moreover, in general, such methods allow for faster computations than general-purpose methods. For the symmetric eigenproblem, e.g., applying implicit QR steps to the full symmetric matrix requires $\mathcal{O}(n^3)$ arithmetic operations per step, while applying an implicit QR step to the similar symmetric tridiagonal matrix requires only $\mathcal{O}(n)$ arithmetic operations, where n is the order of the matrix. The QR method is the method of choice when solving small to medium size dense symmetric eigenproblems. Unfortunately, this approach is not suitable when dealing with large and sparse symmetric matrices as an elimination process can not make full use of the sparsity. The preparatory step of the QR algorithm involves the initial reduction of the (large and sparse) symmetric matrix to tridiagonal form. During this reduction process fill-in will occur such that the original sparsity pattern is destroyed. Moreover, in practise, one often does not have direct access to the large and sparse symmetric matrix A itself, one might only have access to the matrix-vector product Ax for any vector x. In this case, the Lanczos algorithm combined with the QR algorithm is the most suitable algorithm for computing some of the eigenvalues of A. It generates a sequence of symmetric tridiagonal matrices $T_j \in \mathbf{R}^{j \times j}$ with the property that the eigenvalues of T_j are progressively better estimates of A's eigenvalues.

Different structure-preserving methods for solving the symplectic eigenproblem have been proposed. Mehrmann [103] describes a symplectic QZ algorithm. This algorithm has all desirable properties, but its applicability is limited to a special symplectic eigenproblem arising in single input/output optimal control problems. Unfortunately, this method can not be applied in general due to the lacking reduction to symplectic J–Hessenberg form in the general case [5]. In [96], Lin uses the $S + S^{-1}$–transformation in order to solve the symplectic eigenvalue problem. The method cannot be used to compute eigenvectors and/or invariant subspaces. Lin and Wang [99] present a structure-preserving algorithm to compute the stable Lagrangian subspace of a symplectic pencil. Their approach is to determine an isotropic Jordan sub-basis corresponding to the unimodular eigenvalues by using the associated Jordan basis of the $S + S^{-1}$-transformation of the symplectic matrix pencil. The algorithm is structure-preserving, but not stable due to use the use of nonorthogonal transformations. Patel [118] shows that the $S + S^{-1}$–idea can also be used to derive a structure-preserving method for the generalized symplectic eigenvalue problem similar to Van Loan's square-reduced method for the Hamiltonian eigenvalue problem [137]. Unfortunately, similar to Van Loan's square-reduced method, a loss of half of the possible accuracy in the eigenvalues is possible when using this algorithm. Based on the multishift idea presented in [5], Patel also describes a method working on a condensed symplectic pencil using implicit QZ steps to compute the stable deflating subspace of a symplectic pencil [117]. The algorithm is theoretically structure-preserving, but in practical computations this can not (easily) be achieved. Benner, Mehrmann and Xu present in [24, 23] a numerically stable, structure preserving method for computing the eigenvalues and the stable invariant sub-space of Hamiltonian matrices/matrix pencils. Using a Cayley transformation the method can also be used for symplectic matrices/matrix pencils. That is, via the Cayley transformation a symplectic matrix (or matrix pencil) is trans-formed to a Hamiltonian matrix pencil. Eigenvalues and the stable invariant subspace of this Hamiltonian matrix pencil are then computed with the method proposed. As the Cayley transformation preserves invariant subspaces, the sta-ble invariant subspace of the original symplectic problem can be read off from the stable invariant subspace of the Hamiltonian matrix pencil. The eigenvalues of the original symplectic problem are obtained via the inverse Cayley trans-formation. This approach is rather expensive for solving a standard symplectic eigenproblem, as such a problem is converted into a generalized Hamiltonian eigenproblem.

Flaschka, Mehrmann, and Zywietz show in [53] how to construct structure-preserving methods for the symplectic eigenproblem based on the SR method [44, 102]. This method is a QR–like method based on the SR decomposition. Almost every matrix $A \in \mathbf{R}^{2n \times 2n}$ can be decomposed into a product $A = SR$, where S is symplectic and R is J–triangular [47]. The SR algorithm is an iterative algorithm that performs an SR decomposition at each iteration. If B is the current iterate, then a spectral transformation function q is chosen and the SR decomposition of $q(B)$ is formed, if possible: $q(B) = SR$. Now the symplectic factor S is used to perform a similarity transformation on B to yield the next iterate $\hat{B}$: $\hat{B} = S^{-1}BS$. In an initial step, the $2n \times 2n$ symplectic matrix is reduced to a more condensed form, the symplectic J–Hessenberg form, which in general contains $2n^2 + 3n - 1$ nonzero entries. As in the general framework of GR algorithms [142], the SR iteration preserves the symplectic J–Hessenberg form at each step and is supposed to converge to a form from which eigenvalues and invariant subspaces can be read off. A $2n \times 2n$ symplectic J–Hessenberg matrix is determined by $4n - 1$ parameters. The SR algorithm can be modified to work only with these parameters instead of the $2n^2 + 3n - 1$ nonzero matrix elements. Thus only $\mathcal{O}(n)$ arithmetic operations per SR step are needed compared to $\mathcal{O}(n^2)$ arithmetic operations when working on the actual J–Hessenberg matrix. The authors note that the algorithm "..*forces the symplectic structure, but it has the disadvantage that it needs $4n - 1$ terms to be nonzero in each step, which makes it highly numerically unstable. ... Thus, so far, this algorithm is mainly of theoretical value.*" [53, page 186, last paragraph].

Banse and Bunse-Gerstner [15, 13, 14] presented a new condensed form for symplectic matrices, the *symplectic butterfly form*. The $2n \times 2n$ condensed matrix is symplectic, contains $8n - 4$ nonzero entries, and, similar to the symplectic J–Hessenberg form of [53], it is determined by $4n - 1$ parameters. It can be depicted by

For every symplectic matrix M, there exist numerous symplectic matrices S such that $B = S^{-1}MS$ is a symplectic butterfly matrix. The SR algorithm preserves the butterfly form in its iterations. The symplectic structure, which will be destroyed in the numerical process due to roundoff errors, can easily be restored in each iteration for this condensed form. There is reason to believe that an SR algorithm based on the symplectic butterfly form has better numerical properties than one based on the symplectic J–Hessenberg form; see Section 4.1.

Here the symplectic butterfly form is discussed in detail. Structure-preserving SR and SZ algorithms based on the symplectic butterfly form are developed for solving small to medium size dense symplectic eigenproblems. A symplectic Lanczos algorithm based on the symplectic butterfly form is presented that is useful for solving large and sparse symplectic eigenproblems.

The notation and definitions used here are introduced in Chapter 2.1. the basic properties of symplectic matrices and matrix pencils are discussed. Moreover, the general eigenvalue algorithms which are the basis for the structure-preserving ones discussed here are reviewed.

In Chapter 3 the symplectic butterfly form for symplectic matrices and symplectic matrix pencils is discussed in detail. We will show that unreduced symplectic butterfly matrices have properties similar to those of unreduced Hessenberg matrices in the context of the QR algorithm and unreduced J–Hessenberg matrices in the context of the SR algorithm.

The $4n-1$ parameters that determine a symplectic butterfly matrix B cannot be read off of B directly. Computing the parameters can be interpreted as factoring B into the product of two even simpler matrices K and N: $B = K^{-1}N$. The parameters can then be read off of K and N directly. Up to now two different ways of factoring symplectic butterfly matrices have been proposed in the literature [12, 19]. We will introduce these factorizations and consider their drawbacks and advantages.

Chapter 4 deals with the SR and SZ algorithm based on the symplectic butterfly form. First we will revisit the SR algorithm for symplectic butterfly matrices. Such an algorithm was already considered in [13, 19]. In those publications, it is proposed to use a polynomial of the form $p(\lambda) = \prod_{i=1}^{k}(\lambda - \mu_i)$ to drive the SR step, just as in the implicit QR (bulge-chasing) algorithm for upper Hessenberg matrices. Here we will show that it is better to use a Laurent polynomial to drive the SR step. This reduces the size of the bulges that are introduced, thereby decreasing the number of computations required per iteration. It also improves the convergence and stability properties of the algorithm by effectively treating each reciprocal pair of eigenvalues as a unit. Further, the choice of the shifts will be discussed. It will be shown that the convergence rate of the butterfly SR algorithm is typically cubic. The method still suffers from loss of the symplectic structure due to roundoff errors, but the loss of symplecticity is normally less severe than in an implementation using a standard polynomial or than in the implementation of the algorithm proposed by Flaschka, Mehrmann, and Zywietz in [53] based on symplectic J–Hessenberg matrices. Moreover, using the factors K and N of the symplectic butterfly matrix B, one can easily and cheaply restore the symplectic structure of the iterates whenever necessary.

To derive a method that is purely based on the $4n - 1$ parameters that determine B and that thus forces the symplectic structure, one needs to work with the factors K and N. We will take two approaches to derive such an algorithm. The first approach uses the ideas that lead to the development of the unitary Hessenberg QR algorithm [61]. The butterfly SR step works on the butterfly matrix B and transforms it into a butterfly matrix $\widetilde{B}$. A parameterized version of the algorithm will work only on the $4n - 1$ parameters that determine B to derive those of $\widetilde{B}$ without ever forming B or $\widetilde{B}$. This is done by making use of the factorization $B = K^{-1}N$, decomposing K and N into even simpler symplectic matrices, and the observation that most of the transformations applied during the implicit SR step commute with most of the simple factors of K and N. The second algorithm that works only on the parameters is an SZ algorithm for the matrix pencil $K - \lambda N$, whose eigenvalues are the same as those of the symplectic matrix $B = K^{-1}N$.

Numerical experiments for all algorithms proposed are presented. The experiments clearly show: The methods converge, cubic convergence can be observed. The parameterized SR algorithm converges slightly faster than the SR algorithm. The eigenvalues are computed to about the same accuracy. The SZ algorithm is considerably better than the SR algorithm in computing the eigenvalues of a parameterized symplectic matrix/matrix pencil.

Finally, at the end of Chapter 4 two interesting remarks on the butterfly SR algorithm are given. First we prove a connection between the SR and HR algorithm: An iteration of the SR algorithm on a $2n \times 2n$ symplectic butterfly matrix using shifts $\mu_i, \mu_i^{-1}, i = 1, \ldots, k$, is equivalent to an iteration of the HR algorithm on a certain $n \times n$ tridiagonal D–symmetric matrix using shifts $\mu_i + \mu_i^{-1}, i = 1, \ldots, k$. Then we discuss how the problem described in Example 1.2 can be solved using the results obtained so far.

All algorithm discussed in Chapter 4 are based on elimination schemes. Therefore they are not suitable when dealing with large and sparse symplectic eigenproblems. As pointed out before, a structure-preserving symplectic Lanczos algorithm would be a better choice here. Such a symplectic Lanczos method which creates the symplectic butterfly form if no breakdown occurs was derived in [13]. Given $v_1 \in \mathbf{R}^{2n}$ and a symplectic matrix $M \in \mathbf{R}^{2n \times 2n}$, the symplectic Lanczos algorithm produces a matrix $S^{2n,2k} = [v_1, v_2, \ldots, v_k, w_1, w_2, \ldots, w_k] \in \mathbf{R}^{2n \times 2k}$ which satisfies a recursion of the form

$$M S^{2n,2k} = S^{2n,2k} B^{2k,2k} + r_{k+1} e_{2k}^T, \tag{1.0.7}$$

where $B^{2k,2k}$ is a butterfly matrix of order $2k \times 2k$, and the columns of $S^{2n,2k}$ are orthogonal with respect to the indefinite inner product defined by J. The residual r_{k+1} depends on v_{k+1} and w_{k+1}; hence $(S^{2n,2k})^T J r_{k+1} = 0$. Such a symplectic Lanczos method will suffer from the well-known numerical difficulties inherent to any Lanczos method for unsymmetric matrices. In [13], a symplectic look-ahead Lanczos algorithm is presented which overcomes breakdown by giving up the strict butterfly form. Unfortunately, so far there do not exist eigenvalue methods that can make use of that special reduced form. Standard eigenvalue methods as QR or SR algorithms have to be employed resulting in a full symplectic matrix after only a few iteration steps.

A different approach to deal with the numerical difficulties of the Lanczos process is to modify the starting vectors by an implicitly restarted Lanczos process (see the fundamental work in [41, 132]; for the unsymmetric eigenproblem the implicitly restarted Arnoldi method has been implemented very successfully, see [94]). The problems are addressed by fixing the number of steps in the Lanczos process at a prescribed value k which depends upon the required number of approximate eigenvalues. J–orthogonality of the k Lanczos vectors is secured by re-J-orthogonalizing these vectors when necessary. The purpose of the implicit restart is to determine initial vectors such that the associated residual vectors are tiny. Given (1.0.7), an implicit Lanczos restart computes the Lanczos factorization

$$M\breve{S}^{2k} = \breve{S}^{2k}\breve{B}^{2k,2k} + \breve{r}_{k+1}e_{2k}^T$$

which corresponds to the starting vector

$$\breve{v}_1 = q(M)v_1$$

(where $q(M) \in \mathbf{R}^{2n \times 2n}$ is a Laurent polynomial) without having to explicitly restart the Lanczos process with the vector $\breve{v}_1$. Such an implicit restarting mechanism is derived here analogous to the technique introduced in [67, 132].

In Chapter 5 we first derive the symplectic Lanczos method itself. Further, we are concerned with finding conditions for the symplectic Lanczos method terminating prematurely such that an invariant subspace associated with certain desired eigenvalues is obtained. We will also consider the important question of determining stopping criteria. An error analysis of the symplectic Lanczos algorithm in finite-precision arithmetic analogous to the analysis for the unsymmetric Lanczos algorithm presented by Bai [11] will be given. As to be expected, it follows that (under certain assumptions) the computed Lanczos vectors loose J–orthogonality when some Ritz values begin to converge. Further, an implicitly restarted symplectic Lanczos method is derived. Numerical properties of the proposed algorithm are discussed. Finally we present some numerical examples. As expected, they demonstrate that re–J–orthogonalizing is necessary, as the computed symplectic Lanczos vectors loose J–orthogonality when some Ritz values begin to converge. Moreover, the observed behavior of the implicitly restarted symplectic Lanczos algorithm corresponds to the reported behavior of the implicitly restarted Arnoldi method of Sorensen [132].

Chapter 2

PRELIMINARIES

2.1 NOTATIONS, DEFINITIONS, AND BASIC PROPERTIES

We will employ Householder notational convention. Capital and lower case letters denote matrices and vectors, respectively, while lower case Greek letters denote scalars. By $\mathbf{R}^{n \times k}$ we denote the real $n \times k$ matrices, by $\mathbb{C}^{n \times k}$ the complex $n \times k$ matrices. We use $\mathbf{K}$ to denote $\mathbf{R}$ or $\mathbb{C}$. The $n \times n$ identity matrix will be denoted by $I^{n,n}$, and the ith unit vector by e_i; $I^{n,n} = [e_1, e_2, \ldots, e_n]$. Let

$$J^{2n,2n} := \begin{bmatrix} 0 & I^{n,n} \\ -I^{n,n} & 0 \end{bmatrix} \qquad (2.1.1)$$

and $P^{2n,2n}$ be the permutation matrix

$$P^{2n,2n} := [e_1, e_3, \ldots, e_{2n-1}, e_2, e_4, \ldots, e_{2n}] \in \mathbf{R}^{2n \times 2n}. \qquad (2.1.2)$$

If the dimension of $I^{n,n}$, $J^{2n,2n}$, or $P^{2n,2n}$ is clear from the context, we leave off the superscript. We denote by $Z^{n,k}$ the first k columns of a $n \times n$ matrix Z.

Using the permutation matrix P, the matrix J can be permuted to the block diagonal matrix

$$J_P := PJP^T = \text{diag}(\begin{bmatrix} 0 & 1 \\ -1 & 0 \end{bmatrix}, \begin{bmatrix} 0 & 1 \\ -1 & 0 \end{bmatrix}, \ldots, \begin{bmatrix} 0 & 1 \\ -1 & 0 \end{bmatrix}).$$

We define an indefinite inner product by

$$(x, y)_J := x^T J y, \qquad x, y \in \mathbf{R}^{2n \times 2n}. \qquad (2.1.3)$$

Let $A \in \mathbf{K}^{n \times k}$. Then we will denote

- the (i, j)th entry of A by a_{ij}

- the jth row of A by $A_{j,1:k}$

- the jth column of A by $A_{1:n,j}$

- the entries $\ell, \ell + 1, \ldots, m$ of the jth row of A by $A_{j,\ell:m}$

- the entries $\ell, \ell + 1, \ldots, m$ of the jth column of A by $A_{\ell:m,j}$

- the transpose of A by A^T;   if $C = A^T$, then $c_{ij} = a_{ji}$

- the conjugate transpose of A by A^H;   if $C = A^H$, then $c_{ij} = \overline{a_{ji}}$

Corresponding notations will be used for vectors $x \in \mathbf{K}^n$.

Sometimes we partition the matrix $A \in \mathbf{K}^{n \times k}$ to obtain

$$
A = \begin{bmatrix} A_{11} & \cdots & A_{1q} \\ \vdots & & \vdots \\ A_{p1} & \cdots & A_{pq} \end{bmatrix} \begin{matrix} n_1 \\ \vdots \\ n_p \end{matrix} \qquad (2.1.4)
$$
$$
\begin{matrix} k_1 & \cdots & k_q \end{matrix}
$$

where $n_1 + \cdots + n_p = n$, $k_1 + \cdots + k_q = k$ and $A_{ij} \in \mathbf{K}^{n_i \times k_j}$ designates the (i, j) block or submatrix.

Throughout this book we will use the terms eigenvalues, spectrum, and invariant/deflating subspace as defined below.

DEFINITION 2.1 *Let* $A, B \in \mathbf{K}^{n \times n}$, $\alpha, \beta \in \mathbb{C}$.

- $\lambda \in \mathbb{C}$ *is an* eigenvalue *of A if* $\det(A - \lambda I) = 0$.

- $\sigma(A) = \{\lambda \in \mathbb{C} | \det(A - \lambda I) = 0\}$ *is called the spectrum of A.*

- $\mathcal{U} \subset \mathbb{C}^m$ *determines an* invariant subspace *of A with respect to the eigenvalues in* $\Lambda = \{\lambda_i | \lambda_i \in \mathbb{C}, i = 1, \ldots, k\}$ *if there exist* $U \in \mathbb{C}^{m \times k}$, *and* $K \in \mathbb{C}^{k \times k}$ *such that U has full column rank, Λ is the spectrum of K, $AU = UK$ and the columns of U span $\mathcal{U}$. Sometimes we refer to U as the invariant subspace of A.*

- *The* spectral radius *of A is defined by*

$$
\rho(A) = \max\{|\lambda| : \lambda \in \sigma(A)\}.
$$

- *The matrix pencil $A - \lambda B$ is called* regular *if* $\det(A - \lambda B)$ *does not vanish identically for all* $\lambda \in \mathbb{C}$.

- *If $A - \lambda B$ is regular, then $\mu \in \mathbb{C}$ is an eigenvalue of $A - \lambda B$ if $\det(A - \mu B) = 0$.*

- *If $A - \lambda B$ is regular, then (α, β) determines an eigenvalue μ of $A - \lambda B$ if $\det(\alpha A - \beta B) = 0$, and $\beta - \alpha\mu = 0$.*

- *If $A - \lambda B$ is regular, then (α, β) determines an* infinite eigenvalue *of $A - \lambda B$ if $\det(\alpha A - \beta B) = 0$, and $\beta \neq 0, \alpha = 0$.*

- $\sigma(A, B) = \{(\alpha, \beta) \in \mathbb{C}^2 \mid \det(\alpha A - \beta B) = 0\}$ *is called the* generalized spectrum *of $A - \lambda B$. The spectrum of $A - \lambda B$ will be denoted by a set of complex numbers μ where $\mu = \beta/\alpha$ for $(\alpha, \beta) \in \sigma(A, B)$. We will use the convention that $\mu = \infty$ if $\alpha = 0, \beta \neq 0$ (i.e., an infinite eigenvalue will be denoted by ∞).*

- $\mathcal{U} \subset \mathbb{C}^m$ *determines a* deflating subspace *of $A - \lambda B$ with respect to the eigenvalues in $\Lambda = \{\lambda_i \mid \lambda_i \in \mathbb{C}, i = 1, \ldots, k\}$ if there exist $U, V \in \mathbb{C}^{m \times k}$, $K_1, K_2 \in \mathbb{C}^{k \times k}$ such that U has full column rank, Λ is the spectrum of $K_1 - \lambda K_2$, $AU = VK_1$, $BU = VK_2$, and the columns of U span $\mathcal{U}$. Sometimes, we refer to U or (U, V) as the deflating subspace of $A - \lambda B$.*

We will make frequent use of the following notations.

- The *Frobenius norm* for $A \in \mathbf{K}^{n \times n}$ will be denoted by

$$\|A\|_F = \sqrt{\sum_{i=1}^{n} \sum_{j=1}^{n} |a_{ij}|^2}.$$

- The *2–norm* of a vector $x \in \mathbf{K}^n$ will be denoted by

$$\|x\|_2 = (x^H x)^{\frac{1}{2}}.$$

- The corresponding matrix norm for $A \in \mathbf{K}^{n \times n}$, the *spectral norm*, will be denoted by

$$\|A\|_2 = \rho(A^H A)^{\frac{1}{2}}.$$

- The *condition number* of a matrix $A \in \mathbf{K}^{n \times n}$ using the 2–norm is denoted by

$$\kappa_2(A) = \|A\|_2 \|A^{-1}\|_2.$$

- The *trace* of a matrix $A \in \mathbf{K}^{n \times n}$ will be denoted by

$$\text{trace}(A) = \sum_{j=1}^{n} a_{jj} = \sum_{j=1}^{n} \lambda_j,$$

where the λ_j's are the eigenvalues of A.

- The *rank* of a matrix $A \in \mathbf{K}^{m \times n}$ will be denoted by

$$\operatorname{rank}(A) = \dim(\operatorname{ran}(A)),$$

where $\operatorname{ran}(A)$ denotes the *range* of A

$$\operatorname{ran}(A) = \{y \in \mathbf{K}^m : y = Ax \text{ for some } x \in \mathbf{K}^n\}.$$

- Given a collection of vectors $a_1, \ldots, a_n \in \mathbf{K}^m$, the set of all linear combinations of these vectors is a subspace referred to as the span of $\{a_1, \ldots, a_n\}$

$$\operatorname{span}\{a_1, \ldots, a_n\} = \left\{ \sum_{j=1}^{n} \beta_j a_j : \beta_j \in \mathbf{K} \right\}.$$

- The *determinant* of $A \in \mathbf{K}^{n \times n}$ is given by

$$\det(A) = \sum_{j=1}^{n} (-1)^{j+1} a_{1j} \det(A_{1j}).$$

Here a_{1j} denote the entries of A in the first row and A_{1j} is an $(n-1) \times (n-1)$ matrix obtained by deleting the first row and jth column of A.

- The *leading principal submatrix* of order m of a matrix $A \in \mathbf{K}^{n \times n}$ is given by $A_{1:m,1:m}$. It will be denoted by $A^{m,m}$. The *trailing principle submatrix* of order m of a matrix $A \in \mathbf{K}^{n \times n}$ is given by $A_{(n-m+1):n,(n-m+1):n}$.

- With $\operatorname{sign}(a), a \in \mathbf{R}$ we denote the *sign* of a

$$\operatorname{sign}(a) = \begin{cases} 0 & \text{if } a = 0 \\ -1 & \text{if } a < 0 \\ 1 & \text{if } a > 0 \end{cases}$$

We use the following types of matrices and matrix factorization for matrices of size $n \times n$.

DEFINITION 2.2 *Let $A \in \mathbf{K}^{n \times n}, v \in \mathbf{K}^n$.*

- *A is a* diagonal *matrix if $a_{ij} = 0$ for $i \neq j, i, j = 1, \ldots, n$, that is,*

$$A = \boxed{\diagdown}.$$

- *A is an* (upper) Hessenberg matrix *if* $a_{ij} = 0$ *for* $i > j + 1, i, j = 1, \ldots, n$, *that is,*

$$A = \boxed{\diagdown}.$$

- *A is an* unreduced (upper) Hessenberg matrix *if A is an upper Hessenberg matrix with* $a_{i,i-1} \neq 0, i = 2, \ldots, n$.

- *A is an* (upper) triangular matrix *if* $a_{ij} = 0$ *for* $i > j, i, j = 1, \ldots, n$, *that is,*

$$A = \boxed{\diagdown}.$$

- *A is a* strict (upper) triangular matrix *if A is an upper triangular matrix with* $a_{ii} = 0, i = 1, \ldots, n$, *that is,*

$$A = \boxed{\diagdown}.$$

- *A is a* quasi (upper) triangular matrix *if it is a block matrix of the form (2.1.4) with blocks of size* 1×1 *or* 2×2 *and* $A_{ij} = 0$ *for* $i > j, i = 1, \ldots, p, j = 1, \ldots, q$.

- *A is a* tridiagonal matrix *if* $a_{ij} = 0$ *for* $i > j + 1$, *and* $i < j - 1, i, j = 1, \ldots, n$, *that is*

$$A = \boxed{\diagdown}.$$

- *A is an* unreduced tridiagonal matrix *if A is a tridiagonal matrix with* $a_{i,i-1} \neq 0, i = 2, \ldots, n$ *and* $a_{i,i+1} \neq 0, i = 1, \ldots, n - 1$.

- *A* signature matrix *is a diagonal matrix* $D = \text{diag}(d_1, \ldots, d_n)$ *where* $d_i \in \{\pm 1\}$.

- *The matrix A is called D–symmetric if* $(DA)^T = DA$ *where D is a signature matrix.*

- *The* Krylov matrix $K(A, v, j) \in \mathbf{K}^{n \times j}$ *is defined by*

$$K(A, v, j) = [v, Av, \ldots, A^{j-1}v].$$

- *A is an* orthogonal matrix, *if* $\mathbf{K} = \mathbf{R}$ *and* $A^T A = I$.

- *A is a* unitary matrix, *if* $\mathbf{K} = \mathbf{C}$ *and* $A^H A = I$.

- *The* QR factorization *of A is given by* $A = QR$ *where* $Q \in \mathbf{R}^{n \times n}$ *is orthogonal and* $R \in \mathbf{R}^{n \times n}$ *is upper triangular if* $\mathbf{K} = \mathbf{R}$. *If* $\mathbf{K} = \mathbf{C}$, *then* $Q \in \mathbb{C}^{n \times n}$ *is unitary and* $R \in \mathbb{C}^{n \times n}$ *is upper triangular.*

LEMMA 2.1 *A tridiagonal matrix T is D–symmetric for some D if and only if $|t_{i+1,i}| = |t_{i,i+1}|$ for $i = 1, \ldots, n-1$. Every unreduced tridiagonal matrix is similar to a D–symmetric matrix (for some D) by a diagonal similarity with positive main diagonal entries.*

D–symmetric tridiagonal matrices are, e.g., generated by the unsymmetric Lanczos process [84].

Hessenberg matrices play a fundamental role for the analysis of the standard eigenvalue algorithms considered in this book. Hence, let us review some of their most important properties. It is well-known that for any matrix $A \in \mathbf{R}^{n \times n}$ an orthogonal transformation matrix $Q \in \mathbf{R}^{n \times n}$ can be computed such that $Q^T A Q = H$ is of Hessenberg form (see, e.g., [58, Section 7.4.3]). Such a Hessenberg decomposition is not unique.

THEOREM 2.2 (IMPLICIT-Q-THEOREM) *Suppose $Q = [q_1, \ldots, q_n]$ and $V = [v_1, \ldots, v_n]$ are orthogonal matrices with the property that both $Q^T A Q = H$ and $V^T A V = G$ are upper Hessenberg matrices where $A \in \mathbf{R}^{n \times n}$. Let k denote the smallest positive integer for which $h_{k+1,k} = 0$, with the convention that $k = n$ if H is unreduced. If $q_1 = v_1$, then $q_i = \pm v_i$ and $|h_{i,i-1}| = |g_{i,i-1}|$ for $i = 2 : k$. Moreover, if $k < n$, then $g_{k+1,k} = 0$.*

PROOF: See, e.g, [58, Theorem 7.4.2]. $\checkmark$

There is a useful connection between the Hessenberg reduction $Q^T A Q = H$ and the QR factorization of the Krylov matrix $K(A, Q(:,1), n)$.

THEOREM 2.3 *Suppose $Q \in \mathbf{R}^{n \times n}$ is an orthogonal matrix and $A \in \mathbf{R}^{n \times n}$. Let $q_1 = Q e_1$ be the first column of Q. Then $Q^T A Q = H$ is an unreduced upper Hessenberg matrix if and only if $Q^T K(A, q_1, n) = R$ is nonsingular and upper triangular.*

PROOF: See, e.g., [58, Theorem 7.4.3]. $\checkmark$

Thus, there is a correspondence between nonsingular Krylov matrices and orthogonal similarity reductions to unreduced upper Hessenberg form.

The last two results mentioned here concern unreduced upper Hessenberg matrices. The left and right eigenvectors of unreduced upper Hessenberg matrices have the following properties.

THEOREM 2.4 *Suppose $H \in \mathbf{R}^{n \times n}$ is an unreduced upper Hessenberg matrix. If $Hs = \lambda s$ with $s \in \mathbf{K}^n \setminus \{0\}$ and $H^T u = \lambda u$ with $u \in \mathbf{K}^n \setminus \{0\}$, then $e_n^T s \neq 0$ and $e_1^T u \neq 0$.*

PROOF: See, e.g., [92, Lemma 2.1]. $\checkmark$

Moreover, unreduced Hessenberg matrices are *nonderogatory*, that is, each eigenvalue has unit geometric multiplicity.

THEOREM 2.5 *Suppose $H \in \mathbf{R}^{n \times n}$ is an unreduced upper Hessenberg matrix. If λ is an eigenvalue of H, then its geometric multiplicity is one.*

PROOF: See, e.g, [58, Theorem 7.4.4].   √

When there is a repeated eigenvalue, the theorem implies that H has less then n linearly independent eigenvectors. If the eigenvectors of a matrix of order n are not a basis for $\mathbf{R}^n$, then the matrix is called *defective* or *nonsimple*. Hence, if H has a repeated eigenvalue it is a defective matrix. Unreduced Hessenberg matrices reveal even more information about the underlying eigensystem. Parlett [112, 114] provides an abundance of results for Hessenberg matrices.

For matrices of size $2n \times 2n$, we use the following types of matrices and matrix factorization.

DEFINITION 2.3 *Let $A \in \mathbf{R}^{2n \times 2n}$, $A = \begin{bmatrix} A_{11} & A_{12} \\ A_{21} & A_{22} \end{bmatrix}$ where $A_{ij} \in \mathbf{R}^{n \times n}$ for $i, j = 1, 2$. Let $v \in \mathbf{R}^{2n}$. Let J be as in (2.1.1).*

- *A is an* (upper) *J–Hessenberg matrix if A_{11}, A_{21}, A_{22} are upper triangular matrices and A_{12} is an upper Hessenberg matrix, that is*

$$A = \begin{bmatrix} \searchbox & \searchbox \\ \searchbox & \searchbox \end{bmatrix}.$$

- *A is an* unreduced (upper) *J–Hessenberg matrix if A is a J–Hessenberg matrix, A_{21}^{-1} exists, and A_{12} is an unreduced upper Hessenberg matrix.*

- *A is an* (upper) *J–triangular matrix if A_{11}, A_{12}, A_{22} are upper triangular matrices and A_{21} is a strict upper triangular matrix, that is,*

$$A = \begin{bmatrix} \searchbox & \searchbox \\ \ddots\searchbox & \searchbox \end{bmatrix}.$$

- *A is a* lower *J–triangular matrix if A^T is an upper J–triangular matrix.*

- *The* generalized Krylov matrix *$L(A, v, j) \in \mathbf{R}^{2n \times 2j}$ is defined by*

$$L(A, v, j) = [v, A^{-1}v, A^{-2}v, \dots, A^{-(j-1)}v, Av, A^2v, \dots, A^jv].$$

- *A is a* symplectic matrix *if $A^T J A = J$.*

- A is a trivial matrix, *if A is symplectic and J–triangular.*

- *The SR factorization of A is given by $A = SR$ where $S \in \mathbf{R}^{2n \times 2n}$ is symplectic and $R \in \mathbf{R}^{2n \times 2n}$ is J–triangular.*

Symplectic matrices can be viewed as orthogonal with respect to $(\cdot, \cdot)_J$. To emphasize this point of view, symplectic matrices are also called *J–orthogonal*.

The SR decomposition has been first introduced by Della-Dora [43, 44]. In contrast to the QR decomposition it does not always exist (see Theorem 2.7 below); but the set of matrices which can be factorized in this way is dense in $\mathbf{R}^{2n \times 2n}$ [34, 47]. While the QR decomposition is usually considered for matrices in $\mathbf{R}$ and $\mathbb{C}$, the SR decomposition is usually not considered for complex matrices $A \in \mathbb{C}^{2n \times 2n}$. This is due to the fact that the set of matrices $A \in \mathbb{C}^{2n \times 2n}$ which have an SR decomposition $A = SR$, where $S^H JS = J$ or $S^H JS = -J$, is not dense in $\mathbb{C}^{2n \times 2n}$ [34].

The following facts are easy to see.

LEMMA 2.6 *Let $A, B \in \mathbf{R}^{2n \times 2n}$, $A = \begin{bmatrix} A_{11} & A_{12} \\ A_{21} & A_{22} \end{bmatrix}$ where $A_{ij} \in \mathbf{R}^{n \times n}$ for $i, j = 1, 2$. Let P be as in (2.1.2), and J be as in (2.1.1).*

a) *If A is a J–triangular matrix, then PAP^T is an upper triangular matrix.*

b) *If A is an upper J–Hessenberg matrix or an unreduced upper J–Hessenberg matrix, then PAP^T is an upper Hessenberg matrix or an unreduced upper Hessenberg matrix, respectively.*

c) *A is trivial (that is, symplectic and J–triangular) if and only if it has the form*

$$A = \begin{bmatrix} C & F \\ 0 & C^{-1} \end{bmatrix}, \qquad (2.1.5)$$

where $C = \mathrm{diag}(c_1, \ldots, c_n)$, $F = \mathrm{diag}(f_1, \ldots, f_n)$.

d) *If A is a regular J–triangular matrix, then A^{-1} is a J–triangular matrix.*

e) *If A and B are J–triangular matrices, then AB is a J–triangular matrix.*

f) *If A is a J–Hessenberg matrix and B a J–triangular matrix, then AB and BA are J–Hessenberg matrices.*

Almost every matrix A can be decomposed into the product of a symplectic matrix S and a J–triangular matrix R.

THEOREM 2.7 *Let $A \in \mathbf{R}^{2n \times 2n}$ be nonsingular. There exists a symplectic matrix S and a J–triangular matrix R such that $A = SR$ if and only if all leading principal minors of even dimension of $PA^T JAP^T$ are nonzero where*

P as in (2.1.2), and J as in (2.1.1). The set of $2n \times 2n$ *SR decomposable matrices is dense in* $\mathbf{R}^{2n \times 2n}$.

PROOF: See [47, Theorem 11] or [34, Theorem 3.8] for a proof.   ✓

Bunse-Gerstner and Mehrmann [38] present an algorithm for computing the SR decomposition of an arbitrary $2n \times 2n$ matrix A (see also Section 2.2.1). First-order componentwise and normwise perturbation bounds for the SR decomposition can be found in [42, 25].

The following observation will be useful later on.

COROLLARY 2.8 *Let* $A \in \mathbf{R}^{2n \times 2n}$. *Let J be as in (2.1.1).*

a) *If* $A^T = SR$ *is an SR decomposition of* A^T, *then there exists a symplectic matrix W and a lower J–triangular matrix L such that* $A = LW$.

b) *Let*

$$\hat{J} = \begin{bmatrix} & & 1 \\ & \diagup & \\ 1 & & \end{bmatrix} \in \mathbf{R}^{n \times n}, \qquad \tilde{J} = \begin{bmatrix} & \hat{J} \\ -\hat{J} & \end{bmatrix} \in \mathbf{R}^{2n \times 2n}.$$

If $\tilde{J}A\tilde{J} = SR$ *is an SR decomposition of* $\tilde{J}A\tilde{J}$, *then there exists a symplectic matrix* $\tilde{S}$ *and a lower J–triangular matrix* $\tilde{L}$ *such that* $A = \tilde{S}\tilde{L}$.

c) *If* $A^T = \tilde{S}\tilde{L}$ *as in 2. exists, then there exists a symplectic matrix* $\widetilde{W}$ *and an upper J–triangular matrix* $\tilde{R}$ *such that* $A = \tilde{R}\widetilde{W}$.

PROOF:

a) If $A^T = SR$ is an SR decomposition of A^T, then $L = R^T$ is a lower J–triangular matrix and $W = S^T$ is a symplectic matrix, and we have $A = LW$.

b) If $\tilde{J}A\tilde{J} = SR$ is an SR decomposition of $\tilde{J}A\tilde{J}$, then $A = (\tilde{J}S\tilde{J})(\tilde{J}R\tilde{J})$. Let $\tilde{S} = \tilde{J}S\tilde{J}$, $\tilde{S}$ is a symplectic matrix. Let $\tilde{L} = \tilde{J}R\tilde{J}$, $\tilde{L}$ is a lower J–triangular matrix.

c) If $A^T = \tilde{S}\tilde{L}$ as in 2., then we have $A = \tilde{L}^T\tilde{S}^T = \tilde{R}\widetilde{W}$. $\tilde{R} = \tilde{L}^T$ is an upper J–triangular matrix and $\widetilde{W} = \tilde{S}^T$ is a symplectic matrix. This will be called the RS decomposition of A.   ✓

Let $X = \tilde{J}A\tilde{J}$. The SR decomposition of X exists if and only if the leading principal minors of even dimension of PX^TJXP^T are nonzero. We have

$$PX^TJXP^T = P\tilde{J}A^TJA\tilde{J}P^T$$

and

$$P\tilde{J} = [-e_{2n}, -e_{2n-2}, \ldots, -e_2, e_{2n-1}, e_{2n-3}, \ldots, e_1].$$

Let $Y = A^T J A = \begin{bmatrix} Y_{11} & Y_{12} \\ Y_{21} & Y_{22} \end{bmatrix} \in \mathbf{R}^{2n \times 2n}$, where $Y_{ij} \in \mathbf{R}^{n \times n}$. Then the leading principal minors of even dimension of $P X^T J X P^T$ are given by

$$\det \begin{bmatrix} (Y_{11})_{2k+1:2n,2k+1:2n} & (Y_{12})_{2k+1:2n,2k+1:2n} \\ (Y_{21})_{2k+1:2n,2k+1:2n} & (Y_{22})_{2k+1:2n,2k+1:2n} \end{bmatrix}.$$

Hence, the leading principal minors of even dimension of $P(\tilde{J} A \tilde{J})^T J (\tilde{J} A \tilde{J}) P^T$ are just the trailing principal minors of even dimension of $P A^T J A P^T$.

Statements similar to the above have been shown for the QR decomposition; see, e.g., [31, Korollar 2.5.2].

Symplectic matrices may serve to transform a $2n \times 2n$ matrix A to J–Hessenberg form. The relation between this transformation to J–Hessenberg form and the SR factorization is completely analogous to the relation between the unitary similarity reduction to Hessenberg form and the QR factorization, as the following theorem shows.

THEOREM 2.9 (IMPLICIT-S-THEOREM) *Let $A \in \mathbf{R}^{2n \times 2n}$.*

a) *Let $A = SR$ and $A = \tilde{S}\tilde{R}$ be SR factorizations of A. Then there exists a trivial matrix D such that $\tilde{S} = SD^{-1}$ and $\tilde{R} = DR$.*

b) *Suppose $S \in \mathbf{R}^{2n \times 2n}$ is a symplectic matrix. Let $s_1 = Se_1$ be the first column of S. Then $S^{-1}AS$ is an unreduced J–Hessenberg matrix if and only if $S^{-1}K(A, s_1, 2n) = R$ is nonsingular and J–triangular.*

PROOF:

a) See, e.g., [38, Proposition 3.3].

b) See, e.g., [38, Theorem 3.4].   √

The essential uniqueness of the factorization $K(A, Se_1, n) = SR$ tells us that the transforming matrix S for the similarity transformation $S^{-1}AS$ is essentially uniquely determined by its first column. This Implicit-S-Theorem can serve as the basis for the construction of an implicit SR algorithm for J–Hessenberg matrices, just as the Implicit-Q-Theorem (Theorem 2.2) provides a basis for the implicit QR algorithm on upper Hessenberg matrices. In both cases uniqueness depends on the unreduced character of the matrix. While the QR decomposition of a matrix always exists, the SR factorization may not exist. Hence, the reduction to J–Hessenberg form may not exist.

Unreduced J–Hessenberg matrices have the similar properties as unreduced Hessenberg matrices.

THEOREM 2.10 *Suppose $H \in \mathbf{R}^{2n \times 2n}$ is an unreduced J–Hessenberg matrix.*

a) *If $Hs = \lambda s$ with $s \in \mathbf{K}^{2n} \setminus \{0\}$ and $H^T u = \lambda u$ with $u \in \mathbf{K}^{2n} \setminus \{0\}$, then $e_n^T s \neq 0$ and $e_1^T u \neq 0$.*

b) *If λ is an eigenvalue of H, then its geometric multiplicity is one.*

PROOF:

a) As PHP^T is an unreduced Hessenberg matrix (Lemma 2.6), we have from Theorem 2.4:

   If $PHP^T \widehat{s} = \lambda \widehat{s}$ and $PH^T P^T \widehat{u} = \lambda \widehat{u}$, then $e_{2n}^T \widehat{s} \neq 0$ and $e_1^T \widehat{u} \neq 0$.

   With $s = P^T \widehat{s}$ and $u = P^T \widehat{u}$, we obtain the assertion as $e_{2n}^T s = e_{2n}^T P^T \widehat{s} = e_{2n}^T \widehat{s} \neq 0$ and $e_1^T u = e_1^T P^T \widehat{u} = e_1^T \widehat{u} \neq 0$.

b) As PHP^T is an unreduced Hessenberg matrix (Lemma 2.6), the assertion follows from Theorem 2.10. $\checkmark$

2.1.1 SYMPLECTIC MATRICES AND MATRIX PENCILS

Symplectic matrices will be the main topic of this book. Let us recall the definition (see Definition 2.3): A matrix $M \in \mathbf{R}^{2n \times 2n}$ is called *symplectic* (or *J–orthogonal*) if

$$MJM^T = J \qquad (2.1.6)$$

(or equivalently, $M^T JM = J$). A *symplectic matrix pencil* $K - \lambda N$, $K, N \in \mathbf{R}^{2n \times 2n}$ is defined by the property

$$KJK^T = NJN^T, \qquad (2.1.7)$$

where J is as in (2.1.1). In other words, the set S of all symplectic matrices is the set of all matrices that preserve the bilinear form defined by J. It is well-known and easy to show from this definition that S forms a multiplicative group (even more, S is a Lie group). While symplectic matrices are nonsingular ($M^{-1} = JM^T J^T$), a symplectic matrix pencil $K - \lambda N$ is not necessarily regular, i.e., there is no guarantee that $\det(K - \lambda N)$ does not vanish identically for all $\lambda \in \mathbb{C}$. K and N may be nonsingular or singular. Hence (2.1.7) is in general not equivalent to $K^T JK = N^T JN$.

The spectrum of a symplectic matrix pencil/matrix is symmetric with respect to the unit circle. Or, in other words, the eigenvalues of symplectic matrix pencils occur in reciprocal pairs: if $\lambda \neq 0$ is a (generalized finite) eigenvalue, then so is λ^{-1}. Furthermore, if $\lambda = 0$ is an eigenvalue of a symplectic pencil, then so is ∞. Let $y^T \in \mathbf{R}^{2n}\setminus\{0\}$ be a left eigenvector of $K - \lambda N$ corresponding to the eigenvalue λ, then $x = JK^T y$ is a right eigenvector corresponding to λ^{-1}. Hence, for symplectic matrices we have: If λ is an eigenvalue of M with right eigenvector x, then λ^{-1} is an eigenvalue of M with left eigenvector $(Jx)^T$. Further, if $\lambda \in \mathbb{C}$ is an eigenvalue of M (or $K - \lambda N$), then so are $\bar{\lambda}, \lambda^{-1}, \overline{\lambda^{-1}}$.

By definition of the symplectic matrix pencil, we obtain the following result.

LEMMA 2.11 *If $K - \lambda N$ is a symplectic matrix pencil, then $Q(K - \lambda N)S$ is a symplectic matrix pencil for all nonsingular matrices Q and all symplectic matrices S.*

In most applications, conditions are satisfied which guarantee the existence of an n–dimensional deflating (or invariant) subspace corresponding to the eigenvalues of $K - \lambda N$ (or M) inside the open unit disk. This is the subspace one usually wishes to compute. The numerical computation of such a deflating (or invariant) subspace is typically carried out by an iterative procedure like the QZ (or QR) algorithm which transforms $K - \lambda N$ into a generalized Schur form (M into Schur form), from which the deflating (invariant) subspace can be read off. See, e.g., [103, 111, 136]. For symplectic matrix pencils/matrices a special generalized Schur form is known.

THEOREM 2.12

a) *Let M be a $2n \times 2n$ real symplectic matrix. Then there exists a real orthogonal and symplectic matrix Z such that*

$$Z^T M Z = \begin{bmatrix} T_1 & T_2 \\ 0 & T_1^{-T} \end{bmatrix},$$

where T_1 is an $n \times n$ real quasi upper triangular matrix, if and only if every unimodular eigenvalue λ of M has even algebraic multiplicity, say $2k$, and any matrix $X_k \in \mathbb{C}^{2n \times 2k}$ with the property that its columns span a basis of the maximal M–invariant subspace corresponding to λ satisfies that $X_k^H J^{2n,2n} X_k$ is congruent to $J^{2k,2k}$. Moreover, Z can be chosen such that T_1 has only eigenvalues inside the closed unit disk.

b) *Let $K - \lambda N$ be a $2n \times 2n$ real regular symplectic matrix pencil. Then there exists a real orthogonal matrix Q and a real orthogonal and symplectic matrix Z such that*

$$Q^T K Z = \begin{bmatrix} T_1 & T_2 \\ 0 & T_3 \end{bmatrix}, \qquad Q^T N Z = \begin{bmatrix} S_1 & S_2 \\ 0 & S_3 \end{bmatrix},$$

where T_1, S_3^T are quasi-upper triangular matrices, and S_1, T_3^T are upper triangular matrices, if and only if every unimodular eigenvalue λ of $K - \lambda N$

has even algebraic multiplicity, say 2k, and any matrix $X_k \in \mathbb{C}^{2n \times 2k}$ with the property that its columns span a basis of the maximal deflating subspace for $K - \lambda N$ corresponding to λ satisfies that $X_k^H J^{2n,2n} X_k$ is congruent to $J^{2k,2k}$. Moreover, Q and Z can be chosen such that $T_1 - \lambda S_1$ has only eigenvalues inside the closed unit disk.

PROOF: This result was first stated and proved in [97]. A simpler proof is given in [98]. Weaker versions of the theorem assuming that M (or $K - \lambda N$) has no eigenvalues of modulus one can be found, e.g., in [87] or [103].   √

The first n columns of the right transformation matrix Z then span an invariant/deflating subspace corresponding to the eigenvalues inside the closed unit disk. This subspace is unique if no eigenvalues are on the unit circle. The construction of numerical methods to compute these Schur forms using only symplectic and orthogonal transformations is still an open problem. Mehrmann [103] describes a QZ–like algorithm for a special symplectic matrix pencil that arises in the single input or single output discrete time optimal linear quadratic control problem. Patel [117] describes a method working on a condensed symplectic pencil using implicit QZ steps to compute the stable deflating subspace of a symplectic pencil. While the algorithm proposed by Mehrmann works only for a very special case, the more general method proposed by Patel suffers from numerical problems due to the difficulty to preserve the symplectic structure of the problem throughout the iteration.

For a discussion of other symplectic canonical forms (e.g., symplectic Jordan or Kronecker canonical forms) see, e.g., [98, 89, 145].

2.1.2 ELEMENTARY SYMPLECTIC MATRICES

During the course of the discussion of the various algorithms considered here we will use the following elementary symplectic transformations:

- *Symplectic Gauss transformations $L_k = L(k, c, d)$*

$$
L_k =
\left[
\begin{array}{cccc|cccc}
I^{k-2,k-2} & & & & & & & \\
 & c & & & & & d & \\
 & & c & & & d & & \\
 & & & I^{n-k,n-k} & & & & \\
\hline
 & & & & I^{k-2,k-2} & & & \\
 & & & & & c^{-1} & & \\
 & & & & & & c^{-1} & \\
 & & & & & & & I^{n-k,n-k}
\end{array}
\right],
$$

where $c, d \in \mathbf{R}$.

- *Symplectic Gauss transformations type II* $\widetilde{L}_k = \widetilde{L}(k, c, d)$

$$
\widetilde{L}_k = \left[
\begin{array}{c|c}
\begin{matrix} I^{k-1,k-1} & & \\ & c & \\ & & I^{n-k,n-k} \end{matrix} & \begin{matrix} & d & \\ & & \\ & & \end{matrix} \\
\hline
\begin{matrix} & & \\ & & \\ & & \end{matrix} & \begin{matrix} I^{k-1,k-1} & & \\ & c^{-1} & \\ & & I^{n-k,n-k} \end{matrix}
\end{array}
\right],
$$

where $c, d \in \mathbf{R}$.

- *Symplectic Householder transformations* $H_k = H(k, v)$

$$
H_k = \left[
\begin{array}{c|c}
\begin{matrix} I^{k-1,k-1} & \\ & P \end{matrix} & \\
\hline
& \begin{matrix} I^{k-1,k-1} & \\ & P \end{matrix}
\end{array}
\right],
$$

where $P = I^{n-k+1,n-k+1} - 2\frac{vv^T}{v^T v}$, $v \in \mathbf{R}^{n-k+1}$.

- *Symplectic Givens transformations* $G_k = G(k, c, s)$

$$
G_k = \left[
\begin{array}{c|c}
\begin{matrix} I^{k-1,k-1} & & \\ & c & \\ & & I^{n-k,n-k} \end{matrix} & \begin{matrix} & s & \\ & & \\ & & \end{matrix} \\
\hline
\begin{matrix} & -s & \\ & & \\ & & \end{matrix} & \begin{matrix} I^{k-1,k-1} & & \\ & c & \\ & & I^{n-k,n-k} \end{matrix}
\end{array}
\right],
$$

where $c^2 + s^2 = 1$, $c, s \in \mathbf{R}$.

The symplectic Givens and Householder transformations are orthogonal, while the symplectic Gauss transformations are nonorthogonal. It is crucial that the simple structure of these elementary symplectic transformations is exploited when computing matrix products of the form $G_k A$, AG_k, $H_k A$, AH_k, $L_k A$, AL_k, $\widetilde{L}_k A$, and $A\widetilde{L}_k$. Note that only rows k and $n + k$ are affected by the premultiplication $G_k A$, and columns k and $n + k$ by the postmultiplication AG_k. Similar, pre- and postmultiplication by L_k affects only the rows (resp., the columns) $k - 1, k, n + k - 1$ and $n + k$, while pre- and postmultiplication by $\widetilde{L}_k$ affects only the rows (resp., the columns) k and $n + k$. Premultiplication by H_k affects only the rows k to n and $n + k$ to $2n$, while postmultiplication affects the corresponding columns. Further, note that for the symplectic Householder transformations we have, e.g., $PA_{k:n,1:2n} = A_{k:n,1:2n} + vw^T$ where $w = \beta A^T_{k:n,1:2n} v$, $\beta = -2/v^T v$. Thus a symplectic

Householder update involves only matrix-vector multiplications followed by an outer product update. Failure to recognize these points and to treat the elementary symplectic transformations as general matrices increases work by an order of magnitude. The updates never entail the explicit formation of the transformation matrix, only the relevant parameters are computed. Algorithms to compute these parameters of the abovementioned transformations are given here for the sake of completeness (see Table 2.1 – 2.4 (in MATLAB[1]-like notation), see also, e.g., [110, 39]).

REMARK 2.13 a) *An efficient implementation of an update of a $2n \times 2n$ matrix by a symplectic Givens transformation G_k (pre- or postmultiplication) requires $6 \cdot 2n$ flops[2] (see, e.g., [58]).*

b) *An efficient implementation of an update of a $2n \times 2n$ matrix by a symplectic Householder transformation H_k (pre- or postmultiplication) requires $2(4m - 2) \cdot 2n$ flops. Here we assume that $v \in \mathbf{R}^m, m = n - k + 1$ with $v_1 = 1$. The flop count can then be seen as follows: The update, e.g., $H_k A$ requires, as noted above, the computation of $P A_{k:n,1:2n}$ and $P A_{n+k:2n,1:2n}$. Let $\widehat{A} = A_{k:n,1:2n}$ or $\widehat{A} = A_{n+k:2n,1:2n}$. Then*

the computation of	requires
$x^T = v^T \widehat{A}$	$2(m - 1) \cdot 2n$ flops,
$y^T = \beta x^T$	$2n$ flops,
$B = vy^T$	$(m - 1) \cdot 2n$ flops,
$\widehat{A} + B = P\widehat{A}$	$m \cdot 2n$ flops.

Hence, the computation of $P\widehat{A}$ requires $(4m - 2) \cdot 2n$ flops. As for a symplectic Householder update two such computations are needed, it requires $2(4m - 2) \cdot 2n$ flops.

c) *An efficient implementation of an update of a $2n \times 2n$ matrix by a symplectic Gauss transformation L_k (pre- or postmultiplication) requires $8 \cdot 2n$ flops.*

d) *An efficient implementation of an update of a $2n \times 2n$ matrix by a symplectic Gauss transformation type II $\widetilde{L}_k$ (pre- or postmultiplication) requires $4 \cdot 2n$ flops.*

---

[1]MATLAB is a trademark of The MathWorks, Inc.
[2]Following [58], we define each floating point arithmetic operation together with the associated integer indexing as a flop.

Algorithm: Generate Symplectic Givens Matrix

Given scalars a and b compute c and s such that $c^2 + s^2 = 1$ and

$$\begin{bmatrix} c & s \\ -s & c \end{bmatrix} \begin{bmatrix} a \\ b \end{bmatrix} = \begin{bmatrix} r \\ 0 \end{bmatrix}.$$

function $[c, s] = $ **givens**(a, b)
   if $b = 0$
   then $c = 1, s = 0$
   else if $|b| > |a|$
      then $t = a/b$
         $s = -1/\sqrt{1 + t^2}$
         $c = st$
      else $t = b/a$
         $c = 1/\sqrt{1 + t^2}$
         $s = ct$
   end
  end

Table 2.1.   Symplectic Givens Matrix

Algorithm: Generate Symplectic Householder Matrix

Given a column vector $x \in \mathbf{R}^n$ compute v such that $v(1) = 1$, and $y(2 : n) = 0$ for $y = (I - 2vv^T/(v^Tv))x$.

function $v = $ **house**(x)
   $m = \|x\|_2$
   $v = x$
   if $m \neq 0$
   then if $x(1) \geq 0$
      then $b = x(1) + m$
      else $b = x(1) - m$
      end
      $v = (1/b)v$
   end
   $v(1) = 1$

Table 2.2.   Symplectic Householder Matrix

Algorithm: Generate Symplectic Gauss Matrix

Given scalars a and b, where $b = 0$ only if $a = 0$, compute c and d such that

$$\begin{bmatrix} c & & d & \\ & c & d & \\ & & c^{-1} & \\ & & & c^{-1} \end{bmatrix} \begin{bmatrix} \star \\ a \\ b \\ 0 \end{bmatrix} = \begin{bmatrix} \star \\ 0 \\ r \\ 0 \end{bmatrix}.$$

Further the condition number κ of the elimination matrix is computed.

function $[c, d, \kappa] = $ **gauss1**(a, b)
    if $a = 0$
    then $t = 0$
    else $t = -a/b$
    end
    $c = 1/\sqrt[4]{1 + t^2}$
    $d = ct$
    $\kappa = \sqrt{1 + t^2} + |t|$

Table 2.3. Symplectic Gauss Matrix

Algorithm: Generate Symplectic Gauss Matrix Type II

Given scalars a and b, where $b = 0$ only if $a = 0$, compute c and d such that

$$\begin{bmatrix} c & d \\ & c^{-1} \end{bmatrix} \begin{bmatrix} a \\ b \end{bmatrix} = \begin{bmatrix} 0 \\ r \end{bmatrix}.$$

Further the condition number κ of the elimination matrix is computed.

function $[c, d, \kappa] = $ **gauss2**(a, b)
    $[c, d, \kappa] = $ gauss1(a, b)

Table 2.4. Symplectic Gauss Matrix Type II

The Gauss transformations are computed such that among all possible transformations of that form, the one with the minimal condition number is chosen. The following lemma is easy to see.

LEMMA 2.14 *Let $M \in \mathbf{R}^{2n \times 2n}$ and $j, k \in \mathbf{N}$, $1 \leq j \leq 2n$, $1 \leq k \leq n$ given indices.*

a) *Let $[c, s] = \mathbf{givens}(M(k, j), M(k + n, j))$ and $G_k = G(k, c, s)$, then*

$$(G_k M)_{k+n,j} = 0.$$

Further G_k is symplectic and orthogonal.

b) *Let $[c, s] = \mathbf{givens}(M(j, k + n), M(j, k))$ and $G_k = G_k(k, c, s)$, then*

$$(M G_k)_{j,k} = 0.$$

Further G_k is symplectic and orthogonal.

c) *Let $v = \mathbf{house}(M(k : n, j))$ and $H_k = H(k, v)$, then*

$$(H_k M)_{k+1:n,j} = 0.$$

Further H_k is symplectic and orthogonal.

d) *Let $v = \mathbf{house}(M(j, k + n : 2n))$ and $H_k = H(k, v)$, then*

$$(M H_k)_{j,k+1+n:2n} = 0.$$

Further H_k is symplectic and orthogonal.

e) *Let $k > 1$ and $M(k-1+n, j) = 0$ only if $M(k, j) = 0$. Let $L_k = L(k, c, d)$ where $[c, d, \kappa] = \mathbf{gauss1}(M(k, j), M(k - 1 + n, j))$, then*

$$(L_k M)_{k,j} = 0.$$

Further L_k is symplectic with the condition number κ. κ is minimal, that is, there is no corresponding elimination matrix with smaller condition number.

f) *Let $k > 1$ and $M(j, k - 1) = 0$ only if $M(j, k + n) = 0$, $[c, d, \kappa] = \mathbf{gauss1}(M(j, k + n), M(j, k - 1))$ and $L_k = L(k, c^{-1}, d)$, then*

$$(M L_k)_{j,k+n} = 0.$$

Further L_k is symplectic with condition number κ. κ is minimal, that is, there is no corresponding symplectic elimination matrix with smaller condition number.

g) *Let* $M(j,k) = 0$ *only if* $M(j, k + n) = 0$, $[c, d, \kappa] = \textbf{gauss2}(M(j, k + n), M(j,k))$ *and* $\widetilde{L}_k = \widetilde{L}(k, c^{-1}, d)$, *then*

$$(M\widetilde{L}_k)_{j,k+n} = 0.$$

Further $\widetilde{L}_k$ *is symplectic with condition number* κ. κ *is minimal, that is, there is no corresponding symplectic elimination matrix with smaller condition number.*

REMARK 2.15 *Any orthogonal symplectic matrix Q can be expressed as the product of symplectic Givens and symplectic Householder transformations, see [110, Corollary 2.2].*

2.2 EIGENVALUE ALGORITHMS

In order to develop fast and efficient numerical methods for the symplectic eigenproblem one should make use of the rich mathematical structure of the problem in a similar way as it has been successfully done for symmetric/Hermitian and orthogonal/unitary eigenproblems. E.g., for the symmetric eigenproblem, one of the nowadays standard approaches involves first the reduction of the symmetric matrix to symmetric tridiagonal form followed by a sequence of implicit QR steps which preserve this symmetric tridiagonal form, see, e.g., [58]. Such structure-preserving methods are desirable as important properties of the original problem are preserved during the actual computations and are not destroyed by rounding errors. Moreover, in general, such methods allow for faster computations than general-purpose methods. For the symmetric eigenproblem, e.g., applying implicit QR steps to the full symmetric matrix requires $\mathcal{O}(n^3)$ arithmetic operations per step, while applying an implicit QR step to a similar symmetric tridiagonal matrix requires only $\mathcal{O}(n)$ arithmetic operations, where n is the order of the matrix. If the matrix under consideration is large and sparse, the QR method might not be a suitable tool for computing the eigeninformation. In that case, usually the Lanczos method, a technique especially tuned to solve large, sparse eigenproblems should be used.

The eigenvalues and invariant subspaces of symplectic matrices S may be computed by the QR algorithm. But the QR method cannot take advantage of the symplectic structure of S, it will treat S like any arbitrary $2n \times 2n$ matrix. The computed eigenvalues will in general not come in reciprocal pairs, although the exact eigenvalues have this property. Even worse, small perturbations may cause eigenvalues close to the unit circle to cross the unit circle such that the number of true and computed eigenvalues inside the open unit disk may differ.

To preserve the symplectic structure of S, we have to employ similarity transformations with symplectic matrices instead of the transformations with the usual unitary matrices in the QR algorithm. (Recall: The symplectic matrices form a group under multiplication.) In order to ensure numerical stability, it would be best to employ symplectic and orthogonal transformations. Under certain conditions a symplectic matrix M may be reduced to symplectic Hessenberg form

$$U^T M U = \begin{bmatrix} \diagdown & \square \\ 0| & \square \end{bmatrix}$$

using a symplectic and orthogonal transformation matrix U. This form stays invariant under a QR like iteration which uses only symplectic and orthogonal transformations. However, the computation of the initial unreduced symplectic Hessenberg form is not always possible. As shown in [5] the components of the first column of U must satisfy a system of n quadratic equations in $2n$ unknowns. Consequently, such a reduction is not always possible. Hence, more general QR like methods have to be considered in order to derive a structure-preserving QR like eigenvalue method for the symplectic eigenproblem.

General QR like methods, in which the QR decompositions are replaced by other decompositions have been studied by several authors, see, e.g., [43, 142]. The factorizations have to satisfy several conditions to lead to a reasonable computational process. The one that meets most of these requirements for the symplectic eigenproblem is the SR decomposition. This decomposition can serve as a basis for a QR like method, the SR *algorithm*, which works for arbitrary matrices of even dimensions. It preserves the symplectic structure and, as will be seen, allows to develop fast and efficient implementations.

The SR algorithm [43, 44, 38] is a member of the family of GR algorithms [142] for calculating eigenvalues and invariant subspaces of matrices. The oldest member of the family is Rutishauser's LU algorithm [120, 121] and the most widely used is the QR algorithm [54, 79, 144, 138, 58, 139]. The GR algorithm is an iterative procedure that begins with a matrix A whose eigenvalues and invariant subspaces are sought. It produces a sequence of similar matrices (A_i) that (hopefully) converge to a form exposing the eigenvalues. The transforming matrices for the similarity transformations $A_i = G_i^{-1} A_{i-1} G_i$ are obtained from a "GR" decomposition $p_i(A_{i-1}) = G_i R_i$ in which p_i is a polynomial and R_i is upper triangular. The degree of p_i is called the multiplicity of the ith step. If p_i has degree 1, it is a *single step*. If the degree is 2, it is a *double step*, and so on. Writing p_i in factored form $p_i(A) = \alpha_i (A - \mu_1^{(i)} I)(A - \mu_2^{(i)} I) \cdots (A - \mu_{m_i}^{(i)} I)$ we call the roots $\mu_1^{(i)}, \mu_2^{(i)}, \ldots, \mu_{m_i}^{(i)}$ the *shifts* for the ith step. Each step of multiplicity m_i has m_i shifts. A procedure for choosing the p_i is called a *shift strategy* because the choice of p_i implies a certain choice of shifts $\mu_1^{(i)}, \ldots, \mu_{m_i}^{(i)}$. In [142] it is shown that every GR algorithm is a form of a nested subspace iteration in which a change of coordinate system is made at each step. Convergence theorems for the GR algorithm are proved. The theorems guarantee convergence only if the condition numbers of the accumulated transforming matrices $\widehat{G}_i = G_1 G_2 \cdots G_i$ remain bounded throughout the iterations. The global convergence theorem holds for shift strategies that converge – unfortunately, no one has yet been able to devise a practical shift strategy that is guaranteed to converge for all matrices and can be shown to converge rapidly. The local convergence rate for the generalized Rayleigh-quotient strategy is typically quadratic. For matrices having certain types of special structure, it is cubic. In the *generalized Rayleigh-quotient strategy* p_i is chosen to be the characteristic polynomial of the trailing $m_i \times m_i$ principal submatrix of A_{i-1}.

Algorithms in the GR family are usually implemented implicitly, as chasing algorithms. The matrix whose eigenvalues are sought is first reduced to some upper Hessenberg-like form. Then the chasing algorithm is set in motion by a similarity transformation that introduces a bulge in the Hessenberg-like form near the upper left-hand corner of the matrix. A sequence of similarity transformations then chases the bulge downward and to the right, until the Hessenberg-like form is restored. Chasing steps like this are repeated until (hopefully) the matrix converges to a form from which the eigenvalues can be read off. A GR *step* consists of a similarity transformation $X = G^{-1}AG$ where $p(A) = GR$. One can show that G is more or less uniquely determined by its first column (e.g., in the QR step this follows from the Implicit-Q-Theorem). The implicit GR algorithm performs a different similarity transformation $\widetilde{X} = \widetilde{G}^{-1}A\widetilde{G}$, but $\widetilde{G}$ is constructed in such a way that its first column is proportional to the first column of G. It follows from the Implicit-G-Theorem that G and $\widetilde{G}$ are essentially the same, and consequently X and $\widetilde{X}$ are essentially the same. Watkins and Elsner analyze general GR chasing algorithms in [141].

A SZ algorithm is the analogue of the SR algorithm for the generalized eigenproblem, just as the QZ algorithm is the analogue of the QR algorithm for the generalized eigenproblem. Both are instances of the GZ algorithm [143]. The GZ algorithm is an iterative procedure that begins with a regular matrix pencil $A - \lambda B$, where B is nonsingular. It produces a sequence of equivalent matrix pencils $(A_i - \lambda B_i)$ that (hopefully) converge to a form exposing the eigenvalues. The transforming matrices for the equivalence transformations $A_i = G_i^{-1}A_{i-1}Z_i$ and $B_i = G_i^{-1}B_{i-1}Z_i$ are obtained from the GR decompositions $p_i(A_{i-1}B_{i-1}^{-1}) = G_iR_i$ and $p_i(B_{i-1}^{-1}A_{i-1}) = Z_iS_i$ in which p_i is a polynomial, R_i, S_i are upper triangular, and G_i, Z_i are nonsingular. In the special case $B_{i-1} = I$ we have $Z_i = G_i$ and $S_i = R_i$. That is, the algorithm reduces to the generic GR algorithm for the standard eigenvalue problem.

In Section 2.2.1 we will review the SR algorithm for general, real $2n \times 2n$ matrices, while in Section 2.2.2 the SZ algorithm for general, real $2n \times 2n$ matrix pencils is reviewed. Any successful implementation of the SR/SZ (or, a general GR/GZ) algorithm will make use of some elimination scheme to reduce the matrix/matrix pencil under consideration to some upper Hessenberg-like form in each of the iteration steps. In case the matrix under consideration is very large and sparse, such an elimination scheme is often not suitable. Fill-in would increase the memory space needed to store the matrix; sometimes the matrix is even too large to be stored online at all. In Section 2.2.4 we will review the unsymmetric Lanczos method, a well known technique that can be used to solve large, sparse eigenproblems $Ax = \lambda x$. The method involves partial tridiagonalizations of the given matrix. Only matrix-vector multiplications Ay and $A^T y$ are required, the matrix A itself is never altered. Information about A's (extremal) eigenvalues tends to emerge long before the tridiagonalization is complete. This makes the Lanczos algorithm particularly useful in situations where a few of A's (largest) eigenvalues are desired. Finally, in Section 2.2.3 we will briefly review the HR algorithm. The HR algorithm is, like the SR algorithm, a member of the family of GR algorithms. It is useful as it preserves sign-symmetric structure, like the one that arises in the unsymmetric Lanczos algorithm.

2.2.1 SR ALGORITHM

The SR algorithm is based on the SR decomposition. Recall that the SR factorization of a real $2n \times 2n$ matrix A is given by $A = SR$ where $S \in \mathbf{R}^{2n \times 2n}$ is symplectic and $R \in \mathbf{R}^{2n \times 2n}$ is J–triangular. Almost every matrix A can be decomposed into such a product, see Theorem 2.7.

If the SR decomposition exists, then other SR decompositions of A can be built from it by passing trivial factors (2.1.5) back and forth between S and R. That is, if D is a trivial matrix, $\tilde{S} = SD^{-1}$ and $\tilde{R} = DR$, then $A = \tilde{S}\tilde{R}$ is another SR decomposition of A (see Theorem 2.9). If A is nonsingular, then this is the only way to create other SR decompositions. In other words, the SR decomposition is unique up to trivial factors.

In Section 2.1 we have already seen that the relation between the symplectic transformation of a $2n \times 2n$ matrix to J–Hessenberg form and the SR decomposition is completely analogous to the relation between the unitary reduction to upper Hessenberg form and the QR decomposition, see Theorem 2.9.

The SR decomposition $A = SR$ and, therefore, also the reduction to J–Hessenberg form can, in general, not be performed with a symplectic orthogonal matrix S. A necessary and sufficient condition for the existence of such an orthogonal SR decomposition is that A is of the form

$$A = \begin{bmatrix} X & Y \\ -Y & X \end{bmatrix} R$$

where $X, Y \in \mathbf{R}^{n \times n}$, and R is a J–triangular matrix [34]. Hence, for the computation of the SR decomposition (or the reduction to J–Hessenberg form) one has to employ nonorthogonal symplectic transformations.

Bunse-Gerstner and Mehrmann [38] present an algorithm for computing the SR decomposition of an arbitrary $2n \times 2n$ matrix A. The algorithm uses the symplectic Givens transformations G_k, the symplectic Householder transformations H_k, and the symplectic Gauss transformation L_k introduced in Section 2.1.2. Symplectic elimination matrices S_j are determined such that $R = S_{2n} \cdots S_2 S_1 A$ is of J–triangular form. Then $A = SR$ with $S = S_1^{-1} S_2^{-1} \cdots S_{2n}^{-1}$ is an SR decomposition of A. The basic idea of the algorithm can be summarized as follows:

> let $S = I$
> let $R = A$
> for $j = 1$ to n
>   determine a symplectic matrix S_{2j-1} such that the jth column of $S_{2j-1}R$ is of the
>   desired form
>   set $S = SS_{2j-1}^{-1}$, $R = S_{2j-1}R$
>   determine a symplectic matrix S_{2j} such that the $(n+j)$th column of $S_{2j}R$ is of the
>   desired form
>   set $S = SS_{2j}^{-1}$, $R = S_{2j}R$

The entries $n + i$ to $2n$ of the ith column and the entries $n + i + 1$ to $2n$ of the $(n+i)$th column are eliminated using symplectic Givens matrices. The entries $i + 1$ to n of the ith column and the entries $i + 2$ to n of the $(n+i)$th column are eliminated using symplectic Householder matrices. The entry $(n + i + 1)$ of the $(n+i)$th column is eliminated using a symplectic Gauss matrix. This algorithm for computing the SR decomposition of an arbitrary matrix (as given in [38]) can be summarized as given in Table 2.5 (in MATLAB-like notation).

If at any stage $j \in \{1, \ldots, n-1\}$ the algorithm ends because of the stopping condition, then the $2j$th leading principal minor of $PA^T JAP^T$ is zero and A has no SR decomposition (see Theorem 2.7).

All but $(n - 1)$ transformations are orthogonal, which are known to be numerically stable transformations. Applying symplectic Gauss transformations for elimination, problems can arise not only because the algorithm may break down but also in those cases where we are near to such a breakdown. If we eliminate the jth nonzero entry of a vector x with a symplectic Gauss matrix L_j and x_{n-j-1} is very small relative to x_j, then the condition number $\kappa_2(L_j)$, here essentially given by

$$||L_j||_2 = (1 + v^2)^{1/2} + |v|,$$

where $v = -x_j/x_{n-j-1}$, will be very large. A transformation with L_j will then cause dramatic growth of the rounding errors in the result. Here we will always choose the symplectic Gauss matrix among all possible ones with optimal (smallest possible) condition number.

Algorithm: SR Decomposition

Given a $2n \times 2n$ matrix A compute a $2n \times 2n$ symplectic matrix S, and a $2n \times 2n$ J–triangular matrix R such that $A = SR$.

$S = I^{2n,2n}$
for $j = 1 : n$
    for $k = n : -1 : j$
        compute G_k such that $(G_k A)_{k+n,j} = 0$
        $A = G_k A$
        $S = SG_k^T$
    end
    if $j < n$
    then compute H_j such that $(H_j A)_{j+1:n,j} = 0$.
        $A = H_j A$
        $S = SH_j^T$
        for $k = n : -1 : j + 1$
            compute G_k such that $(G_k A)_{k+n,j+n} = 0$.
            $A = G_k A$
            $S = SG_k^T$
        end
        if $j < n - 1$
        then compute H_{j+1} such that $(H_{j+1} A)_{j+2:n,j+n} = 0$.
            $A = H_{j+1} A$
            $S = SH_{j+1}^T$
        end
        if $A(j + n, j + n) = 0$ and $A(j + 1, j + n) \neq 0$
        then stop, SR decomposition does not exist
        end
        compute L_{j+1} such that $(L_{j+1} A)_{j+1,j+n} = 0$.
        $A = L_{j+1} A$
        $S = SL_{j+1}^{-1}$
    end
end

Table 2.5. SR Decomposition

Using the SR decomposition the SR algorithm for an arbitrary $2n \times 2n$ matrix A is given as

```
let A₀ = A
for k = 1, 2, ...
    choose a shift polynomial pₖ
    compute the SR decomposition pₖ(Aₖ₋₁) = SₖRₖ
    compute Aₖ = Sₖ⁻¹Aₖ₋₁Sₖ
```

The SR decomposition of $p_k(A_{k-1})$ might not exist. As the set of the matrices, for which the SR decomposition does not exist, is of measure zero (Theorem 2.7), the polynomial p_k is discarded and an implicit SR step with a random shift is performed as proposed in [38] in context of the Hamiltonian SR algorithm. For an actual implementation this might be realized by checking the condition number of the Gauss transformation L_k needed in each step and performing an exceptional step if it exceeds a given tolerance.

REMARK 2.16 *How does a small perturbation of A influence the SR step? Will the SR step on $A + E$, where E is an error matrix with small norm, yield a transformed matrix $S(A + E)S^{-1}$ close to SAS^{-1}? How does finite-precision arithmetic influence the SR step? As in the SR algorithm nonorthogonal symplectic similarity transformations are employed, a backward error analysis would yield*

$$S(A + E)S^{-1} = SAS^{-1} + G$$

where $\|G\|_2 = \|SES^{-1}\|_2 \leq \kappa_2(S)\|E\|_2$. The condition number $\kappa_2(S)$ can be arbitrarily large. The QR algorithm does not have this problem. As in the QR algorithm only unitary similarity transformation are employed, an error analysis yields

$$Q(A + E)Q^T = QAQ^T + F$$

where $\|F\|_2 = \|QEQ^T\|_2 = \|E\|_2$. First-order componentwise and norm-wise perturbation bounds for the SR decomposition can be found in [42, 25].

The shift polynomials p_k are usually chosen according to the generalized Rayleigh-quotient strategy modified for the situation given here, that is p_k is chosen to be the characteristic polynomial of the trailing $m_i \times m_i$ submatrix of $PA_{i-1}P^T$ where P as in (2.1.2). A convergence proof can be deduced from the corresponding proof of convergence for general GR algorithms in [142].

THEOREM 2.17 *Let $A_0 \in \mathbf{R}^{2n \times 2n}$, and let p be a polynomial. Let λ_1, ..., λ_{2n} denote the eigenvalues of A_0, ordered so that $|p(\lambda_1)| \geq |p(\lambda_2)| \geq \ldots \geq |p(\lambda_{2n})|$. Suppose k is a positive integer less than $2n$ such that $|p(\lambda_k)| \geq |p(\lambda_{k+1})|$, let $\rho = |p(\lambda_{k+1})|/|p(\lambda_k)|$, and let (p_i) be a sequence of polynomials such that $p_i \to p$ and $p_i(\lambda_j) \neq 0$ for $j = 1, \ldots, k$ and all i. Let U be the invariant subspace of PA_0P^T associated with $\lambda_{k+1}, \ldots, \lambda_{2n}$, and suppose* span$\{e_1, \ldots, e_k\} \cap \mathcal{U} = \{0\}$ (P as in (2.1.2)). Let (A_i) be the sequence of iterates of the SR algorithm using these p_i, starting from A_0. If there exists a constant $\widehat{\kappa}$ such that the cumulative transformation matrices $\widehat{S}_i = S_1 S_2 \cdots S_i$ all satisfy $\kappa_2(\widehat{S}_i) \leq \widehat{\kappa}$, then (PA_iP^T) tends to block upper triangular form, in the following sense. Write

$$PA_iP^T = \left[\begin{array}{cc} X_{11}^{(i)} & X_{12}^{(i)} \\ X_{21}^{(i)} & X_{22}^{(i)} \end{array} \right],$$

where $X_{11}^{(i)} \in \mathbf{R}^{k \times k}$. Then for every $\widehat{\rho}$ satisfying $\rho < \widehat{\rho} < 1$ there exists a constant C such that $\| X_{21}^{(i)} \|_2 \leq C\widehat{\rho}^{\,i}$ for all i.

PROOF: See Theorem 6.2 in [142]. $\checkmark$

The condition $p_i(\lambda_j) \neq 0$ for $j = 1, \ldots, k$ may occasionally be violated. If $p_i(\lambda_j) = 0$, then $p_i(A_i)$ is singular. It can be shown that in this case, the eigenvalue λ_j can be deflated from the problem after the ith iteration. The theorem further implies that the eigenvalues of $X_{11}^{(i)}$ and $X_{22}^{(i)}$ converge to $\lambda_1, \ldots, \lambda_k$ and $\lambda_{k+1}, \ldots, \lambda_{2n}$, respectively.

REMARK 2.18 The condition span$\{e_1, \ldots, e_k\} \cap \mathcal{U} = \{0\}$ is automatically satisfied for all unreduced J–Hessenberg matrices. Suppose $x \in$ span$\{e_1, \ldots, e_k\}$ is nonzero. Let its last nonzero component be $x_r, r \leq k$. If A_0 has unreduced J–Hessenberg form, then PA_0P^T is an unreduced upper Hessenberg matrix. The last nonzero component of $PA_0P^T x$ is its $(r+1)$st, the last nonzero component of $PA_0^2P^T x$ is its $(r+2)$nd, and so on. It follows that $x, A_0x, A_0^2x, \ldots, A_0^m x$ are linearly independent, where $m = 2n - k$. Therefore the smallest invariant subspace of PA_0P^T that contains x has dimension at least $m + 1$. Since $\mathcal{U}$ is invariant under PA_0P^T and has dimension m, it follows that $x \notin \mathcal{U}$. Thus span$\{e_1, \ldots, e_k\} \cap \mathcal{U} = \{0\}$.

The following theorem indicates quadratic and cubic converge under certain circumstances.

THEOREM 2.19 Let $A_0 \in \mathbf{R}^{2n \times 2n}$ have distinct eigenvalues. Let (A_i) be the sequence generated by the SR algorithm starting from A_0, using the generalized Rayleigh-quotient shift strategy with polynomials of degree m. Suppose there is a constant $\widehat{\kappa}$ such that $\kappa_2(\widehat{S}_i) \leq \widehat{\kappa}$ for all i, and the PA_iP^T converge to block triangular form, in the sense described in Theorem 2.17, with $k = 2n - m$. Then the convergence is quadratic. Moreover, suppose that each of the iterates

$$PA_iP^T = \left[\begin{array}{cc} X_{11}^{(i)} & X_{12}^{(i)} \\ X_{21}^{(i)} & X_{22}^{(i)} \end{array} \right]$$

satisfies $||X_{12}^{(i)}|| = ||X_{21}^{(i)}||$ *for some fixed norm* $||\cdot||$. *Then the iterates converge cubically if they converge.*

PROOF: See Theorems 6.3 and 6.5 in [142]. $\checkmark$

The most glaring shortcoming associated with the above algorithm is that each step requires a full SR decomposition costing $\mathcal{O}((2n)^3)$ flops. Fortunately, the amount of work per iteration can be reduced by an order of magnitude if we first reduce the full matrix A to J–Hessenberg form as the SR algorithm preserves the J–Hessenberg form: If $p_k(A_{k-1})$ is nonsingular and $p_k(A_{k-1}) = S_k R_k$, then R_k is nonsingular as S_k is symplectic. Therefore,

$$
\begin{aligned}
A_k &= S_k^{-1} A_{k-1} S_k \\
&= R_k p_k(A_{k-1})^{-1} A_{k-1} p_k(A_{k-1}) R_k^{-1} \\
&= R_k A_{k-1} R_k^{-1}
\end{aligned}
$$

because $p_k(A_{k-1})$ and A_{k-1} commute. If A_{k-1} is of J–Hessenberg form, then so is A_k as A_k is a product of a J–Hessenberg matrix A_{k-1} and J-triangular matrices R_k, and R_k^{-1}. For singular $p_k(A_{k-1})$ one has to check the special form of S_k to see that A_k is of desired form if A_{k-1} is of J–Hessenberg form. In this case the problem can be split into two problems of smaller dimensions: If $\text{rank}(p_k(A_{k-1})) = 2n - 2\nu = 2j$, then the problem splits into a problem of size $2j \times 2j$ with J–Hessenberg form and a problem of size $2\nu \times 2\nu$ whose eigenvalues are exactly the shifts that are eigenvalues of A_{k-1} (that is, that are eigenvalues of A), see, e.g., [141, Section 4]. The SR decomposition of a J–Hessenberg matrix requires only $\mathcal{O}((2n)^2)$ flops to calculate as compared to $\mathcal{O}((2n)^3)$ flops for the SR decomposition of a full $2n \times 2n$ matrix. Hence, as the initial reduction to J–Hessenberg form is an $\mathcal{O}((2n)^3)$ process, a reasonable implementation of the SR algorithm should first reduce A to J–Hessenberg form.

Because of the essential uniqueness of the reduction to J–Hessenberg form, the SR algorithm can be performed without explicitly computing the decompositions $p_k(A_{k-1}) = S_k R_k$. In complete analogy to the GR algorithm, we can perform the SR step implicitly:

compute a symplectic matrix $\widetilde{S}_k$ such that $\widetilde{S}_k^{-1} p_k(A_{k-1}) e_1 = \alpha e_1$ for some $\alpha \in \mathbf{R}$
set $\widehat{A}_k = \widetilde{S}_k^{-1} A_{k-1} \widetilde{S}_k$
compute a symplectic matrix $\widehat{S}_k$ such that $\widehat{S}_k^{-1} \widehat{A}_k \widehat{S}_k$ is of J–Hessenberg form

The resulting J–Hessenberg matrix

$$
\widehat{S}_k^{-1} \widehat{A}_k \widehat{S}_k
$$

is essentially the same as

$$
S_k^{-1} A_{k-1} S_k,
$$

since $\widehat{S}_k = DS_k$ for some trivial matrix D (2.1.5).

Applying the first transformation $\widetilde{S}_k$ to the J–Hessenberg matrix A_{k-1} yields a matrix with almost J–Hessenberg form having a small bulge, that is there will be some additional entries in the upper left hand corner of each $n \times n$ block of $\widetilde{S}_k^{-1} A_{k-1} \widetilde{S}_k$. The remaining implicit transformations (that is, the computation of $\widehat{S}_k$) perform a bulge chasing sweep down the diagonal to restore the J–Hessenberg form.

Bunse-Gerstner and Mehrmann present in [38] an algorithm for reducing an arbitrary matrix to J–Hessenberg form. Depending on the size of the bulge in $\widetilde{S}_k^{-1} A_{k-1} \widetilde{S}_k$, the algorithm can be greatly simplified to reduce $\widetilde{S}_k^{-1} A_{k-1} \widetilde{S}_k$ to J–Hessenberg form. The algorithm uses the symplectic Givens transformations G_k, the symplectic Householder transformations H_k, and the symplectic Gauss transformations L_k introduced in Section 2.1.2. The basic idea of the algorithm can be summarized as follows:

> for $j = 1$ to n
> > determine a symplectic matrix S such that the jth column of $S^{-1}A$ is of the desired form
> > set $A = S^{-1}AS$
> > determine a symplectic matrix S such that the $(n + j)$th column of $S^{-1}A$ is of the desired form
> > set $A = S^{-1}AS$

In order to compute a symplectic matrix S such that the jth column of $S^{-1}A$ is of the desired form the following actions are taken. The entries $n+j$ to $2n$ of the jth column are eliminated using symplectic Givens matrices. The entries $j + 2$ to n of the jth column are eliminated using symplectic Householder matrices. The entry $(j + 1)$ of the jth column is eliminated using a symplectic Gauss matrix. Similar, in order to compute a symplectic matrix S such that the $(n + j)$th column of $S^{-1}A$ is of the desired form the following actions are taken. The entries $n + j$ to $2n$ of the $(n + j)$th column are eliminated using symplectic Givens matrices. The entries $j + 2$ to n of the $(n + j)$th column are eliminated using symplectic Householder matrices. This algorithm for computing the reduction of an arbitrary matrix to J–Hessenberg form (as given in [38]) can be summarized as given in Table 2.6 (in MATLAB-like notation).

REMARK 2.20 *The algorithm for reducing a $2n \times 2n$ arbitrary matrix to J–Hessenberg form as given in Table 2.6 requires about $47n^3$ flops. If the transformation matrix S is required, then $28n^3$ flops have to be added. This flop count is based on the fact that $n^2 - n$ symplectic Givens transformations, $n - 1$ symplectic Gauss transformations and 2 symplectic Householder transformation with $v \in \mathbf{R}^j$ for each $j = 2, \ldots, n - 1$ are used. The successively generated zeros in A are taken into account.*

Algorithm: Reduction to J–Hessenberg Form

Given a $2n \times 2n$ arbitrary matrix A compute its reduction to J–Hessenberg form. A will be overwritten by its J–Hessenberg form.

for $j = 1 : n - 1$
    for $k = n : -1 : j + 1$
        compute G_k such that $(G_k A)_{k+n,j} = 0$
        $A = G_k A G_k^T$
    end
    if $j < n - 1$
    then compute H_j such that $(H_j A)_{j+2:n,j} = 0$
        $A = H_j A H_j^T$
    end
    if $A(j+1,j) \neq 0$ and $A(n+j, n+j) = 0$
    then stop, reduction not possible
    end
    compute L_{j+1} such that $(L_{j+1} A)_{j+1,j} = 0$
    $A = L_{j+1} A L_{j+1}^{-1}$
    for $k = n : -1 : j + 1$
        compute G_k such that $(G_k A)_{n+k,n+j} = 0$
        $A = G_k A G_k^T$
    end
    if $j < n - 1$
    then compute H_j such that $(H_j A)_{j+2:n,n+j} = 0$
        $A = H_j A H_j^T$
    end
end

Table 2.6.   Reduction to J–Hessenberg Form

As in the reduction to J–Hessenberg form and in the SR algorithm only symplectic similarity transformations are employed, the J–Hessenberg form based SR algorithm preserves the symplectic structure. That is, if A is of symplectic J–Hessenberg form, then all iterates A_i of the SR algorithm are of symplectic J–Hessenberg form.

2.2.2   *SZ* ALGORITHM

The SZ algorithm is the analogue of the SR algorithm for the generalized eigenproblem, just as the QZ algorithm is the analogue of the QR algorithm for the generalized eigenproblem. Both are instances of the GZ algorithm [143].

The SZ algorithm is an iterative procedure that begins with a regular matrix pencil $A - \lambda B$, where B is nonsingular. It produces a sequence of equivalent matrix pencils $(A_i - \lambda B_i)$ that (hopefully) converge to a form exposing the eigenvalues. The transforming matrices for the equivalence transformations $A_i = S_i^{-1} A_{i-1} Z_i$ and $B_i = S_i^{-1} B_{i-1} Z_i$ are obtained from the SR decompositions $p_i(A_{i-1} B_{i-1}^{-1}) = S_i R_i$ and $p_i(B_{i-1}^{-1} A_{i-1}) = Z_i T_i$ in which p_i is a polynomial, R_i, T_i are J–triangular, and S_i, Z_i are symplectic. In the special case $B_{i-1} = I$ we have $Z_i = S_i$ and $T_i = R_i$. That is, the algorithm reduces to the generic SR algorithm for the standard eigenvalue problem.

If the given pencil is singular, the staircase algorithm of Van Dooren [135] can be used to remove the singular part (see also [45]). This algorithm also removes the infinite eigenvalue and its associated structure (which may be present if B is singular) and the zero eigenvalue and its associated structure (which may be present if A is singular). What is left is a regular pencil for which both A and B are nonsingular. Hence, for the discussion of the generic GZ algorithm Watkins and Elsner in [143] assume that the pencil is regular with a nonsingular B.

Using the results given in the last section, it follows immediately that both of the sequences $P(B_{i-1}^{-1} A_{i-1}) P^T$ and $P(A_{i-1} B_{i-1}^{-1}) P^T$ generated by the above described algorithm converge to (block) upper triangular form in the sense of Theorem 2.17, provided that the condition numbers of the accumulated transforming matrices $\widehat{S}_i = S_1 \cdots S_i$ and $\widehat{Z}_i = Z_1 \cdots Z_i$ remain bounded and the shifts converge as $i \to \infty$. More relevant is a statement about the convergence of the sequences (A_i) and (B_i) separately, since these are the matrices with which one actually works.

THEOREM 2.21 *Let $A_0, B_0 \in \mathbf{R}^{n \times n}$, and let p be a polynomial. Let λ_1, ..., λ_n denote the eigenvalues of $A_0 - \lambda B_0$, ordered so that $|p(\lambda_1)| \geq |p(\lambda_2)| \geq \ldots \geq |p(\lambda_n)|$. Suppose k is a positive integer less than n such that $|p(\lambda_k)| \geq |p(\lambda_{k+1})|$, let $\rho = |p(\lambda_{k+1})|/|p(\lambda_k)|$, and let (p_i) be a sequence of polynomials such that $p_i \to p$ and $p_i(\lambda_j) \neq 0$ for $j = 1, \ldots, k$ and all i. Let (U_d, U_r) be the deflating subspace of $A_0 - \lambda B_0$ associated with $\lambda_{k+1}, \ldots, \lambda_n$. Suppose $\text{span}\{e_1, \ldots, e_k\} \cap \text{ran}(U_d) = \{0\}$ and $\text{span}\{e_1, \ldots, e_k\} \cap \text{ran}(U_r) = \{0\}$. Let $(A_i - \lambda B_i)$ be the sequence of iterates of the SZ algorithm using these p_i, starting from $A_0 - \lambda B_0$. If there exists a constant $\widehat{\kappa}$ such that the cumulative transformation matrices $\widehat{S}_i = S_1 S_2 \cdots S_i$ and $\widehat{Z}_i = Z_1 \cdots Z_i$ all satisfy $\kappa_2(\widehat{S}_i) \leq \widehat{\kappa}$ and $\kappa_2(\widehat{Z}_i) \leq \widehat{\kappa}$, then $P(A_i - \lambda B_i) P^T$*

tends to block upper triangular form, in the following sense. Let C_i denote either A_i or B_i and write C_i as

$$PC_iP^T = \begin{bmatrix} X_{11}^{(i)} & X_{12}^{(i)} \\ X_{21}^{(i)} & X_{22}^{(i)} \end{bmatrix},$$

where $X_{11}^{(i)} \in \mathbf{R}^{k \times k}$. Then for every $\widehat{\rho}$ satisfying $\rho < \widehat{\rho} < 1$ there exists a constant M such that

$$\frac{\|X_{21}^{(i)}\|_2}{\|C_i\|_2} \leq M\widehat{\rho}^{\,i} \text{ for all } i.$$

PROOF: See Theorem 3.2 in [143]. ✓

Comments similar to those given after Theorem 2.17 apply here. The condition $p_i(\lambda_j) \neq 0$ may occasionally be violated. In this case the eigenvalue λ_i can be deflated from the problem. The eigenvalues of $A_{11}^{(i)} - \lambda B_{11}^{(i)}$ and $A_{22}^{(i)} - \lambda B_{22}^{(i)}$ converge to $\lambda_1, \ldots, \lambda_k$ and $\lambda_{k+1}, \ldots, \lambda_n$, respectively (here $A_{jj}^{(i)}$ denotes the (j, j) block of PA_iP^T and $B_{jj}^{(i)}$ denotes the (j, j) block of PB_iP^T).

If A_0 is an unreduced upper Hessenberg matrix and B_0 an upper triangular matrix, then the subspace conditions $\text{span}\{e_1, \ldots, e_k\} \cap \text{ran}(U_d) = \{0\}$ and $\text{span}\{e_1, \ldots, e_k\} \cap \text{ran}(U_r) = \{0\}$ are satisfied for all k. The argument is similar to that of Remark 2.18.

A natural way to choose the shift polynomials is to let p_i be the characteristic polynomial of the $m_i \times m_i$ lower right-hand corner pencil $A_{22}^{(i)} - \lambda B_{22}^{(i)}$. This will be called the *generalized Rayleigh-quotient shift strategy*. If the GZ algorithm converges under the conditions of Theorem 2.21, and generalized Rayleigh-quotient shifts with $m_i = n - k$ are used, the asymptotic convergence rate will be quadratic, provided that $A_0B_0^{-1}$ is simple (or nondefective).

Algorithms in the GZ family are usually implemented implicitly similar to the implicit implementation of a GR algorithm. Watkins and Elsner [143] analyze the implementation of GZ algorithms assuming that A_0 is upper Hessenberg and B_0 upper triangular. This implies that all iterates are of these forms as well. Their approach is much more involved than the usual approach used for deriving the QZ algorithm which invokes the Implicit-Q-Theorem (Theorem 2.2), but it gives a much clearer picture of the relationship between the explicit and implicit versions of the algorithm. In the following we only briefly review their results in the context of the SZ algorithm.

Using the SR decomposition the explicit SZ algorithm for an arbitrary $2n \times 2n$ matrix pencil $A - \lambda B$ is given as

let $A_0 = A, B_0 = B$
for $k = 1, 2, \ldots$

choose a shift polynomial p_k
compute the SR decomposition $p_k(A_{k-1}B_{k-1}^{-1}) = S_k R_k$
compute the SR decomposition $p_k(B_{k-1}^{-1}A_{k-1}) = T_k V_k$
compute $A_k = S_k^{-1}A_{k-1}T_k$ and $B_k = S_k^{-1}B_{k-1}T_k$

The shift polynomials p_k are usually chosen according to the generalized Rayleigh-quotient strategy modified for the situation given here, that is p_k is chosen to be the characteristic polynomial of the trailing $m_i \times m_i$ submatrix of $P(A_{i-1} - \lambda B_{i-1})P^T$ where P as in (2.1.2).

The SR decomposition of $p_k(A_{k-1}B_{k-1}^{-1})$ or $p_k(B_{k-1}^{-1}A_{k-1})$ might not exist. As the set of matrices, for which the SR decomposition does not exist, is of measure zero (Theorem 2.7), the polynomial p_k is discarded and an implicit SR shift with a random shift is performed as proposed in [38] in context of the Hamiltonian SR algorithm. For an actual implementation this might be realized by checking the condition number of the Gauss transformations needed to compute the SR decomposition.

The most glaring shortcoming associated with the above algorithm is that each step requires two full SR decomposition costing $\mathcal{O}((2n)^3)$ flops. As seen in the last section, the SR algorithm preserves the J–Hessenberg form. Hence it seems to be desirable to reduce A and B to forms such that AB^{-1} and $B^{-1}A$ are of J–Hessenberg form. Such a reduction has to be carried out using only symplectic transformations. As will be shown next, it is possible to simultaneously reduce A to J–Hessenberg form and B to J–triangular form using symplectic transformations. Then AB^{-1} and $B^{-1}A$ are of J–Hessenberg form and the amount of work per iteration of the SZ algorithm is reduced by an order of magnitude.

It is well-known (see, e.g., [58, Section 7.74]) that a matrix pencil can be reduced in a finite number of steps, using only orthogonal transformations, to a matrix pencil of the form

$$\boxed{\diagdown} - \lambda \boxed{\diagdown}.$$

Consider $A_P = PAP^T$ and $B_P = PBP^T$ (P as in (2.1.2)) and assume that we have transformed $A_P - \lambda B_P$ to the above form

$$Q_P^T(A_P - \lambda B_P)V_P = H - \lambda R$$

where H is of Hessenberg form, R is upper triangular and Q_P and V_P are orthogonal. Let $Q = P^T Q_P P, V = P^T V_P P$, $H_P = PHP^T$, and $R_P = PRP^T$. Then

$$Q^T(A - \lambda B)V = H_P - \lambda R_P$$

where H_P is of J–Hessenberg form and R_P of J–triangular form. Assume that the SR decomposition $V = S_V R_V$ and the RS decomposition $Q^T = R_Q S_Q$

exist. Then from $R_Q S_Q (A - \lambda B) S_V R_V = H_P - \lambda R_P$ we obtain

$$S_Q(A - \lambda B)S_V = R_Q^{-1}H_P R_V^{-1} - \lambda R_Q^{-1}R_P R_V^{-1} =: \widetilde{H} - \lambda \widetilde{R}.$$

As R_V, R_Q and R_P are of J–triangular form, and H_P is of J–Hessenberg form, using Lemma 2.6 yields that $\widetilde{H}$ is of J–Hessenberg form and $\widetilde{R}$ of J–triangular form. Hence almost all matrix pencils $A - \lambda B$ can be reduced to a pencil of the form

$$\begin{bmatrix} \searrow & \searrow \\ \searrow & \searrow \end{bmatrix} - \lambda \begin{bmatrix} \searrow & \searrow \\ \cdot_\circ & \searrow \end{bmatrix} \qquad (2.2.8)$$

using symplectic transformations. Moreover, $\widetilde{H}\widetilde{R}^{-1}$ and $\widetilde{R}^{-1}\widetilde{H}$ are of J–Hessenberg form. Therefore, the amount of work per iteration in the SZ algorithm can be reduced by an order of magnitude if we first reduce the full matrix A to J–Hessenberg form and the full matrix B to J–triangular form. Then AB^{-1} and $B^{-1}A$ are of J–Hessenberg form.

As in the SR algorithm we would like to avoid the actual forming of the matrices $p(AB^{-1})$ and $p(B^{-1}A)$. The basic idea of an implicit SZ algorithm can be given as follows:

> simultaneously reduce A to J–Hessenberg form A_0 and B to J–triangular form B_0
> > using symplectic transformations
> for $k = 1, 2, \ldots$
> > choose a shift polynomial p_k
> > compute a symplectic matrix $\check{S}_k$ such that $\check{S}_1^{-k}p_k(A_{k-1}B_{k-1}^{-1})e_1 = \alpha e_1$ for some $\alpha \in \mathbf{R}$
> > set $\widehat{A}_k = \check{S}_k^{-1}A_{k-1}$, $\widehat{B}_k = \check{S}_k^{-1}B_{k-1}$
> > compute symplectic matrices $\widehat{S}_k$ and $\widehat{T}_k$ such that $A_k = \widehat{S}_k^{-1}\widehat{A}_k\widehat{T}_k$ is of J–Hessenberg form and $B_k = \widehat{S}_k^{-1}\widehat{B}_k\widehat{T}_k$ is of J–triangular form

Although we have not presented an explicit algorithm for reducing $A - \lambda B$ to the form (2.2.8), it follows from the previous discussion that $\widetilde{S}_k = \widehat{S}_k\check{S}_k$ has the same first column as $\check{S}_k$. By the way the initial transformation is determined, we can apply the Implicit-S-Theorem (Theorem 2.9) and assert that $A_k B_k^{-1} = \widetilde{S}_k^{-1}A_{k-1}B_{k-1}^{-1}\widetilde{S}_k$ is essentially the same matrix that we would obtain by applying the SR iteration to $A_{k-1}B_{k-1}^{-1}$ directly. Hence there exists a J–triangular matrix $\widetilde{R}_k$ such that $p_k(A_{k-1}B_{k-1}^{-1}) = \widetilde{S}_k\widetilde{R}_k$. Define $\widehat{V}_k = \widehat{T}_k^{-1}p_k(B_{k-1}^{-1}A_{k-1})$, then $p_k(B_{k-1}^{-1}A_{k-1}) = \widehat{T}_k\widehat{V}_k$. We conclude

$$\begin{aligned} B_k\widehat{V}_k &= B_k\widehat{T}_k^{-1}p_k(B_{k-1}^{-1}A_{k-1}) \\ &= \widetilde{S}_k^{-1}B_{k-1}p_k(B_{k-1}^{-1}A_{k-1}) \\ &= \widetilde{S}_k^{-1}p_k(A_{k-1}B_{k-1}^{-1})B_{k-1} \\ &= \widetilde{R}_k B_{k-1}. \end{aligned}$$

As $\widetilde{R}_k, B_{k-1}$ and B_k are J–triangular, $\widehat{V}_k$ is J–triangular and $p_k(B_{k-1}^{-1}A_{k-1}) = \widehat{T}_k\widehat{V}_k$ is an SR decomposition of $p_k(B_{k-1}^{-1}A_{k-1})$.

Applying the first transformation $\widehat{S}_k$ to A and B yields matrices of almost J–Hessenberg (J–triangular, resp.) form having a small bulge, that is there will be some additional entries in the upper left hand corner of each $n \times n$ block of $\widehat{S}_k^{-1}A$ ($\widehat{S}_k^{-1}B$, resp.). The remaining implicit transformations perform a bulge chasing sweep to restore the J–Hessenberg and J–triangular forms.

We refrain from presenting an algorithm for reducing an arbitrary matrix pencil to a pencil of the form (2.2.8) as such an algorithm will not be used for our further discussion.

2.2.3 HR ALGORITHM

The HR algorithm [29, 33] is just like the SR algorithm a member of the family of GR algorithms [142] for calculating eigenvalues and invariant subspaces of matrices. Unlike the SR algorithm, the HR algorithm deals with matrices in $\mathbf{R}^{n \times n}$. Before we briefly introduce the HR algorithm, we recall some definitions from Definition 2.2. A signature matrix is a diagonal matrix $D = \text{diag}(d_1, \ldots, d_n)$ such that each $d_i \in \{1, -1\}$. Given a signature matrix D, we say that a matrix $A \in \mathbf{R}^{n \times n}$ is D–symmetric if $(DA)^T = DA$. Moreover, from Lemma 2.1 we know that a tridiagonal matrix T is D–symmetric for some D if and only if $|t_{i+1,i}| = |t_{i,i+1}|$ for $i = 1, \ldots, n-1$. Every unreduced tridiagonal matrix is similar to a D–symmetric matrix (for some D) by a diagonal similarity with positive main diagonal entries. D–symmetric tridiagonal matrices are generated by the unsymmetric Lanczos process (see Section 2.2.4), for example.

Almost every $A \in \mathbf{R}^{n \times n}$ has an HR *decomposition*

$$A = HU,$$

in which U is upper triangular, and H satisfies the hyperbolic property

$$H^T D H = \widehat{D},$$

where $\widehat{D}$ is another signature matrix [33]. For nonsingular A the HR decomposition is unique up to a signature matrix. We can make it unique by insisting that the upper triangular factor U satisfies $u_{ii} > 0$, $i = 1, \ldots, n$. The HR *algorithm* [29, 33] is an iterative process based on the HR decomposition. Choose a spectral transformation function p for which $p(A)$ is nonsingular, and form the HR decomposition of $p(A)$, if possible:

$$p(A) = HU.$$

Then use H to perform a similarity transformation on A to get the next iterate:

$$\widehat{A} = H^{-1}AH.$$

The HR algorithm has the following structure preservation property: If A is D–symmetric and $H^T DH = \widehat{D}$, then $\widehat{A}$ is $\widehat{D}$–symmetric. If A is also tridiagonal, then so is $\widehat{A}$. For a detailed discussion see [33, 29]. See also [32, 100].

2.2.4 LANCZOS ALGORITHM

In 1950, Lanczos [84] proposed a method for the successive reduction of a given, in general non-Hermitian, $n \times n$ matrix A to tridiagonal form. The Lanczos process generates a sequence of $m \times m$ tridiagonal matrices $T^{m,m}, m = 1, 2, ...$, which, in a certain sense, approximate A. Furthermore, in exact arithmetic and if no breakdown occurs, the Lanczos method terminates after at most $l \leq n$ steps with $T^{l,l}$ a tridiagonal matrix that represents the restriction of A or A^T to an A–invariant or A^T–invariant subspace of $\mathbb{C}^n$, respectively. In particular, all the eigenvalues of $T^{l,l}$ are also eigenvalues of A and, in addition, the method produces basis vectors for the A–invariant or A^T–invariant subspace found.

The Lanczos tridiagonalization is essentially the Gram-Schmidt bi-ortho-gonalization method for generating bi-orthogonal bases for a pair of Krylov subspaces

$$\{q, Aq, A^2 q, A^3 q, ...\} \quad \text{and} \quad \{p^T, p^T A, p^T A^2, p^T A^3, ...\}$$

$(p, q \in \mathbf{R}^n$ arbitrary). Applying the two-sided Gram-Schmidt process to the vectors $\{A^k q\}_{k \geq 0}$ and $\{p^T A^k\}_{k \geq 0}$, one arrives at a three term recurrence relation which, when $k = n$, represents a similarity transformation of the matrix A to tridiagonal form. The three term recurrence relation produces a sequence of vectors which can be viewed as forming the rows and columns, respectively, of rectangular matrices, $P^{n,k}$ and $Q^{n,k}$, such that after n steps, $P^{n,n}$ and $Q^{n,n}$ are $n \times n$ matrices with $Q^{n,n} = (P^{n,n})^{-1}$ and $P^{n,n} AQ^{n,n}$ is tridiagonal. At each step, an orthogonalization is performed, which requires a division by the inner product of (multiples of) the vectors produced at the previous step. Thus the algorithm suffers from breakdown and instability if any of these inner products is zero or nearly zero.

Let us be more precise. Given $p_1, q_1 \in \mathbf{R}^n$ and an unsymmetric matrix $A \in \mathbf{R}^{n \times n}$, the standard unsymmetric Lanczos algorithm produces matrices $P^{n,k} = [p_1, ..., p_k] \in \mathbf{R}^{n \times k}$ and $Q^{n,k} = [q_1, ..., q_k] \in \mathbf{R}^{n \times k}$ which satisfy the recursive identities

$$\begin{align} AP^{n,k} &= P^{n,k} T^{k,k} + \beta_{k+1} p_{k+1} e_k^T, & (2.2.9) \\ A^T Q^{n,k} &= Q^{n,k} (T^{k,k})^T + \gamma_{k+1} q_{k+1} e_k^T, & (2.2.10) \end{align}$$

where

$$
T^{k,k} = \begin{bmatrix}
\alpha_1 & \gamma_2 & & \\
\beta_2 & \ddots & \ddots & \\
& \ddots & \ddots & \gamma_k \\
& & \beta_k & \alpha_k
\end{bmatrix}
$$

is a truncated reduction of A. Generally, the elements β_j and γ_j are chosen so that $|\beta_j| = |\gamma_j|$ and $(Q^{n,k})^T P^{n,k} = I^{k,k}$ (bi-orthogonality). One pleasing result of this bi-orthogonality condition is that multiplying (2.2.9) on the left by $(Q^{n,k})^T$ yields the relationship $(Q^{n,k})^T A P^{n,k} = T^{k,k}$.

Encountering a zero $\beta_{k+1} p_{k+1}$ or $\gamma_{k+1} q_{k+1}$ in the Lanczos iteration is a welcome event in that it signals the computation of an exact invariant subspace. If $\beta_{k+1} p_{k+1} = 0$, then the iteration terminates and span$\{p_1, \ldots, p_k\}$ is an invariant subspace for A. If $\gamma_{k+1} q_{k+1} = 0$, then the iteration also terminates and span$\{q_1, \ldots, q_k\}$ is an invariant subspace for A^T. If neither condition is true and $q_{k+1}^T p_{k+1} = 0$, then the tridiagonalization process ends without any invariant subspace information. This is called a *serious breakdown*. However, an exact zero or even a small $\beta_{k+1} p_{k+1}$ or $\gamma_{k+1} q_{k+1}$ is a rarity in practice. Nevertheless, the extremal eigenvalues of $T^{k,k}$ turn out to be surprisingly good approximations to A's extremal eigenvalues. Hence, the successive tridiagonalization by the Lanczos algorithm combined with a suitable method for computing the eigenvalues of the resulting tridiagonal matrices is an appropriate iterative method for solving large and sparse eigenproblem, if only some of the eigenvalues are sought. As the resulting tridiagonal matrices are sign-symmetric, the HR or the LU algorithm are appropriate QR like methods for computing the eigenvalues and invariant subspaces, as they preserve the special structure.

Yet, in practice, there are a number of difficulties associated with the Lanczos algorithm. At each step of the unsymmetric Lanczos tridiagonalization, an orthogonalization is performed, which requires a division by the inner product of (multiples of) the vectors produced at the previous step. Thus the algorithm suffers from breakdown and instability if any of these inner products is zero or close to zero. It is known [74] that vectors q_1 and p_1 exist so that the Lanczos process with these as starting vectors does not encounter breakdown. However, determining these vectors requires knowledge of the minimal polynomial of A. Further, there are no theoretical results showing that p_1 and q_1 can be chosen such that small inner products can be avoided. Thus, no algorithm for successfully choosing p_1 and q_1 at the start of the computation yet exists.

In theory, the three-term recurrences in (2.2.9) and (2.2.10) are sufficient to guarantee $(Q^{n,k})^T P^{n,k} = I^{k,k}$. It is known [108] that bi-orthogonality will in fact be lost when at least one of the eigenvalues of $T^{k,k}$ converges to an eigenvalue of A. In order to overcome this problem, re-bi-orthogonalization of the vectors q_j and p_j is necessary. Different strategies have been proposed for this, e.g., complete or selective re-bi-orthogonalization. For a more detailed discussion on the various aspects of the difficulties of the Lanczos method in the context of computing some eigenvalues of large and sparse matrices see, e.g., [123] and the references therein.

It is possible to modify the Lanczos process so that it skips over exact breakdowns. A complete treatment of the modified Lanczos algorithm and its intimate connection with orthogonal polynomials and Padé approximation was presented by Gutknecht [68, 69]. Taylor [134] and Parlett, Taylor, and Liu [116] were the first to propose a look-ahead Lanczos algorithm that skips over breakdowns and near-breakdowns. The price paid is that the resulting matrix is no longer tridiagonal, but has a small bulge in the tridiagonal form to mark each occurrence of a (near) breakdown. Freund, Gutknecht, and Nachtigal presented in [57] a look-ahead Lanczos code that can handle look-ahead steps of any length.

A different approach to deal with the inherent difficulties of the Lanczos process is to modify the starting vectors by an implicitly restarted Lanczos process (see the fundamental work in [41, 132]; for the unsymmetric eigenproblem the implicitly restarted Arnoldi method has been implemented very successfully, see [94]). The problems are addressed by fixing the number of steps in the Lanczos process at a prescribed value k which is dependent on the required number of approximate eigenvalues. J–orthogonality of the k Lanczos vectors is secured by re–J–orthogonalizing these vectors when necessary. The purpose of the implicit restart is to determine initial vectors such that the associated residual vectors are tiny. Given that $P^{n,k}$ and $Q^{n,k}$ from (2.2.9) and (2.2.10) are known, an implicit Lanczos restart computes the Lanczos factorization

$$A\widetilde{P}^{n,k} \;=\; \widetilde{P}^{n,k}\widetilde{T}^{k,k} + \widetilde{r}_k e_k^T, \tag{2.2.11}$$

$$A^T\widetilde{Q}^{n,k} \;=\; \widetilde{Q}^{n,k}(\widetilde{T}^{k,k})^T + \widetilde{q}_k e_k^T, \tag{2.2.12}$$

which corresponds to the starting vectors

$$\widetilde{p}_1 = \rho_p(A - \mu I)p_1, \qquad \widetilde{q}_1 = \rho_q(A^T - \mu I)q_1, \tag{2.2.13}$$

without explicitly restarting the Lanczos process with the vectors in (2.2.13). For a detailed derivation see [67] and the related work in [41, 132].

Chapter 3

THE BUTTERFLY FORM FOR SYMPLECTIC MATRICES AND MATRIX PENCILS

The SR algorithm preserves the symplectic structure of a matrix during its iterations. Applying an implicit SR step to a full symplectic matrix requires $\mathcal{O}(n^3)$ arithmetic operations implying a computational cost of $\mathcal{O}(n^4)$ for the SR algorithm. Here we will consider reduced symplectic matrix forms which are preserved by the SR algorithm and allow for a faster implementation of the implicit SR step. As already explained in Section 2.2.1, the amount of work per iteration can be reduced by an order of magnitude if first the full matrix is reduced to J–Hessenberg form since the SR algorithm preserves the J–Hessenberg form. This approach has been considered in [53]. There it is also noted that symplectic J–Hessenberg matrices can be characterized by $4n - 1$ parameters. The SR algorithm can be rewritten to work only with these parameters instead of the $2n^2 + 3n - 1$ nonzero matrix elements of a symplectic J–Hessenberg matrix. This results in $\mathcal{O}(n)$ arithmetic operations per implicit SR step. Here we will consider a different condensed form, the symplectic butterfly form, introduced by Banse and Bunse-Gerstner in [15, 13, 12, 14]. This form is preserved under the SR iteration and also allows the implementation of an implicit SR step in $\mathcal{O}(n)$ arithmetic operations. Numerical experiments show that the SR algorithm based on the symplectic butterfly form has better numerical properties than one based on the symplectic J–Hessenberg form.

The reduced symplectic forms considered here can be motivated analogously to the Schur parameter form of a unitary matrix [3, 61]. Any unitary matrix H can be transformed to a unitary upper Hessenberg matrix $\widetilde{H}$ by a unitary similarity transformation. Then premultiplying $\widetilde{H}$ successively by suitable unitary matrices

$$G_k^H = \text{diag}(I^{k-1,k-1}, \begin{bmatrix} -\gamma_k & \sigma_k \\ \sigma_k & \overline{\gamma_k} \end{bmatrix}, I^{n-k-1,n-k-1}),$$

$|\gamma_k|^2 + \sigma_k^2 = 1$, for $k = 1, \ldots, n$, the subdiagonal elements of $\widetilde{H}$ can be eliminated such that $G_n^H \cdots G_1^H \widetilde{H}$ is an upper triangular matrix with 1's on the diagonal. Since $G_n^H \cdots G_1^H \widetilde{H}$ is still unitary, it has to be the identity, that is, $\widetilde{H} = G_1 \cdots G_n$. Hence, $\widetilde{H}$ is uniquely determined by the n complex parameters γ_k. This can be exploited to derive fast algorithms for the unitary eigenproblem: a QR algorithm for unitary Hessenberg matrices [61, 64, 65], a divide and conquer algorithm [6, 62, 63], a bisection algorithm [37], an Arnoldi algorithm [60, 75, 36]. See also [2, 4, 28, 35, 48, 122].

Recalling that symplectic matrices are J–orthogonal, one is lead to try to factorize a symplectic matrix having a J–Hessenberg form into the product of J–orthogonal elementary matrices. Motivated by this idea, Banse and Bunse-Gerstner [15, 13, 12, 14] present a new condensed form for symplectic matrices which can be depicted as a symplectic matrix of the following form:

The $2n \times 2n$ condensed matrix is symplectic, contains $8n - 4$ nonzero entries, and is determined by $4n - 1$ parameters. This condensed form, called *symplectic butterfly form*, is no longer of J–Hessenberg form. But as observed in [13], the SR iteration preserves the symplectic butterfly form. The symplectic structure, which will be destroyed in the numerical process due to roundoff errors, can easily be restored in each iteration for this condensed form. There is reason to believe that an SR algorithm based on the symplectic butterfly form has better numerical properties than one based on the symplectic J–Hessenberg form; see Section 4.1.

Section 3.1 briefly summarizes the results on parameterized symplectic J–Hessenberg matrices presented in [53]. In Section 3.2 we will introduce symplectic butterfly matrices. Further, we will show that unreduced symplectic butterfly matrices have properties similar to those of unreduced J–Hessenberg matrices in the context of the SR algorithm. The $4n - 1$ parameters that determine a symplectic butterfly matrix B cannot be read off of B directly. Computing the parameters can be interpreted as factoring B into the product of two even simpler matrices K and N: $B = K^{-1}N$. The parameters can then be read off of K and N directly. Up to now two different ways of factoring symplectic butterfly matrices have been proposed in the literature [12, 19]. We will introduce these factorizations and consider their drawbacks and advantages.

The reduction of a symplectic matrix to butterfly form is unique up to a trivial factor. Making use of this fact, in Section 3.3 a canonical form for symplectic butterfly matrices is introduced which is helpful for certain theoretical discussions. Finally, in Section 3.4 condensed forms for symplectic matrix pencils are presented which are closely related to the symplectic butterfly form of a symplectic matrix.

Some of the results discussed in this chapter appeared in [19, 20, 21].

3.1 PARAMETERIZED SYMPLECTIC J–HESSEN-BERG FORM

In [53], Flaschka, Mehrmann and Zywietz consider symplectic J–Hessenberg matrices. These are symplectic matrices of the form

$$M = \begin{bmatrix} \diagdown & \diagdown \\ \diagdown & \diagdown \end{bmatrix} = \begin{bmatrix} M_{11} & M_{12} \\ M_{21} & M_{22} \end{bmatrix}, \qquad (3.1.1)$$

where $M_{11}, M_{21}, M_{22} \in \mathbf{R}^{n \times n}$ are upper triangular matrices and $M_{12} \in \mathbf{R}^{n \times n}$ is an upper Hessenberg matrix. As discussed in Section 2.2.1, the SR iteration preserves this form at each step and is supposed to converge to a form from which the eigenvalues can be read off. In the context of the SR algorithm, unreduced J–Hessenberg matrices have properties similar to those of unreduced Hessenberg matrices in the context of the QR algorithm. Like the QR step, an efficient implementation of the SR step for J–Hessenberg matrices requires $\mathcal{O}(n^2)$ arithmetic operations; hence no gain in efficiency is obtained compared to the standard Hessenberg QR algorithm. Further, the authors report the loss of the symplectic structure due to roundoff errors after only a few SR steps. As a symplectic J–Hessenberg matrix looks like a general J–Hessenberg matrix, it is not easy to check and to guarantee that the structure is kept invariant in the presence of roundoff errors.

In [53] it is also shown that any symplectic J–Hessenberg matrix is determined uniquely by $4n - 1$ parameters: the diagonal elements of M_{11}, the diagonal and subdiagonal elements of M_{12} and the diagonal elements of M_{21} (assuming that M_{11}^{-1} exists). In order to find such a parameterization of symplectic J–Hessenberg matrices the authors need the following proposition.

PROPOSITION 3.1 *Let* $M = \begin{bmatrix} M_{11} & M_{12} \\ M_{21} & M_{22} \end{bmatrix} \in \mathbf{R}^{2n \times 2n}$ *be a symplectic matrix.*

a) *If* M_{11} *is invertible, then* M *has the decomposition*

$$M = \begin{bmatrix} I & 0 \\ M_{21}M_{11}^{-1} & I \end{bmatrix} \begin{bmatrix} M_{11} & 0 \\ 0 & M_{11}^{-T} \end{bmatrix} \begin{bmatrix} I & M_{11}^{-1}M_{12} \\ 0 & I \end{bmatrix}$$

where $M_{21}M_{11}^{-1}$ *and* $M_{11}^{-1}M_{12}$ *are symmetric. All three factors are symplectic.*

b) *If* M_{22} *is invertible, then* M *has the decomposition*

$$M = \begin{bmatrix} I & M_{12}M_{22}^{-1} \\ 0 & I \end{bmatrix} \begin{bmatrix} M_{22}^{-T} & 0 \\ 0 & M_{22} \end{bmatrix} \begin{bmatrix} I & 0 \\ M_{22}^{-1}M_{21} & I \end{bmatrix}$$

where $M_{12}M_{22}^{-1}$ and $M_{22}^{-1}M_{21}$ are symmetric. All three factors are symplectic.

PROOF: See [103].   ✓

Making use of this three factor decomposition of symplectic matrices, the authors show that a symplectic J–Hessenberg matrix depends uniquely on $4n - 1$ parameters.

THEOREM 3.2 *Let $M = \begin{bmatrix} M_{11} & M_{12} \\ M_{21} & M_{22} \end{bmatrix} \in \mathbf{R}^{2n \times 2n}$ be a symplectic matrix. Let M_{11} or M_{22} be invertible. Then M depends uniquely on $4n - 1$ parameters.*

PROOF: See [53, Theorem 2.7].   ✓

One possible parameterization of a symplectic J–Hessenberg matrix (assuming that M_{11} is regular) is given by the diagonal elements of M_{11}, the diagonal and subdiagonal elements of M_{12} and the diagonal elements of M_{21}.
Further, the following important result is proved in [53].

THEOREM 3.3 *Let $M = \begin{bmatrix} M_{11} & M_{12} \\ M_{21} & M_{22} \end{bmatrix} \in \mathbf{R}^{2n \times 2n}$ be a symplectic J–Hessenberg matrix. Then for $1 \le i \le j \le n$,*

$$\begin{bmatrix} (M_{11})_{i:j,i:j} & (M_{12})_{i:j,i:j} \\ (M_{21})_{i:j,i:j} & (M_{22})_{i:j,i:j} \end{bmatrix}$$

is also a symplectic J–Hessenberg matrix, where $(M_{pq})_{i:j,i:j}$ is the submatrix of M_{pq} consisting of the rows and columns $i, i + 1, \ldots, j$.

PROOF: See [53, Theorem 2.10].   ✓

The SR algorithm can be modified to work only with the $4n - 1$ parameters instead of the $2n^2 + 3n - 1$ nonzero matrix elements. Thus only $\mathcal{O}(n)$ arithmetic operations per SR step are needed compared to $\mathcal{O}(n^2)$ arithmetic operations when working on the actual J–Hessenberg matrix. In [53], the authors give the details of such an SR step for the single shift case. They note that the algorithm "...*forces the symplectic structure, but it has the disadvantage that it needs $4n - 1$ terms to be nonzero in each step, which makes it highly numerically unstable. ... Thus, so far, this algorithm is mainly of theoretical value.*" [53, page 186, last paragraph].

3.2   THE SYMPLECTIC BUTTERFLY FORM

Recently, Banse and Bunse-Gerstner [15, 13, 12, 14] presented a new condensed form for symplectic matrices. The $2n \times 2n$ condensed matrix is symplectic, contains $8n - 4$ nonzero entries, and is determined by $4n - 1$ parameters.

This condensed form, called *symplectic butterfly form*, is defined as a symplectic matrix of the following form:

$$\begin{bmatrix} B_{11} & B_{12} \\ B_{21} & B_{22} \end{bmatrix} = \begin{bmatrix} \diagdown & \diagdown\!\diagdown \\ \diagdown & \diagdown\!\diagdown \end{bmatrix}, \tag{3.2.2}$$

where $B_{ij} \in \mathbf{R}^{n \times n}$, B_{11} and B_{21} are diagonal, and B_{12} and B_{22} are tridiagonal.

The symplectic butterfly form can be motivated analogously to the Schur parameter form of a unitary matrix. Let us assume that the symplectic matrix M can be reduced to J–Hessenberg form

$$S_1^{-1} M S_1 = \begin{bmatrix} \diagdown\!\!\search & \diagdown \\ \diagdown & \diagdown \end{bmatrix}, \tag{3.2.3}$$

and further that we can successively eliminate the element $m_{k+n,k}$ and the elements $m_{k,k+n}$ and $m_{k+1,k+n}$ for $k = 1, \ldots, n$. The $(k + n, k)$th element is eliminated by premultiplying a suitable symplectic elimination matrix U_k that in addition normalizes the (k, k)th element to 1. The elements $(k, k + n)$ and $(k + 1, k + n)$ are eliminated by premultiplying a suitable symplectic matrix V_k^{-1}, that is

$$V_n^{-1} U_n \cdots V_1^{-1} U_1 M = \begin{bmatrix} \diagdown & \diagdown \\ \diagdown & \diagdown \end{bmatrix}.$$

Since M, U_k, V_k^{-1} are symplectic, their product is symplectic and hence the above matrix has to be the identity. Thus

$$M = U_1^{-1} V_1 \cdots U_n^{-1} V_n. \tag{3.2.4}$$

One possible choice of suitable symplectic U_k and V_k, $k = 1, \ldots, n$, is

$$U_k = \left[\begin{array}{ccc|ccc} I^{k-1,k-1} & & & & & \\ & a_k & & & & \\ & & I^{n-k,n-k} & & & \\ \hline & & & I^{k-1,k-1} & & \\ & b_k & & & a_k^{-1} & \\ & & & & & I^{n-k,n-k} \end{array} \right],$$

$$
V_k \;=\; \left[\begin{array}{ccc|ccc}
I^a & & & & & \\
& 1 & & & c_k & d_{k+1} \\
& & 1 & & d_{k+1} & \\
& & & I^b & & \\
\hline
& & & I^a & & \\
& & & & 1 & \\
& & & & & 1 \\
& & & & & & I^b
\end{array}\right],
$$

where $I^a = I^{k-1,k-1}$ and $I^b = I^{n-k-1,n-k-1}$, and

$$
V_n \;=\; \left[\begin{array}{cc|cc}
I^{n-1,n-1} & & & \\
& 1 & & c_n \\
\hline
& & I^{n-1,n-1} & \\
& & & 1
\end{array}\right].
$$

Because of their special structure most of the U_k and V_k commute:

$$
\begin{array}{rcll}
U_k^{-1}U_m^{-1} &=& U_m^{-1}U_{k.}^{-1} & \text{for all } k, m, \\
V_k V_m &=& V_m V_k & \text{for all } k, m, \\
U_q^{-1}V_p &=& V_p U_q^{-1} & \text{for all } q < p \text{ or } q \geq p + 2.
\end{array} \qquad (3.2.5)
$$

Similar to the rearrangement of the factors of the Schur parameter form of a unitary upper Hessenberg matrix $H = G_1 G_2 \cdots G_n$ to obtain a Schur parameter pencil $G_o - \lambda G_e$ with $G_o = G_1 G_3 G_5 \cdots$, $G_e = G_2 G_4 \cdots$ in [35], the commuting properties (3.2.5) can be used to rearrange the factors in (3.2.4) to obtain a symplectic matrix S_2 such that

$$
\begin{aligned}
S_2^{-1} M S_2 &= U_1^{-1} \cdots U_n^{-1} V_n \cdots V_1 \\[4pt]
&= \begin{bmatrix} \diagdown \\ & \diagdown \end{bmatrix}\begin{bmatrix} I & \diagdown \\ & I \end{bmatrix} \\[4pt]
&= \begin{bmatrix} \diagdown & \diagdown \\ & \diagdown \end{bmatrix}.
\end{aligned} \qquad (3.2.6)
$$

Thus, $S_2^{-1} M S_2$ is in butterfly form. Combining (3.2.3) and (3.2.6), we have found a symplectic matrix $S = S_1 S_2 \in \mathbf{R}^{2n \times 2n}$ that transforms a symplectic matrix M into butterfly form.

Instead of U_k, e.g., a symplectic Givens rotation G_k can be used (see [15]) yielding a slightly different form of the factorization

$$
\begin{bmatrix} \diagdown & \diagdown \\ & \diagdown \end{bmatrix}\begin{bmatrix} \diagdown & \diagdown \\ & \diagdown \end{bmatrix} = \begin{bmatrix} \diagdown & \diagdown \\ & \diagdown \end{bmatrix}.
$$

In order to state existence and uniqueness theorems for the reduction of a symplectic matrix to butterfly form, let us introduce unreduced butterfly matrices. An *unreduced butterfly matrix* is a butterfly matrix $B = \begin{bmatrix} B_{11} & B_{12} \\ B_{21} & B_{22} \end{bmatrix}$ in which the tridiagonal matrix B_{22} is unreduced. Unreduced butterfly matrices play a role analogous to that of unreduced Hessenberg/J–Hessenberg matrices in the standard QR/SR theory. They have the following property.

LEMMA 3.4 *If B is an unreduced butterfly matrix, then B_{21} is nonsingular and B can be factored as*

$$
\begin{bmatrix} B_{11} & B_{12} \\ B_{21} & B_{22} \end{bmatrix} = \begin{bmatrix} B_{21}^{-1} & B_{11} \\ 0 & B_{21} \end{bmatrix} \begin{bmatrix} 0 & -I \\ I & B_{21}^{-1}B_{22} \end{bmatrix} \tag{3.2.7}
$$

$$
= \begin{bmatrix} \diagdown & \diagdown \\ 0 & \diagdown \end{bmatrix} \begin{bmatrix} 0 & -I \\ I & \diagdown \end{bmatrix}.
$$

This factorization is unique. Note that $B_{21}^{-1}B_{22}$ is symmetric.

PROOF: The fact that B is symplectic implies $B_{11}B_{22} - B_{21}B_{12} = I$. Assume that B_{21} is singular, that is $(B_{21})_{jj} = 0$ for some j. Then the jth row of $B_{11}B_{22} - B_{21}B_{12} = I$ gives

$$(B_{11})_{jj}(B_{22})_{j,j-1} = 0, \quad (B_{11})_{jj}(B_{22})_{jj} = 1, \quad (B_{11})_{jj}(B_{22})_{j,j+1} = 0.$$

This can only happen for $(B_{11})_{jj} \neq 0$, $(B_{22})_{jj} \neq 0$, and $(B_{22})_{j,j-1} = (B_{22})_{j,j+1} = 0$, but B_{22} is unreduced. Hence, B_{21} has to be nonsingular if B_{22} is unreduced. Thus, for an unreduced butterfly matrix we obtain

$$
\begin{bmatrix} B_{21} & -B_{11} \\ 0 & B_{21}^{-1} \end{bmatrix} \begin{bmatrix} B_{11} & B_{12} \\ B_{21} & B_{22} \end{bmatrix} = \begin{bmatrix} 0 & -I \\ I & B_{21}^{-1}B_{22} \end{bmatrix}.
$$

As both matrices on the left are symplectic, their product is symplectic and hence $B_{21}^{-1}B_{22}$ has to be a symmetric tridiagonal matrix. Thus

$$
\begin{bmatrix} B_{11} & B_{12} \\ B_{21} & B_{22} \end{bmatrix} = \begin{bmatrix} B_{21}^{-1} & B_{11} \\ 0 & B_{21} \end{bmatrix} \begin{bmatrix} 0 & -I \\ I & B_{21}^{-1}B_{22} \end{bmatrix}
$$

$$
= \begin{bmatrix} \diagdown & \diagdown \\ 0 & \diagdown \end{bmatrix} \begin{bmatrix} 0 & -I \\ I & \diagdown \end{bmatrix}.
$$

The uniqueness of this factorization follows from the choice of signs of the identities. $\checkmark$

We will frequently make use of this decomposition and will denote it by

$$
K_u^{-1} := \begin{bmatrix} B_{21}^{-1} & B_{11} \\ 0 & B_{21} \end{bmatrix} = \left[\begin{array}{ccc|ccc} a_1^{-1} & & & b_1 & & \\ & \ddots & & & \ddots & \\ & & a_n^{-1} & & & b_n \\ \hline & & & a_1 & & \\ & & & & \ddots & \\ & & & & & a_n \end{array} \right] , \quad (3.2.8)
$$

$$
N_u := \begin{bmatrix} 0 & -I \\ I & B_{21}^{-1} B_{22} \end{bmatrix}
$$

$$
= \left[\begin{array}{ccc|ccc} & & & -1 & & \\ & & & & \ddots & \\ & & & & & -1 \\ \hline 1 & & & c_1 & d_2 & \\ & \ddots & & d_2 & \ddots & \ddots \\ & & 1 & & \ddots & \ddots & d_n \\ & & & & & d_n & c_n \end{array} \right] . \quad (3.2.9)
$$

Hence $B = K_u^{-1} N_u$ is given by

$$
\left[\begin{array}{cccc|cccc} b_1 & & & & b_1 c_1 - a_1^{-1} & b_1 d_2 & & \\ & \ddots & & & & b_2 d_2 & \ddots & \ddots \\ & & \ddots & & & & \ddots & \ddots & b_{n-1} d_n \\ & & & b_n & & & b_n d_n & b_n c_n - a_n^{-1} \\ \hline a_1 & & & & a_1 c_1 & a_1 d_2 & & \\ & \ddots & & & & a_2 d_2 & \ddots & \ddots \\ & & \ddots & & & & \ddots & \ddots & a_{n-1} d_n \\ & & & a_n & & & a_n d_n & a_n c_n \end{array} \right] . \quad (3.2.10)
$$

From (3.2.8) – (3.2.10) we obtain the following corollary.

COROLLARY 3.5 *Any unreduced butterfly matrix $B \in \mathbf{R}^{2n \times 2n}$ can be represented by $4n - 1$ parameters $a_1, \ldots, a_n, d_2, \ldots, d_n \in \mathbf{R} \setminus \{0\}, b_1, \ldots, b_n, c_1, \ldots, c_n \in \mathbf{R}$. Of these, $2n - 1$ parameters have to be nonzero.*

REMARK 3.6 *We will have deflation if $d_j = 0$ for some j. Then the eigenproblem can be split into two smaller ones with symplectic butterfly matrices.*

Moreover, any submatrix of B (3.2.10) of the form

$$B_{k:\ell} := \left[\begin{array}{cccc|cccc} b_k & & & & b_kc_k - a_k^{-1} & b_kd_{k+1} & & \\ & \ddots & & & & b_{k+1}d_{k+1} & \ddots & \ddots \\ & & \ddots & & & & \ddots & \ddots & b_{\ell-1}d_\ell \\ & & & b_\ell & & & & b_\ell d_\ell & b_\ell c_\ell - a_\ell^{-1} \\ \hline a_k & & & & a_kc_k & & a_kd_{k+1} & \\ & \ddots & & & & a_{k+1}d_{k+1} & \ddots & \ddots \\ & & \ddots & & & & \ddots & \ddots & a_{\ell-1}d_\ell \\ & & & a_\ell & & & & a_\ell d_\ell & a_\ell c_\ell \end{array}\right]$$

is a symplectic butterfly matrix. If B is unreduced, then so is $B_{k:\ell}$.

Recall the definition of a generalized Krylov matrix (Definition 2.3)

$$L(A, v, j) = [v, A^{-1}v, A^{-2}v, \ldots, A^{-(j-1)}v, Av, A^2v, \ldots, A^jv]$$

for $A \in \mathbf{R}^{2n \times 2n}, v \in \mathbf{R}^{2n}$. Now we can state the uniqueness and existence theorem for the reduction to butterfly form.

THEOREM 3.7 (IMPLICIT-S-THEOREM) *Let M and S be $2n \times 2n$ symplectic matrices and denote by s_1 the first column of S.*

a) *Let $L(M, s_1, n)$ be nonsingular. If $L(M, s_1, n) = SR$ is an SR decomposition, then $S^{-1}MS$ is an unreduced butterfly matrix.*

b) *If $S^{-1}MS = B$ is a symplectic butterfly matrix, then $L(M, s_1, n)$ has an SR decomposition $L(M, s_1, n) = SR$. If B is unreduced, then R is nonsingular.*

c) *Let $S, \widetilde{S} \in \mathbf{R}^{2n \times 2n}$ be symplectic matrices such that $S^{-1}MS = B$ and $\widetilde{S}^{-1}M\widetilde{S} = \widetilde{B}$ are butterfly matrices. Then there exists a trivial matrix D (2.1.5) such that $S = \widetilde{S}D$ and $B = D\widetilde{B}D^{-1}$.*

PROOF:

a) Since $L(M, s_1, n)$ is nonsingular and $L(M, s_1, n) = SR$, R is nonsingular as S is symplectic. Furthermore, as $L(M, s_1, n)$ is nonsingular, its columns span a basis of $\mathbf{R}^{2n}$. Hence, there exist $c_1, \ldots, c_{2n}, d_1, \ldots, d_{2n} \in \mathbf{R}$ such that

$$M^{n+1}s_1 = c_1s_1 + c_2M^{-1}s_1 + \cdots + c_nM^{-(n-1)}s_1$$
$$+ c_{n+1}Ms_1 + \cdots + c_{2n}M^ns_1$$

and

$$
\begin{aligned}
M^{-n}s_1 \;=\;\; & d_1 s_1 + d_2 M^{-1}s_1 + \cdots + d_n M^{-(n-1)}s_1 \\
& + d_{n+1}M s_1 + \cdots + d_{2n}M^n s_1.
\end{aligned}
$$

Let

$$
C := \begin{bmatrix}
0 & 1 & & & & & & c_1 \\
& \ddots & \ddots & & & & & \vdots \\
& & \ddots & 1 & & & & \vdots \\
& & & 0 & & & & c_n \\
\hline
1 & & & & 0 & & & c_{n+1} \\
& & & & 1 & \ddots & & \vdots \\
& & & & & \ddots & 0 & \vdots \\
& & & & & & 1 & c_{2n}
\end{bmatrix},
$$

$$
D := \begin{bmatrix}
0 & & & d_1 & 1 & & & \\
1 & \ddots & & \vdots & & & & \\
& \ddots & 0 & \vdots & & & & \\
& & 1 & d_n & & & & \\
\hline
& & & d_{n+1} & 0 & 1 & & \\
& & & \vdots & & \ddots & \ddots & \\
& & & \vdots & & & \ddots & 1 \\
& & & d_{2n} & & & & 0
\end{bmatrix}.
$$

Then

$$
\begin{aligned}
ML(M, s_1, n) &= \\
&= [M s_1, s_1, M^{-1}s_1, \ldots, M^{-(n-2)}s_1 \,|\, M^2 s_1, \ldots, M^{n+1}s_1] \\
&= L(M, s_1, n)C.
\end{aligned}
$$

Using $L(M, s_1, n) = SR$, we obtain $MSR = SRC$, that is $S^{-1}MS = RCR^{-1}$. Due to the special form of R and C, RCR^{-1} and therefore $S^{-1}MS$ must have the form

$$
S^{-1}MS = \begin{bmatrix} \searrow & \searrow \\ \searrow & \searrow \end{bmatrix}.
$$

Similar, we can show that $M^{-1}L(M, s_1, n) = L(M, s_1, n)D$, hence $M^{-1}SR = SRD$ and $M^{-1} = SRDR^{-1}S^{-1}$. As M and S are symplectic, we have $M^T = J^{-1}M^{-1}J$ and $S^T J^{-1}S = -S^T JS = -J$ and therefore

$$
\begin{aligned}
(S^{-1}MS)^T &= S^T M^T S^{-T} = S^T J^{-1}SRDR^{-1}S^{-1}JS^{-T} \\
&= -JRDR^{-1}(-J)^{-1} = -JRDR^{-1}J.
\end{aligned}
$$

Due to the special form of J, R and D we obtain

$$
(S^{-1}MS)^T = \begin{bmatrix} \searrow & \searrow \\ \searrow & \searrow \end{bmatrix}.
$$

Comparing the two forms for $S^{-1}MS$, it follows that $S^{-1}MS$ is a butterfly matrix. Taking a closer look at RCR^{-1} shows that the subdiagonal elements of the $(2,2)$ block are given by quotients of diagonal entries of R

$$
(RCR^{-1})_{n+i+1,n+i} = \frac{r_{n+i+1,n+i+1}}{r_{n+i,n+i}}.
$$

The subdiagonal elements of the $(2,2)$ block of $-JRDR^{-1}J$ are also given by quotients of diagonal entries of R

$$
(-JRDR^{-1}J)_{n+i+1,n+i} = \frac{r_{i+1,i+1}}{r_{i,i}}.
$$

Hence the $(2,2)$ block of the butterfly matrix $S^{-1}MS$ is an unreduced tridiagonal matrix.

b) From $S^{-1}MS = B$ we obtain $M = SBS^{-1}$ and $M^{-1} = SB^{-1}S^{-1}$. Hence, for $i = 1,\ldots,n$ and $j = 1,\ldots,n-1$ we have $S^{-1}M^i s_1 = S^{-1}SB^i S^{-1}s_1 = B^i e_1$ and $S^{-1}M^{-j}s_1 = (B^{-1})^j e_1$. As the inverse of $B = \begin{bmatrix} B_{11} & B_{12} \\ B_{21} & B_{22} \end{bmatrix}$ is given by $\begin{bmatrix} B_{22}^T & -B_{21} \\ -B_{12}^T & B_{11} \end{bmatrix}$, B^{-1} is of the form

$$
B^{-1} = \begin{bmatrix} \searrow & \searrow \\ \searrow & \searrow \end{bmatrix}.
$$

Using induction, it is easy to see that in $B^i e_1$, the components $i+1$ to n and $i+1+n$ to $2n$ and in $(B^{-1})^j e_1$ the components $j+2$ to n and $j+1+n$ to $2n$ are zero. Hence for $R := S^{-1}L(M, s_1, n)$ we have

$$
R = [S^{-1}s_1, S^{-1}M^{-1}s_1, \ldots, S^{-1}M^{-(n-1)}s_1 | S^{-1}Ms_1, \ldots, S^{-1}M^n s_1]
$$

is a J–triangular matrix. Furthermore for $i = 1, \ldots, n$

$$
\begin{aligned}
r_{ii} &= (S^{-1}M^{-(i-1)}s_1)_i \\
&= ((B^{-1})^{i-1}e_1)_{i,1} \\
&= -b_{1,n+1}b_{n+1,n+2}b_{n+2,n+3}\cdots b_{n+i-1,n+i}, \\
r_{n+i,n+i} &= (S^{-1}M^i s_1)_{n+i} \\
&= (B^i e_1)_{n+i,1} \\
&= b_{1,n+1}b_{n+2,n+1}b_{n+3,n+2}\cdots b_{n+i,n+i-1}.
\end{aligned}
$$

Hence, if B is unreduced, then R is nonsingular.

c) If $S^{-1}MS = B$ and $\widetilde{S}^{-1}M\widetilde{S} = \widetilde{B}$ are butterfly matrices, then using b) we have the SR factorizations $L(M, s_1, n) = SR$ and $L(M, \lambda s_1, n) = \widetilde{S}\widetilde{R}$. But $L(M, \lambda s_1, n) = \lambda L(M, s_1, n)$, hence $SR = \widetilde{S}(\frac{1}{\lambda}\widetilde{R})$. Using Theorem 2.7 there is a trivial matrix D (2.1.5) such that $S = \widetilde{S}D$ and $B = S^{-1}MS = D^{-1}\widetilde{S}^{-1}M\widetilde{S}D = D^{-1}\widetilde{B}D$.   ✓

REMARK 3.8 *A weaker version of Theorem 3.7 not involving unreduced butterfly matrices was first stated and proved as Theorem 3.6 in [13].*

REMARK 3.9 *In the SR decomposition (and correspondingly in the symplectic similarity transformation to butterfly form) we still have a certain degree of freedom. E.g., any unreduced butterfly matrix is similar to an unreduced butterfly matrix with $b_i = 1$ and $|a_i| = 1$ for $i = 1, \ldots, n$ and $\mathrm{sign}(a_i) = \mathrm{sign}(d_i)$ for $i = 2, \ldots, n$ (this follows from Theorem 3.7 c)).*

Given M, the matrix $L(M, s_1, n)$ is determined by the first column of S. The essential uniqueness of the factorization $L(M, s_1, n) = SR$ tells us that the transforming matrix S for the similarity transformation $B = S^{-1}MS$ is essentially uniquely determined by its first column. This Implicit-S-Theorem can serve as the basis for the construction of an implicit SR algorithm for butterfly matrices, just as the Implicit-Q-Theorem (Theorem 2.2) provides a basis for the implicit QR algorithm on upper Hessenberg matrices. In both cases uniqueness depends on the unreduced character of the matrix.

The next result is well-known for Hessenberg matrices (see Theorem 2.5) and J–Hessenberg matrices (see Theorem 2.10) and will turn out to be essential when examining the properties of the SR algorithm based on the butterfly form.

LEMMA 3.10 *If λ is an eigenvalue of an unreduced symplectic butterfly matrix $B \in \mathbf{R}^{2n \times 2n}$, then its geometric multiplicity is one.*

PROOF: Since B is symplectic, B is nonsingular and its eigenvalues are nonzero. For any $\lambda \in \mathbb{C}$ we have $\mathrm{rank}(B - \lambda I) \geq 2n - 1$ because the first

$2n - 1$ columns of $B - \lambda I$ are linear independent. This can be seen by looking at the permuted expression $B_P - \lambda I = PBP^T - \lambda I =$

$$
\begin{bmatrix}
b_1 & b_1 c_1 - a_1^{-1} & 0 & b_1 d_2 & & & & \\
a_1 & a_1 c_1 & 0 & a_1 d_2 & & & & \\
0 & b_2 d_2 & & & & & & \\
0 & a_2 d_2 & \ddots & & \ddots & & & \\
& & & & & 0 & b_{n-1} d_n & \\
& & \ddots & & \ddots & 0 & a_{n-1} d_n & \\
& & & 0 & b_n d_n & b_n & b_n c_n - a_n^{-1} \\
& & & 0 & a_n d_n & a_n & a_n c_n
\end{bmatrix} - \lambda I
$$

where P as in (2.1.2). Obviously, the first two columns of the above matrix are linear independent as B is unreduced. We can not express the third column as a linear combination of the first two columns:

$$
\begin{bmatrix} 0 \\ 0 \\ b_2 - \lambda \\ a_2 \end{bmatrix} = \beta_1 \begin{bmatrix} b_1 - \lambda \\ a_1 \\ 0 \\ 0 \end{bmatrix} + \beta_2 \begin{bmatrix} b_1 c_1 - a_1^{-1} \\ a_1 c_1 - \lambda \\ b_2 d_2 \\ a_2 d_2 \end{bmatrix}.
$$

From the fourth row we obtain $\beta_2 = d_2^{-1}$. With this the third row yields

$$
b_2 - \lambda = b_2.
$$

As λ is an eigenvalue of B and is therefore nonzero, this equation can not hold. Hence the first three columns are linear independent. Similarly, we can see that the first $2n - 1$ columns are linear independent.

Hence, the eigenspace are one-dimensional. $\qquad \checkmark$

Analogous to the unreduced Hessenberg/J–Hessenberg case (Theorems 2.4 and 2.10), the right eigenvectors of unreduced butterfly matrices have the following property.

LEMMA 3.11 *Suppose that $B \in \mathbf{R}^{2n \times 2n}$ is an unreduced butterfly matrix as in (3.2.10). If $Bx = \lambda x$ with $x \neq 0$ then $e_{2k}^T x \neq 0$.*

PROOF: The proof is by induction on the size of B. As usual, the entries of the eigenvector x will be denoted by x_i; $x = [x_1, x_2, \ldots, x_{2n}]^T$.

Suppose that $n = 2$. The second and fourth row of $Bx = \lambda x$ yield

$$b_2 x_2 + b_2 d_2 x_3 + (b_2 c_2 - a_2^{-1}) x_4 = \lambda x_2, \qquad (3.2.11)$$

$$a_2 x_2 + a_2 d_2 x_3 + a_2 c_2 x_4 = \lambda x_4. \qquad (3.2.12)$$

Since B is unreduced, we know that $a_2 \neq 0$ and $d_2 \neq 0$. If $x_4 = 0$ then from (3.2.12) we obtain

$$x_2 + d_2 x_3 = 0, \qquad (3.2.13)$$

while (3.2.11) gives $b_2(x_2 + d_2 x_3) = \lambda x_2$. Using (3.2.13) we obtain $x_2 = 0$, and further $x_3 = 0$. The third row of $Bx = \lambda x$ gives

$$a_1 x_1 + a_1 c_1 x_3 + a_1 d_2 x_4 = \lambda x_3.$$

As B is unreduced, $a_1 \neq 0$. Using $x_2 = x_3 = x_4 = 0$, we obtain $x_1 = 0$. Thus $x = 0$ which is a contradiction since $x \neq 0$ by assumption.

Assume that the lemma is true for matrices of order $2(n-1)$. Let $B^{2n,2n} \in \mathbf{R}^{2n \times 2n}$ be an unreduced butterfly matrix. For simplicity we will consider the permuted equation $B_P^{2n,2n} x_P = \lambda x_P$ where $B_P^{2n,2n} = P B^{2n,2n} P^T$ and $x_P = Px$. Partition $B_P^{2n,2n}, x_P$ as

$$B_P^{2n,2n} = \left[\begin{array}{c|cc} B_P^{2(n-1),2(n-1)} & 0 & d_n(b_{n-1}e_{2n-3} + a_{n-1}e_{2n-2}) \\ \hline b_n d_n e_{2n-2}^T & b_n & b_n c_n - a_n^{-1} \\ a_n d_n e_{2n-2}^T & a_n & a_n c_n \end{array} \right],$$

$$x_P = \left[\begin{array}{c} y \\ \widetilde{x}_{2n-1} \\ \widetilde{x}_{2n} \end{array} \right],$$

where $B_P^{2(n-1),2(n-1)} \in \mathbf{R}^{(2n-2) \times (2n-2)}$ is an unreduced butterfly matrix and $y \in \mathbf{R}^{2n-2}$. Suppose $x_{2n} = \widetilde{x}_{2n} = 0$. This implies

$$d_n y_{2n-2} + \widetilde{x}_{2n-1} = 0 \qquad (3.2.14)$$

since $a_n \neq 0$ as $B^{2n,2n}$ is unreduced. Further we have

$$b_n(d_n y_{2n-2} + \widetilde{x}_{2n-1}) = \lambda \widetilde{x}_{2n-1}.$$

Hence, using (3.2.14) $\widetilde{x}_{2n-1} = 0$. This implies $B_P^{2(n-1),2(n-1)} y = \lambda y$. Using $\widetilde{x}_{2n-1} = \widetilde{x}_{2n} = 0$ we further obtain from (3.2.14) $y_{2n-2} = 0$. This is a contradiction, because by induction hypothesis $e_{2n-2}^T y \neq 0$. $\qquad \checkmark$

REMARK 3.12 *If $Bx = \lambda x$, then $(Jx)^T$ is the left eigenvector of B corresponding to λ^{-1}: $(Jx)^T B = \lambda^{-1}(Jx)^T$. Let y be the right eigenvector of B corresponding to λ^{-1}: $By = \lambda^{-1}y$, then $(Jy)^T B = \lambda(Jy)^T$. From Lemma 3.11 it follows that $e_{2n}^T y \neq 0$, hence the nth component of the left eigenvector of B corresponding to λ is $\neq 0$.*

As mentioned before, the symplectic butterfly form was introduced by Banse and Bunse-Gerstner in [15, 13, 12, 14]. They took a slightly different point of view in order to argue that a butterfly matrix can be represented by $4n - 1$ parameters. A *strict butterfly form* is introduced in which the upper left diagonal matrix B_{11} of the butterfly form is nonsingular. Then, using similar arguments as above, since

$$\begin{bmatrix} B_{11}^{-1} & 0 \\ -B_{21} & B_{11} \end{bmatrix} \begin{bmatrix} B_{11} & B_{12} \\ B_{21} & B_{22} \end{bmatrix} = \begin{bmatrix} I & V \\ 0 & I \end{bmatrix}$$

and since $V = B_{11}^{-1} B_{12}$ is a symmetric tridiagonal matrix (same argument as used above), one obtains

$$\begin{bmatrix} B_{11} & B_{12} \\ B_{21} & B_{22} \end{bmatrix} = \begin{bmatrix} B_{11}^{-1} & 0 \\ B_{21} & B_{11} \end{bmatrix} \begin{bmatrix} I & V \\ 0 & I \end{bmatrix}$$

$$= \begin{bmatrix} \diagdown & 0 \\ \diagdown & \diagdown \end{bmatrix} \begin{bmatrix} I & \diagdown\diagdown \\ 0 & I \end{bmatrix}. \tag{3.2.15}$$

Hence $4n - 1$ parameters that determine the symplectic matrix can be read off directly. Obviously, n of these parameters have to be nonzero (the diagonal elements of B_{11}). If any of the $n-1$ subdiagonal elements of V is zero, deflation can take place; that is, the problem can be split into at least two problems of smaller dimension, but with the same symplectic butterfly structure.

This decomposition was introduced because of its close resemblance to symplectic matrix pencils that appear naturally in control problems. These pencils are typically of the form

$$K - \lambda N = \begin{bmatrix} F & 0 \\ H & I \end{bmatrix} - \lambda \begin{bmatrix} I & -G \\ 0 & F^T \end{bmatrix}, \quad F, G = G^T, H = H^T \in \mathbf{R}^{n \times n}.$$

(Note: For $F \neq I$, K and N are not symplectic.) Assuming that K and N are nonsingular (that is, F is nonsingular), we can rewrite the above equation

$$\begin{bmatrix} I & 0 \\ 0 & F^{-T} \end{bmatrix} (K - \lambda N) =: \tilde{K} - \lambda \tilde{N}$$

$$= \begin{bmatrix} F & 0 \\ F^{-T}H & F^{-T} \end{bmatrix} - \lambda \begin{bmatrix} I & -G \\ 0 & I \end{bmatrix}.$$

(Note: $\tilde{K}$ and $\tilde{N}$ are symplectic matrices.) Solving this generalized eigenproblem is equivalent to solving the eigenproblem for the symplectic matrix

$$M := \tilde{K}^{-1} \tilde{N} = \begin{bmatrix} F^{-1} & 0 \\ -HF^{-1} & F^T \end{bmatrix} \begin{bmatrix} I & -G \\ 0 & I \end{bmatrix}.$$

Obviously, not every unreduced butterfly matrix B is a strict butterfly matrix, but B can be turned into a strict one by a similarity transformation with a trivial matrix D (2.1.5). Numerous choices of D will work. Thus, it is practically true that every unreduced butterfly matrix is strict. The converse is false. There are strict butterfly matrices that are not similar to any unreduced butterfly matrix. In particular, Lemma 3.10 and Lemma 3.11 do not hold for strict butterfly matrices as can be seen by the next example.

EXAMPLE 3.13 *Let*

$$
B = \left[\begin{array}{cc|cc}
1 & 0 & 0 & 1 \\
0 & 1 & 1 & 0 \\
\hline
0 & 0 & 1 & 0 \\
0 & 0 & 0 & 1
\end{array}\right].
$$

Then B is a strict symplectic butterfly matrix that it is not unreduced. It is easy to see that the spectrum of B is given by $\{1, 1\}$ with geometric multiplicities two. The vector e_1 is an eigenvector of B to the eigenvalue 1; obviously $e_4^T e_1 = 0$.

Because not every strict butterfly matrix is unreduced, the class of strict butterfly matrices lacks the theoretical basis for an implicit SR algorithm. If one wishes to build an algorithm based on the decomposition (3.2.15), one is obliged to restrict oneself to unreduced, strict butterfly matrices. The following considerations show that this is not a serious restriction.

REMARK 3.14 *If both B_{11} and B_{21} are nonsingular, then the matrices $V = B_{11}^{-1} B_{12}$ and $T = B_{21}^{-1} B_{22}$ are related by $V = T - B_{11}^{-1} B_{21}^{-1}$. Thus, the off-diagonal entries of V and T are the same. It follows that corresponding off-diagonal entries of B_{12} and B_{22} are either zero or nonzero together. In connection with the decomposition (3.2.15), this implies that whenever B is not unreduced, V will also be reducible, and we can split the eigenvalue problem into smaller ones.*

This relationship breaks down, however, if B_{21} is singular. Consider, for example, the class of matrices

$$
B = \left[\begin{array}{cc|cc}
1 & 0 & a & g \\
0 & 1 & g & c \\
\hline
1 & 0 & 1+a & g \\
0 & 0 & 0 & 1
\end{array}\right]
$$

with $g \neq 0$. These are strict butterfly matrices for which B_{12} is unreduced but B_{22} is not. Notice that the $(2, 2)$ and $(4, 4)$ entries are eigenvalues and can be deflated from the problem.

In general, if B_{21} is singular, a deflation (and usually a splitting) is possible. If $(B_{21})_{i,i} = 0$, then $(B_{11})_{i,i}$ must be nonzero, since B is nonsingular. This forces $(B_{22})_{i,i-1} = (B_{22})_{i,i+1} = 0$, because $B_{11}^T B_{22} - B_{21}^T B_{12} = I$. It follows *that $(B_{11})_{i,i}$ and $(B_{22})_{i,i}$ are a reciprocal pair of eigenvalues, which can be deflated from the problem. Unless $i = 1$ or $i = n$, the remaining problem can be split into two smaller problems.*

Banse presents in [13] an algorithm to reduce an arbitrary symplectic matrix to butterfly form. The algorithm uses the symplectic Givens transformations G_k, the symplectic Householder transformations H_k, and the symplectic Gauss transformation L_k introduced in Section 2.1.2. Zeros in the rows of M will be introduced by applying one of the above mentioned transformations from the right, while zeros in the columns will be introduced by applying the transformations from the left. Of course, in order to perform a similarity transformation, the inverse of each transformation applied from the right/left has to be applied from the left/right as well. The basic idea of the algorithm can be summarized as follows

    for $j = 1$ to n
        bring the jth column of M into the desired form
        bring the $(n + j)$th row of M into the desired form

The remaining rows and columns in M that are not explicitly touched during the process will be in the desired form due to the symplectic structure. For an 8×8 symplectic matrix, the elimination process can be summarized as in the scheme given in Table 3.1.

$$
\begin{bmatrix}
\star & \hat{O}_6 & \hat{O}_7 & \hat{O}_8 & \star & \star & \hat{O}_9 & \hat{O}_9 \\
5, L^{pr} & \star & \hat{O}_{14} & \hat{O}_{15} & \star & \star & \star & \hat{O}_{16} \\
4, H^{pr} & 13, L^{pr} & \star & \hat{O}_{19} & \hat{O}_{13} & \star & \star & \star \\
4, H^{pr} & 12, H^{pr} & 18, L^{pr} & \star & \hat{O}_{12} & \hat{O}_{18} & \star & \star \\
\star & 6, G^{po} & 7, G^{po} & 8, G^{po} & \star & \star & 9, H^{po} & 9, H^{po} \\
3, G^{pr} & \star & 14, G^{po} & 15, G^{po} & \star & \star & \star & 16, H^{po} \\
2, G^{pr} & 11, G^{pr} & \star & 19, G^{po} & \hat{O}_{11} & \star & \star & \star \\
1, G^{pr} & 10, G^{pr} & 17, G^{pr} & \star & \hat{O}_{10} & \hat{O}_{17} & \star & \star
\end{bmatrix}
$$

Table 3.1.   Elimination Scheme for a Symplectic Matrix

In Table 3.1, the capital letter indicates the type of elimination matrix used to eliminate the entry (G used for a symplectic Givens, H for a symplectic Householder, and L for a symplectic Gauss transformation). The upper index indicates whether the elimination is done by pre- or postmultiplication. The numbers indicate the order in which the entries are annihilated. A zero that is not created by explicit elimination but because of the symplectic structure, is denoted by $\hat{O}$. Its index indicates which transformation causes this zero. E.g., if after step 6 the first column of M is denoted by $m_1 = [m_{11}\ 0\ 0\ 0\ m_{41}\ 0\ 0\ 0]^T, m_{41} \neq 0$, and the second column by $m_2 = [m_{12}\ \star\ \star\ \star\ 0\ \star\ \star\ \star]^T$, then as M is symplectic throughout the whole reduction process, from $M^T J M = J$ we have $0 = m_1^T J m_2 = -m_{41} m_{12}$ and therefore $m_{12} = 0$.

As can be seen from this scheme, pivoting can be incorporated in the reduction process in order to increase numerical stability of the process. Instead of bringing the $(n + j)$th row into the desired form, one can just as well bring the jth row into the desired form.

The algorithm for reducing an arbitrary symplectic matrix to butterfly form as given in [13] can be summarized as given in Table 3.2 (in MATLAB-like notation). Note that pivoting is incorporated in order to increase numerical stability.

REMARK 3.15 *a)* *The algorithm for reducing a $2n \times 2n$ symplectic matrix M to butterfly form as given in Table 3.2 requires about $\frac{76}{3}n^3 - 14n^2$ flops. If the transformation matrix S is required, then $28n^3$ flops have to be added. This flop count is based on the fact that $n^2 - n$ symplectic Givens transformations, $n - 1$ symplectic Gauss transformations and 2 symplectic Householder transformation with $v \in \mathbf{R}^j$ for each $j = 2, \ldots, n - 1$ are used. Moreover, when updating M only during the first iteration step the transformations have to be applied to all columns and rows. During the second iteration step, the transformations have to be applied only to columns $2, \ldots, 2n$ and rows $2, \ldots, n$ and $n+2, \ldots, 2n$. Similar, during the third iteration step, the transformations have to be applied only to columns $2, \ldots, n$ and $n + 1, \ldots, 2n$ and rows $3, \ldots, n$ and $n + 3, \ldots, 2n$; and so on.*

b) *The reduction of a symplectic matrix to butterfly form is cheaper than the reduction to J–Hessenberg form (see Table 2.6, and Remark 2.20), although more zeros are generated here. It is also slightly cheaper than the reduction to Hessenberg form (used as a preparatory step in the QR algorithm), which requires about $\frac{80}{3}n^3$ flops, see [58]. But the accumulation of the transformation matrix requires only about $11n^3$ flops in that case.*

c) *All transformation matrices used in the algorithm for reducing a symplectic matrix to butterfly form have as first column a multiple of e_1. Hence, the algorithm as given in Table 3.2 determines a symplectic matrix S with first column λe_1 which transforms the symplectic matrix M to butterfly form if such a reduction exists.*

3.3   A CANONICAL FORM FOR SYMPLECTIC BUTTERFLY MATRICES

We have noted that the reduction to symplectic butterfly form is not quite uniquely determined; it is determined up to a similarity transformation by a trivial (i.e., symplectic and J–triangular) matrix. For some of the following

---

Algorithm: Reduction to Butterfly Form

Given a $2n \times 2n$ symplectic matrix M compute its reduction to butterfly form. M will be overwritten by its butterfly form.

for $j = 1 : n - 1$
    for $k = n : -1 : j + 1$
        compute G_k such that $(G_k M)_{k+n,j} = 0$
        $M = G_k M G_k^T$
    end
    if $j < n - 1$
    then compute H_j such that $(H_j M)_{j+2:n,j} = 0$
        $M = H_j M H_j^T$
    end
    if $M_{j+1,j} \neq 0$ and $M_{n+j,n+j} = 0$
    then stop, reduction not possible
    end
    compute L_{j+1} such that $(L_{j+1} M)_{j+1,j} = 0$
    $M = L_{j+1} M L_{j+1}^{-1}$
    if $|M_{j,j}| > |M_{j+n,j}|$
    then $p = j + n$
    else $p = j$
    end
    for $k = n : -1 : j + 1$
        compute G_k such that $(M G_k)_{p,k} = 0$
        $M = G_k^T M G_k$
    end
    if $j < n - 1$
    then compute H_j such that $(M H_j)_{p,j+2+n:2n} = 0$
        $M = H_j^T M H_j$
    end
end

Table 3.2.   Reduction to Butterfly Form

discussions it is of interest to develop a canonical form for butterfly matrices under similarity transformations by trivial matrices. We restrict our attention to unreduced symplectic butterfly matrices, since every butterfly matrix can be decomposed into two or more smaller unreduced ones.

In Remark 3.9, it was already observed that any unreduced symplectic butterfly matrix is similar to an unreduced butterfly matrix with

$$
\begin{aligned}
b_i &= 1 && \text{for } i = 1, \ldots, n, \\
|a_i| &= 1 && \text{for } i = 1, \ldots, n, \\
\operatorname{sign}(a_i) &= \operatorname{sign}(d_i) && \text{for } i = 2, \ldots, n.
\end{aligned}
$$

This canonical form is unique. In the following we will define a different canonical form that is only of theoretical interest, while the one mentioned above may be used in actual computations.

THEOREM 3.16 *Let $\widetilde{B}$ be an unreduced symplectic butterfly matrix. Then there exists a symplectic J–triangular matrix X such that $B = X^{-1} \widetilde{B} X$ has the canonical form*

$$
B = \begin{bmatrix} 0 & -D \\ D & T \end{bmatrix}, \tag{3.3.16}
$$

where D is a signature matrix, and T is a D–symmetric, unreduced tridiagonal matrix. D is uniquely determined, T is determined up to a similarity transformation by a signature matrix, and X is unique up to multiplication by a signature matrix of the form $\operatorname{diag}(C, C)$. The eigenvalues of T are $\lambda_i + \lambda_i^{-1}$, $i = 1, \ldots, n$, where $\lambda_i, \lambda_i^{-1}, i = 1, \ldots, n$ are the eigenvalues of B.

PROOF: We are motivated by the decomposition (3.2.7), in which the nonsingular matrix B_{21} is used as a pivot to eliminate B_{11}. We now seek a similarity transformation that achieves a similar end. Let

$$
X = \begin{bmatrix} Y^{-1} & -F \\ 0 & Y \end{bmatrix} \tag{3.3.17}
$$

be a trivial matrix. We shall determine conditions on Y and F under which the desired canonical form is realized. Focusing on the first block column of the similarity transformation $B = X^{-1} \widetilde{B} X$, we have

$$
\begin{bmatrix} B_{11} \\ B_{21} \end{bmatrix} = \begin{bmatrix} Y & F \\ 0 & Y^{-1} \end{bmatrix} \begin{bmatrix} \widetilde{B}_{11} \\ \widetilde{B}_{21} \end{bmatrix} Y^{-1} = \begin{bmatrix} Y\widetilde{B}_{11} + F\widetilde{B}_{21} \\ Y^{-1}\widetilde{B}_{21} \end{bmatrix} Y^{-1}.
$$

We see that $B_{11} = 0$ if and only if $Y\widetilde{B}_{11} + F\widetilde{B}_{21} = 0$, which implies $F = -Y\widetilde{B}_{11}\widetilde{B}_{21}^{-1}$. Thus F is uniquely determined, once Y has been chosen. We have $B_{21} = Y^{-1}\widetilde{B}_{21}Y^{-1}$, which shows that B_{21} and $\widetilde{B}_{21}$ must have the same inertia. Thus the best we can do is to take $B_{21} = D = \operatorname{sign}(\widetilde{B}_{21})$, which is achieved by choosing $Y = |\widetilde{B}_{21}|^{1/2}$.

In summary, we should take X as in (3.3.17), where

$$
Y = |\widetilde{B}_{21}|^{1/2} \quad \text{and} \quad F = -Y\widetilde{B}_{11}\widetilde{B}_{21}^{-1}.
$$

The resulting B has the desired form. The only aspect of the computation that is not completely straightforward is showing that $B_{12} = -D$. However, this becomes easy when one applies the following fact: If B is a symplectic matrix with $B_{11} = 0$, then $B_{12} = -B_{21}^{-1}$. The D–symmetry of T is also an easy consequence of the symplectic structure of B.

The uniqueness statements are easily verified.

If $\begin{bmatrix} x \\ y \end{bmatrix}$ is an eigenvector of B with eigenvalue λ, then $y \neq 0$, and $Ty = (\lambda + \lambda^{-1})y$. $\checkmark$

REMARK 3.17 a) *The canonical form could be made unique by insisting that T's subdiagonal entries be positive: $t_{i+1,i} > 0$, $i = 1, \ldots, n-1$.*

b) *The decomposition (3.2.7) of the canonical form B is*

$$B = \begin{bmatrix} 0 & -D \\ D & T \end{bmatrix} = \begin{bmatrix} D & 0 \\ 0 & D \end{bmatrix}\begin{bmatrix} 0 & -I \\ I & DT \end{bmatrix}.$$

c) *Theorem 3.16 is a theoretical result. From the standpoint of numerical stability, it might not be advisable to transform a symplectic butterfly matrix into canonical form. In the process, the spectral information λ, λ^{-1} is condensed into T as $\lambda + \lambda^{-1}$. The original information can be recovered via the inverse transformation*

$$\nu \to \left(\frac{\nu}{2}\right) \pm \sqrt{\left(\frac{\nu}{2}\right)^2 - 1}.$$

However, eigenvalues near ± 1 will be resolved poorly because this map is not Lipschitz continuous at $\nu = \pm 2$.

The behavior is similar to that of Van Loan's method for the Hamiltonian eigenvalue problem [137]. One may lose up to half of the significant digits as compared to the standard QR algorithm. For instance, try to compute the eigenvalues of the symplectic matrix

$$S = G^T \begin{bmatrix} 1+\delta & 0 \\ 0 & \frac{1}{1+\delta} \end{bmatrix} G,$$

where G is a randomly generated Givens rotation and δ is less than the square root of the machine precision, once by applying the QR algorithm to S and once to $S + S^{-1}$ followed by the inverse transformation given above.

d) *Since*

$$B^{-1} = JB^T J^T = \begin{bmatrix} T^T & D \\ -D & 0 \end{bmatrix},$$

we have $B + B^{-1} = \text{diag}(T^T, T)$. Thus, forming T is equivalent to adding B^{-1} to B. The transformation $S \to S + S^{-1}$ was used in similar fashion in [96, 118] to compute the eigenvalues of a symplectic pencil.

e) In the proof we have shown that the eigenvectors of T can obtained from those of B. It is also possible to recover the eigenvectors of B from those of T: If $Ty = (\lambda + \lambda^{-1})y$ $(y \neq 0)$, then

$$\begin{bmatrix} -\lambda^{-1}Dy \\ y \end{bmatrix} \quad and \quad \begin{bmatrix} -\lambda Dy \\ y \end{bmatrix}$$

are eigenvectors of B associated with λ and λ^{-1}, respectively.

3.4   REDUCED SYMPLECTIC MATRIX PENCILS

Based on the results given in Section 3.2, one can easily derive condensed forms for symplectic matrix pencils. For this consider the factorization $B = K^{-1}N$ (3.2.15) or (3.2.7) of an unreduced symplectic butterfly matrix B. The eigenvalue problem $K^{-1}Nx = \lambda x$ is equivalent to $(\lambda K - N)x = 0$ and $(K - \lambda N)x = 0$ because of the symmetry of the spectrum. In the latter equations the $4n - 1$ parameters are given directly. For the decomposition (3.2.15) we obtain

$$K_s - \lambda N_s = \begin{bmatrix} B_{11}^{-1} & 0 \\ -B_{21} & B_{11} \end{bmatrix} - \lambda \begin{bmatrix} I & V \\ 0 & I \end{bmatrix} \tag{3.4.18}$$

while for (3.2.7) we obtain

$$K_u - \lambda N_u = \begin{bmatrix} B_{21} & -B_{11} \\ 0 & B_{21}^{-1} \end{bmatrix} - \lambda \begin{bmatrix} 0 & -I \\ I & T \end{bmatrix}. \tag{3.4.19}$$

Here, the symmetric tridiagonal matrix $B_{21}^{-1}B_{22}$ is denoted by T, while V denotes $B_{11}^{-1}B_{12}$ as before. As noted in Remark 3.14, if B_{11} and B_{21} are both nonsingular, V and T are related by $V = T - B_{11}^{-1}B_{21}^{-1}$.

It is well-known (Lemma 2.11) that if $K - \lambda N$ is a symplectic matrix pencil, $Q \in \mathbf{R}^{2n \times 2n}$ is nonsingular, and $S \in \mathbf{R}^{2n \times 2n}$ is symplectic, then $Q(K - \lambda N)S$ is a symplectic matrix pencil and the eigenproblems $K - \lambda N$ and $QKS - \lambda QNS$ are equivalent. Obviously the eigenproblems $K_s - \lambda N_s$ and $K_u - \lambda N_u$ are equivalent if B_{11} and B_{21} are both nonsingular: $Q(K_s - \lambda N_s) = K_u - \lambda N_u$ where

$$Q = \begin{bmatrix} 0 & -I \\ I & B_{11}^{-1}B_{21}^{-1} \end{bmatrix}.$$

Hence, if x_s is a right eigenvector of $K_s - \lambda N_s$, then $x_u = x_s$ is a right eigenvector of $K_u - \lambda N_u$. If y_u is a left eigenvector of $K_u - \lambda N_u$, then $y_s = Q^T y_u$ is a left eigenvector of $K_s - \lambda N_s$.

Which of these two equivalent eigenproblems should be preferred in terms of accuracy of the computed eigenvalues? As a measure of the sensitivity of a simple eigenvalue of the generalized eigenproblem $A - \lambda C$, one usually considers the reciprocal of

$$\frac{\sqrt{(y^H A x)^2 + (y^H C x)^2}}{||x||_2 \, ||y||_2} \tag{3.4.20}$$

as the condition number, where x is the right eigenvector, y the left eigenvector corresponding to the same eigenvalue μ. If the expression (3.4.20) is small, one says that the eigenvalue μ is ill conditioned. Let λ be an eigenvalue of B, x_u and y_u the corresponding right and left eigenvectors of $K_u - \lambda N_u$, and $x_s = x_u$ and $y_s = Q^T y_u$ the corresponding right and left eigenvectors of $K_s - \lambda N_s$. Simple algebraic manipulations show

$$||x_s||_2 = ||x_u||_2,$$
$$|y_s^H K_s x_s| = |y_u^H K_u x_u|,$$
$$|y_s^H N_s x_s| = |y_u^H N_u x_u|,$$

while $||y_s||_2 = ||Q^T y_u||_2$. Therefore, the expressions for the eigenvalue condition number differs only in the 2–norm of the respective left eigenvector. Tests in MATLAB indicate that the pencil $K_s - \lambda N_s$ resolves eigenvalues near 1 better then the pencil $K_u - \lambda N_u$, while $K_u - \lambda N_u$ resolves eigenvalues near $\sqrt{-1}$ better. For other eigenvalues both pencils show the same behavior. Hence, from this short analysis there is no indication whether to prefer one of the pencils because of better numerical behavior.

In [13] an elimination process for computing the reduced matrix pencil form (3.4.18) of a symplectic matrix pencil (in which both matrices are symplectic) is given. Based on this reduction process, an SZ algorithm for computing the eigenvalues of symplectic matrix pencils of the form (3.4.18) can be developed. The SZ algorithm is the analogue of the SR algorithm for the generalized eigenproblem, see Section 2.2.2. As the algorithm works on the factors of the butterfly matrix, it works directly on the $4n - 1$ parameters that determine a symplectic butterfly matrix.

An elimination process for computing the reduced matrix pencil of the form (3.4.19) of a symplectic matrix pencil (in which both matrices are symplectic) is given below. Based on this reduction process, an SZ algorithm for computing the eigenvalues of symplectic matrix pencils of the form (3.4.19) is developed in Section 4.3. It turns out that the SZ algorithm for the pencil (3.4.19) requires slightly fewer operations than the SZ algorithm for the pencil (3.4.18); see Section 4.3 for details.

The algorithm to reduce a symplectic matrix pencil $K - \lambda N$, where K and N are symplectic, to the reduced matrix pencil form (3.4.19), uses the symplectic

Givens transformations G_k, the symplectic Householder transformations H_k, and the symplectic Gauss transformations $L_k, \tilde{L}_k$ introduced in Section 2.1.2. In this elimination process, zeros in the rows of K and N will be introduced by applying one of the above mentioned transformations from the right, while zeros in the columns will be introduced by applying the transformations from the left.

The basic idea of the algorithm can be summarized as follows:

bring the first column of N into the desired form
for $j = 1$ to n
    bring the jth row of K into the desired form
    bring the jth column of K into the desired form
    bring the $(n + j)$th column of N into the desired form
    bring the jth row of N into the desired form

The remaining rows and columns in K and N that are not explicitly touched during the process will be in the desired form due to the symplectic structure. For an 8×8 symplectic matrix pencil, the elimination process can be summarized as in the scheme given in Table 3.3.

$$\begin{bmatrix}
\star & 8,G^{po} & 7,G^{po} & 6,G^{po} & \star & 10,L^{po} & 9,H^{po} & 9,H^{po} \\
13,G^{pr} & \star & 26,G^{po} & 25,G^{po} & \hat{0}_{13} & \star & 28,L^{po} & 27,H^{po} \\
12,G^{pr} & 30,G^{pr} & \star & 40,G^{po} & \hat{0}_{12} & \hat{0}_{30} & \star & 41,L^{po} \\
11,G^{pr} & 29,G^{pr} & 42,G^{pr} & \star & \hat{0}_{11} & \hat{0}_{29} & \hat{0}_{42} & \star \\
16,\tilde{L}^{pr} & \hat{0}_{15} & \hat{0}_{15} & \hat{0}_{15} & \star & \hat{0}_{15} & \hat{0}_{15} & \hat{0}_{15} \\
15,L^{pr} & 33,\tilde{L}^{pr} & \hat{0}_{32} & \hat{0}_{32} & \hat{0}_{15} & \star & \hat{0}_{32} & \hat{0}_{32} \\
14,H^{pr} & 32,L^{pr} & 44,\tilde{L}^{pr} & \hat{0}_{43} & \hat{0}_{14} & \hat{0}_{32} & \star & \hat{0}_{43} \\
14,H^{pr} & 31,H^{pr} & 43,L^{pr} & 47,\tilde{L}^{pr} & \hat{0}_{14} & \hat{0}_{31} & \hat{0}_{43} & \star
\end{bmatrix}$$

$$-\lambda \begin{bmatrix}
4,G^{pr} & \hat{0}_5 & \hat{0}_5 & \hat{0}_5 & \star & \hat{0}_5 & \hat{0}_5 & \hat{0}_5 \\
3,G^{pr} & 23,G^{po} & 22,G^{po} & 21,G^{po} & 19,G^{pr} & \star & 24,H^{po} & 24,H^{po} \\
2,G^{pr} & \hat{0}_{24} & 38,G^{po} & 37,G^{po} & 18,G^{pr} & 35,G^{pr} & \star & 39,H^{po} \\
1,G^{pr} & \hat{0}_{24} & \hat{0}_{39} & 46,G^{po} & 17,G^{pr} & 34,G^{pr} & 45,G^{pr} & \star \\
\star & \hat{0}_{23} & \hat{0}_{22} & \hat{0}_{21} & \star & \star & \hat{0}_{24} & \hat{0}_{24} \\
5,H^{pr} & \star & \hat{0}_{38} & \hat{0}_{37} & \star & \star & \star & \hat{0}_{39} \\
5,H^{pr} & \hat{0}_{24} & \star & \hat{0}_{46} & 20,H^{pr} & \star & \star & \star \\
5,H^{pr} & \hat{0}_{24} & \hat{0}_{39} & \star & 20,H^{pr} & 36,H^{pr} & \star & \star
\end{bmatrix}$$

Table 3.3.   Elimination Scheme for a Symplectic Matrix Pencil

As before, the capital letter indicates the type of elimination matrix used to eliminate the entry (G used for a symplectic Givens, H for a symplectic Householder, and $L, \tilde{L}$ for the symplectic Gauss transformation). The upper index indicates whether the elimination is done by pre- or postmultiplication. The numbers indicate the order in which the entries are annihilated. A zero that is not created by explicit elimination but because of the symplectic structure, is denoted by $\hat{0}$. Its index indicates which transformation causes this zero. E.g., if after step 5 the first column of N is denoted by $[0\,0\,0\,0\,n_{51}\,0\,0\,0]^T$, $n_{51} \neq 0$, and the first row by $[0 \star \star \star n_{15} \star \star \star]$, then as N is symplectic throughout the whole reduction process, from $N^T J N = J$ we have $e_1^T (N^T J N) = e_1^T J = e_{n+1}$, or in other words,

$$n_{51}[0 \, -n_{12} \, -n_{13} \, -n_{14} \, -n_{15} \, -n_{16} \, -n_{17} \, -n_{18}] = [0\,0\,0\,0\,1\,0\,0\,0].$$

Hence, the entries of the first row of N have to be zero, only the $(n+1, 1)$ entry is nonzero.

As can be seen from this scheme, similar to the reduction of a symplectic matrix to butterfly form, pivoting can be incorporated in the reduction process in order to make it more stable. E.g., in the process as described above the jth column of K will be brought into the desired form. Due to symplecticity, the $(n + j)$th column of K will then be of desired form as well. One could just as well attack the $(n + j)$th column of K, the jth column will then be of desired form due to symplecticity. Or, instead of bringing the jth row of N into the desired form, one can just as well bring its $(n + j)$th row into the desired form.

In Table 3.4 an algorithm for reducing a symplectic matrix pencil $K - \lambda N$, where K and N are both symplectic, to a reduced pencil of the form (3.4.19) is given. The process will be called *reduction to butterfly pencil*. In order to keep the presentation as simple as possible, no pivoting is introduced here, but should be used in an actual implementation. This algorithm can be used to derive a bulge chasing process for an SZ step.

REMARK 3.18 *a) A careful implementation of this process as a bulge chasing process in an implicit SZ step will just work with the $4n - 1$ parameters and some additional variables instead of with the matrices K and N. See Section 4.3 for some details.*

b) As discussed before, pivoting can be incorporated in the reduction process in order to make it more stable.

c) The algorithm for reducing a $2n \times 2n$ symplectic matrix pencil $K - \lambda N$ to butterfly form as given in Table 3.4 requires about $\frac{224}{3}n^3 - 16n^2$ flops. If the transformation matrices S and Z are required, then $28n^3 + 8n^2$ flops and $28n^3 - 28n^2$ flops, respectively, have to be added. This flop count is based on the fact that n^2 symplectic Givens transformations, $n - 1$ symplectic Gauss transformations, n symplectic Gauss transformations type II and 2 symplectic Householder transformation with $v \in \mathbf{R}^j$ for each $j = 2, \ldots, n - 1$ are used. Moreover, the successively generated zeros in K and N are taken into account.

Algorithm: Reduction to Butterfly Pencil

Given a symplectic matrix pencil $K - \lambda N$, where $K, N \in \mathbf{R}^{2n \times 2n}$ are both symplectic matrices, the following algorithm computes symplectic matrices S and Z such that $S(K - \lambda N)Z$ is a symplectic pencil of the form (3.4.19). K is overwritten by SKZ and N by SNZ.

$Z = I^{2n,2n}$; $S = I^{2n,2n}$;
for $k = n : -1 : 1$
    compute G such that $(GN)_{k,1} = 0$.
    $N = GN$; $K = GK$; $S = GS$;
end
compute H such that $(HN)_{n+2:2n,1} = 0$.
$N = HN$; $K = HK$; $S = HS$;
for $j = 1 : n$
    if $j > 1$
        for $k = n : -1 : j$
            compute G such that $(NG)_{j,k} = 0$.
            $N = NG$; $K = KG$; $Z = ZG$;
        end
    end
    if $j < n$
        if $j > 1$
            compute H such that $(NH)_{j,j+n+1:2n} = 0$.
            $N = NH$; $K = KH$; $Z = ZH$;
        end
        for $k = n : -1 : j + 1$
            compute G such that $(KG)_{j,k} = 0$.
            $N = NG$; $K = KG$; $Z = ZG$;
        end
    end
    if $j < n - 1$
        compute H such that $(KH)_{j,j+2+n:2n} = 0$.
        $N = NH$; $K = KH$; $Z = ZH$;
    end
    if $j < n$
        compute L such that $(KL)_{j,j+n+1} = 0$. Transformation might not exist!!
        $N = NL$; $K = KL$; $Z = ZL$;
        for $k = n : -1 : j + 1$
            compute G such that $(GK)_{k,j} = 0$.
            $N = GN$; $K = GK$; $S = GS$;
        end
    end
    if $j < n - 1$
        compute H such that $(HK)_{j+2+n:2n,j} = 0$.
        $N = HN$; $K = HK$; $S = HS$;
    end
    if $j < n$
        compute L such that $(LK)_{j+1+n,j} = 0$. Transformation might not exist!!
        $N = LN$; $K = LK$; $S = LS$;
    end
    compute $\widetilde{L}$ such that $(\widetilde{L}K)_{j+n,j} = 0$. Transformation might not exist!!
    $N = \widetilde{L}N$; $K = \widetilde{L}K$; $S = \widetilde{L}S$;
    if $j < n$
        for $k = n : -1 : j + 1$
            compute G such that $(GN)_{k,j+n} = 0$.
            $N = GN$; $K = GK$; $S = GS$;
        end
    end
    if $j < n - 1$
        compute H such that $(HN)_{j+2+n:2n,j+n} = 0$.
        $N = HN$; $K = HK$; $S = HS$;
    end
end

Table 3.4.   Reduction to Butterfly Pencil

The reduction of a $2n \times 2n$ symplectic matrix pencil to butterfly form is slightly more expensive than the reduction of such a pencil to Hessenberg-triangular form used as a preparatory step for the QZ algorithm; the flop count for that reduction is given in [58] as $64n^3$ for the matrix pencil plus $32n^2$ flops for the transformation matrix Q and $24n^3$ flops for the transformation matrix Z.

d) *The use of symplectic transformations throughout the reduction process assures that the factors K and N remain symplectic separately. If the objective is only to preserve the symplectic property of the pencil ($KJK^T = NJN^T$), one has greater latitude in the choice of transformations. Only the right-hand (Z) transformations need to be symplectic; the left (S) transforms can be more general as long as they are regular.*

3.4.1   ARBITRARY SYMPLECTIC MATRIX PENCILS

We would like to stress once more that the presented algorithm for reducing a symplectic matrix pencil to a butterfly pencil works only on symplectic matrix pencils $K - \lambda N$ where K and N are symplectic. Hence, the algorithm can not be applied to general symplectic matrix pencils $K - \lambda N$, where $KJK^T = NJN^T \neq J$. But, as will be seen in the subsequent sections, the given algorithm is very useful as the building block of a bulge chasing process for an implicit SZ step. If a symplectic matrix/matrix pencil is given in parameterized form, then one should not form the corresponding butterfly matrix, but compute the eigenvalues via an SZ algorithm based on the above reduction process.

But how should one treat a general symplectic matrix pencil $K - \lambda N$, where $KJK^T = NJN^T \neq J$? What is a good reduced form for such a pencil and how can it be computed efficiently?

If K is nonsingular, then $K^{-1}N$ is a symplectic matrix. Hence, the results of Section 3.2 can be applied to $K^{-1}N$. Assume that the symplectic matrix S transforms $K^{-1}N$ to unreduced butterfly form: $S^{-1}K^{-1}NS = B$. The symplectic butterfly matrix B can be decomposed into the product $K_s^{-1}N_s$ as in (3.2.15) or into the product $K_u^{-1}N_u$ as in (3.2.7). Instead of considering the symplectic eigenproblem $Bx = \lambda x$, the generalized symplectic eigenproblem $(\lambda K_v - N_v)x = 0$, or, equivalently, $(K_v - \lambda N_v)x = 0$ can be considered, for $v \in \{s, u\}$. Hence, one can reduce a general symplectic matrix pencil either to the form (3.4.18) or to the form (3.4.19). An algorithm similar to the one above could be devised to compute these reductions. That is, one could develop an algorithm to compute a symplectic matrix Z and a regular matrix S such that $S(K - \lambda N)Z$ is of the desired form.

But what if K is singular? Then the just given derivation will fail. Can we still reduce $K - \lambda N$ to the form $K_v - \lambda N_v$, $v = s$ or $v = u$? Moreover, at least for symplectic matrix pencils that appear naturally in control problems the reduced form $K_v - \lambda N_v$ does not seem to be the appropriate one. These matrix pencils are typically of the form

$$K - \lambda N = \begin{bmatrix} F & 0 \\ -H & I \end{bmatrix} - \lambda \begin{bmatrix} I & G \\ 0 & F^T \end{bmatrix} \tag{3.4.21}$$

with G, H symmetric. K and N themselves are not symplectic, just the pencil is symplectic: $KJK^T = NJN^T \neq J$. A reduction to the form $K_v - \lambda N_v$ implies that at one point during the reduction the nonsymplectic matrices K and N have to be transformed into symplectic ones as K_v and N_v are symplectic. This implies the implicit inversion of the matrix F. Or in other words, we first have to transform $K - \lambda N$ into a symplectic pencil with symplectic matrices, e.g.,

$$\begin{bmatrix} F^{-1} & 0 \\ 0 & I \end{bmatrix} (K - \lambda N) = \begin{bmatrix} I & 0 \\ H & I \end{bmatrix} - \lambda \begin{bmatrix} F^{-1} & -F^{-1}G \\ 0 & F^T \end{bmatrix} \tag{3.4.22}$$

$$\begin{bmatrix} I & 0 \\ 0 & F^{-T} \end{bmatrix} (K - \lambda N) = \begin{bmatrix} F & 0 \\ F^{-T}H & F^{-T} \end{bmatrix} - \lambda \begin{bmatrix} I & 0 \\ -G & I \end{bmatrix} \tag{3.4.23}$$

then we can apply the discussed derivation. If F is singular, then the above transformation is not possible. The pencil $K - \lambda N$ has at least one zero and one infinite eigenvalue.

Mehrmann proposes in [104, Algorithm 15.16] the following algorithm to deflate zero and infinite eigenvalues of $K - \lambda N$: Assume that rank$(F) = k$. First, use the QR decomposition with column pivoting [58] to determine an orthogonal matrix $V \in \mathbf{R}^{n \times n}$, an upper triangular matrix $U \in \mathbf{R}^{n \times n}$ and a permutation matrix $P \in \mathbf{R}^{n \times n}$ such that

$$PF = UV = [0 \ U_2] \begin{bmatrix} V_1 \\ V_2 \end{bmatrix},$$

where $U_2 \in \mathbf{R}^{n \times n-k}$ and $V_2 \in \mathbf{R}^{n-k \times n}$ have full rank. Then form

$$T = \begin{bmatrix} V^T & 0 \\ -HV^T & V^T \end{bmatrix}, \quad \text{and} \quad \tilde{T} = \begin{bmatrix} V(I+GH)^{-1} & 0 \\ U^T PH(I+GH)^{-1} & V \end{bmatrix}.$$

Now as $FV^T = P^T U = P^T[0 \ U_2]$, we obtain

$$\tilde{T}(K - \lambda N)T = \tilde{K} - \lambda \tilde{N}$$

where

$$\widetilde{K} = \begin{bmatrix} V(I+GH)^{-1}P^TU & 0 \\ U^TPH(I+GH)^{-1}P^TU & I \end{bmatrix}$$

$$= \left[\begin{array}{cc|cc} 0 & \widetilde{F}_1 & 0 & 0 \\ 0 & \widetilde{F} & 0 & 0 \\ \hline 0 & 0 & I & 0 \\ 0 & \widetilde{H} & 0 & I \end{array} \right],$$

$$\widetilde{N} = \begin{bmatrix} I & -V(I+GH)^{-1}GV^T \\ 0 & U^TP(-H(I+GH)^{-1}G+I)V^T \end{bmatrix}$$

$$= \left[\begin{array}{cc|cc} I & 0 & \widetilde{G}_{11} & \widetilde{G}_{12} \\ 0 & I & \widetilde{G}_{21} & \widetilde{G} \\ \hline 0 & 0 & 0 & 0 \\ 0 & 0 & \widetilde{F}_1^T & \widetilde{F}^T \end{array} \right],$$

and $\widetilde{F}_1, \widetilde{G}_{12} \in \mathbf{R}^{k \times n-k}$, $\widetilde{F}, \widetilde{H}, \widetilde{G} \in \mathbf{R}^{n-k \times n-k}$. The first k columns of T span the right deflating subspace of $K - \lambda N$ corresponding to k zero eigenvalues and the rows $n + 1, n + 2, \ldots, n + k$ of $\widetilde{T}$ span the left deflating subspace corresponding to k infinity eigenvalues. We may therefore delete rows and columns $1, 2, \ldots, k, n + 1, n + 2, \ldots, n + k$ and proceed with the reduced pencil $\begin{bmatrix} \widetilde{F} & 0 \\ \widetilde{H} & I \end{bmatrix} - \lambda \begin{bmatrix} I & \widetilde{G} \\ 0 & \widetilde{F} \end{bmatrix}$. There is no guarantee that $\widetilde{F}$ is nonsingular. Hence, the procedure described above has to be repeated until the resulting symplectic matrix pencil has no more zero and infinity eigenvalues and $\widetilde{F}$ is nonsingular. Note that neither the rank of F nor the number of zero eigenvalues of F determine the number of zero and infinity eigenvalues of the symplectic pencil. For instance, in Example 10 of [22], $n = 6$, $\text{rank}(F) = 5$, F has three zero eigenvalues and $K - \lambda N$ as in (3.4.21) has two zero and infinite eigenvalues each.

All the computation in this algorithm can be carried out in a numerically reliable way. The solution of the linear systems with $I + GH$ is well-conditioned, since H and G are symmetric positive semidefinite. If after the first iteration, $\widetilde{F}$ is nonsingular, then this process requires $(11n^2 + 6rn + 2r^2)n$ flops; the initial QR decomposition in order to check the rank of F costs $\frac{4}{3}n^3$ flops. Note that this initial decomposition is always computed. Hence, in case F has full rank, when forming the symplectic pencils in (3.4.22) or (3.4.23), the QR decomposition of the F matrix should be used when computing $F^{-T}Q$ and F^{-T} (F^{-1} and $F^{-1}G$) instead of computing an LU decomposition of F.

Assume that $K - \lambda N$ has ℓ zero and ℓ infinite eigenvalues. In the resulting symplectic pencil $K' - \lambda N'$ of dimension $2(n-\ell) \times 2(n-\ell)$, K' is nonsingular. Hence we can build the symplectic matrix pencil (3.4.22) or (3.4.23) and transform it to a butterfly pencil $K'_u - \lambda N'_u$. Thus, $K' - \lambda N'$ is similar to the symplectic butterfly pencil $K'_u - \lambda N'_u$. Adding k rows and columns of zeros to each block of K'_u and N'_u, and appropriate entries on the diagonals, we can expand the symplectic butterfly pencil $K'_u - \lambda N'_u$ to a symplectic butterfly pencil $\widehat{K}_u - \lambda \widehat{N}_u$ of dimension $2n \times 2n$ that is equivalent to $K - \lambda N$.

But even if F is invertible, from a numerically point of view it might not be a good idea to invert F as F can be close to a singular matrix. Hence, the above described approach does not seem to be necessarily a numerically reasonable one.

Just from inspection, one would like to reduce the pencil (3.4.21) to a pencil of the form

$$
K_r - \lambda N_r = \begin{bmatrix} D_1 & 0 \\ D_2 & I \end{bmatrix} - \lambda \begin{bmatrix} I & W \\ 0 & D_1 \end{bmatrix} = \begin{bmatrix} \diagdown & 0 \\ \diagdown & I \end{bmatrix} - \lambda \begin{bmatrix} I & \diagdown \\ 0 & \diagdown \end{bmatrix}
$$

where D_1, D_2 are diagonal matrices and W is a symmetric tridiagonal. Such a reduction exists, at least if F is invertible. This follows from the fact that in that case $K - \lambda N$ can be reduced to the form $K_s - \lambda N_s$. The above form can be obtained from $K_s - \lambda N_s$ by premultiplication with $\begin{bmatrix} I & 0 \\ 0 & B_{11}^{-1} \end{bmatrix}$.

It would be nice to preserve the special structure of K and N during such a reduction process. In particular, it would be desirable to keep the zero and identity blocks unchanged during the reduction process and to treat the $(1,1)$ block of K and the $(2,2)$ block of N (that is, F and F^T) alike. Treating F and F^T alike implies that only transformations of the form

$$
S = \begin{bmatrix} X & 0 \\ 0 & Y \end{bmatrix}, \qquad Z = \begin{bmatrix} U & 0 \\ 0 & U^{-T} \end{bmatrix}
$$

can be applied, as otherwise the F and H block in K and the F and G block in N are mixed. It is easily seen that such a transformation guarantees that the zeros blocks stay zero. Further, we have

$$
\begin{aligned}
XU &= I &\implies& X = U^{-1}, \\
YU^{-T} &= I &\implies& Y = U^T, \\
XFU &= D_1 &\implies& U^{-1}FU = \begin{bmatrix} \diagdown \end{bmatrix} \in \mathbf{R}^{n \times n}, \\
XGU^{-T} &= W &\implies& U^{-1}GU^{-T} = \begin{bmatrix} \diagdown\diagdown \end{bmatrix}, \\
-YHU &= D_2 &\implies& -U^{-T}HU = \begin{bmatrix} \diagdown \end{bmatrix}.
\end{aligned}
$$

A matrix $A \in \mathbf{R}^{n \times n}$ is diagonalizable by a matrix $X \in \mathbf{C}^{n \times n}$ if and only if A is not defective. Hence not every matrix A is diagonalizable. Here we ask for even more: $F \in \mathbf{R}^{n \times n}$ should be diagonalized by a real $n \times n$ matrix U and that same matrix U should transform the G and the H block to the desired form. This will only be possible for certain special cases, but not in general. Therefore, more general transformation matrices S and Z have to be used, the blocks in K and N have to be mixed in the course of the reduction. We have to allow some fill-in in the zero and identity blocks in the intermediate transformed matrices. The special structure of K and N can not be kept during the reduction process. While in the beginning one can make use of the fact that

$$K^T J K = N^T J N = \begin{bmatrix} 0 & F^T \\ -F & 0 \end{bmatrix},$$

as soon as fill-in in the zero and identity blocks occur, there are no zero blocks in $K^T J K = N^T J N$ anymore. There seems to be no advantage in concentrating our attention on symplectic matrix pencils arising in control problems. One can just as well derive an algorithm for reducing a general symplectic matrix pencil to the form $K_r - \lambda N_r$. It is clear from the above that such a reduction is possible if K and N are nonsingular. What can be done if K and N are singular, is still an open research problem.

Chapter 4

BUTTERFLY SR AND SZ ALGORITHMS

Once the reduction of a symplectic matrix to butterfly form is achieved, the SR algorithm (see Section 2.2.1) is a suitable tool for computing the eigenvalues/eigenvectors of a symplectic matrix. As will be seen in this chapter, the SR algorithm preserves the symplectic butterfly form in its iterations and can be rewritten in a parameterized form that works with $4n - 1$ parameters instead of the $(2n)^2$ matrix elements in each iteration. Such an algorithm was already considered in [13, 19]. In those publications, it is proposed to use a polynomial of the form $p(\lambda) = \prod_{i=1}^{k}(\lambda - \mu_i)$ to drive the SR step, just as in the implicit QR (bulge-chasing) algorithm for upper Hessenberg matrices. Here we will show that it is better to use a Laurent polynomial to drive the SR step. A natural way to choose the spectral transformation function p_j in the butterfly SR algorithm is

- single shift: $p_1(B) = B - \mu I$ for $\mu \in \mathbf{R}$;

- double shift: $p_2(B) = (B - \mu I)(B - \overline{\mu} I)$ for $\mu \in \mathbf{C}$, or
  $$p_2(B) = (B - \mu I)(B - \frac{1}{\mu}I) \text{ for } \mu \in \mathbf{R};$$

- quadruple shift: $p_4(B) = (B - \mu I)(B - \overline{\mu} I)(B - \frac{1}{\mu}I)(B - \frac{1}{\overline{\mu}}I)$, for $\mu \in \mathbf{C}$.

In particular the double shift for $\mu \in \mathbf{R}$, or $\mu \in i\mathbf{R}$, and the quadruple shift for $\mu \in \mathbf{C}$ make use of the symmetries of the spectrum of symplectic matrices. But, as will be seen in Section 4.1.1, a better choice is a Laurent polynomial $q_2(\lambda) = p_2(\lambda)\lambda^{-1}$ or $q_4(\lambda) = p_4(\lambda)\lambda^{-2}$. Each of these is a function in $\lambda + \lambda^{-1}$. For example,

$$\begin{aligned}
q_4(\lambda) &= (\lambda + \lambda^{-1})^2 - (\mu + \mu^{-1} + \overline{\mu} + \overline{\mu^{-1}})(\lambda + \lambda^{-1}) \\
&\quad + (\mu + \mu^{-1})(\overline{\mu} + \overline{\mu^{-1}}) - 2.
\end{aligned}$$

85

This reduces the size of the bulges that are introduced, thereby decreasing the number of computations required per iteration. It also improves the convergence and stability properties of the algorithm by effectively treating each reciprocal pair of eigenvalues as a unit. The method still suffers from loss of the symplectic structure due to roundoff errors, but the loss of symplecticity is normally less severe than in an implementation using a standard polynomial, because less arithmetic is done and the similarity transformations are generally better conditioned. Moreover, using the factors K and N of the symplectic butterfly matrix B, one can easily and cheaply restore the symplectic structure of the iterates whenever necessary.

The butterfly SR algorithm works on the butterfly matrix B and transforms it into a butterfly matrix $\widetilde{B}$ which decouples into simple 2×2 or 4×4 symplectic eigenproblems. Making use of the factorization of B into $K^{-1}N$ as in (3.2.15) or (3.2.7), we will develop a parameterized SR algorithm for computing the eigeninformation of a parameterized symplectic butterfly matrix. The algorithm will work only on the $4n - 1$ parameters that determine the symplectic butterfly matrix B similar to the approach used for the development of the unitary Hessenberg QR algorithm [61]. It computes the parameters which determine the matrix $\widetilde{B}$ without ever forming B or $\widetilde{B}$. This is done by decomposing K and N into even simpler symplectic matrices, and by making use of the observation that most of the transformations applied to $B = K^{-1}N$ during the implicit SR step commute with most of the simple factors of K and N.

Finally we will develop a second algorithm that works only on the parameters. We have seen that B can be factored into $B = K^{-1}N$ as in (3.2.15) or (3.2.7). The eigenvalue problem $K^{-1}Nx = \lambda x$ is equivalent to $(\lambda K - N)x = 0$ and $(K - \lambda N)x = 0$ because of the symmetry of the spectrum. In the latter equations the $4n - 1$ parameters are given directly. The idea here is that instead of considering the eigenproblem for B, we can just as well consider the generalized eigenproblem $(K - \lambda N)x = 0$. An SZ algorithm will be developed to solve these generalized eigenproblems. The SZ algorithm is the analogue of the SR algorithm for the generalized eigenproblem, just as the QZ algorithm is the analogue of the QR algorithm for the generalized eigenproblem, see Section 2.2.2.

The butterfly SR algorithm is discussed in Section 4.1. In particular, the use of Laurent polynomials as shift polynomials, and the choice of the shifts will be discussed. It will be shown that the convergence rate of the butterfly SR algorithm is typically cubic. Section 4.2 deals with the parameterized butterfly SR algorithm. The butterfly SZ algorithm is presented in Section 4.3. Numerical experiments with the proposed algorithms are described in Section 4.4. The experiments clearly show: The methods did always converge, cubic convergence can be observed. The parameterized SR algorithm converges slightly faster than the SR algorithm. The eigenvalues are computed to about the same accuracy. The SZ algorithm is considerably better than the SR algorithm in computing the eigenvalues of a parameterized symplectic matrix/matrix pencil. The number of (quadruple-shift) iterations needed for convergence for each eigenvalue is about $2/3$.

Finally, two interesting comments on the butterfly algorithm are given in Section 4.5. The first comment is on a connection between the butterfly SR algorithm and the HR algorithm. The second comment is on how one of the problems that motivated us to study the symplectic SR algorithm, the problem of solving a discrete-time algebraic Riccati equation, can be solved using the results obtained.

Some of the results discussed in this chapter appeared in [19, 21, 49, 52].

4.1 THE BUTTERFLY SR ALGORITHM

In Section 2.2.1, the SR algorithm for general $2n \times 2n$ real matrices was reviewed. The SR algorithm is a symplectic QR–like method for solving eigenvalue problems based on the SR decomposition. The QR decomposition and the orthogonal similarity transformation to upper Hessenberg form in the QR process are replaced by the SR decomposition and the symplectic similarity reduction to J–Hessenberg form. Unfortunately, a symplectic matrix in butterfly form is not a J–Hessenberg matrix so that we can not simply use the results given in Section 2.2.1 for computing the eigenvalues of a symplectic butterfly matrix. But, as we will see in this section, an SR step preserves the butterfly form. If B is an unreduced symplectic butterfly matrix, $p(B)$ a polynomial such that $p(B) \in \mathbf{R}^{2n \times 2n}$, $p(B) = SR$, and R is invertible, then $S^{-1}BS$ is a symplectic butterfly matrix again. This was already noted and proved in [13], but no results for singular $p(B)$ are given there. Here we will show that, as to be expected, singular $p(B)$ are desirable (that is at least one shift is an eigenvalue of B), as they allow the problem to be deflated after one step.

First, we need to introduce some notation. Let $p(B)$ be a polynomial such that $p(B) \in \mathbf{R}^{2n \times 2n}$. Write $p(B)$ in factored form

$$p(B) := (B - \lambda_1 I^{2n,2n})(B - \lambda_2 I^{2n,2n}) \cdots (B - \lambda_k I^{2n,2n}). \qquad (4.1.1)$$

From $p(B) \in \mathbf{R}^{2n \times 2n}$ it follows that if $\mu \in \mathbb{C}$, and $\mu \in \{\lambda_1, \ldots, \lambda_k\}$, then $\bar{\mu} \in \{\lambda_1, \ldots, \lambda_k\}$. $p(B)$ is singular if and only if at least one of the shifts λ_i is an eigenvalue of B. Such a shift will be called *perfect shift*. Let ν denote the number of shifts that are equal to eigenvalues of B. Here we count a repeated shift according to its multiplicity as a zero of p, except that the number of times we count it must not exceed its algebraic multiplicity as an eigenvalue of B.

LEMMA 4.1 *Let $B \in \mathbf{R}^{2n \times 2n}$ be an unreduced symplectic butterfly matrix. The rank of $p(B)$ in (4.1.1) is $2n - \nu$ with ν as defined above.*

PROOF: Since B is an unreduced butterfly matrix, its eigenspaces are one-dimensional by Lemma 3.10. Hence, we can use the same arguments as in the proof of Lemma 4.4 in [141] in order to prove the statement of this lemma. $\sqrt{}$

In the following we will consider only the case that rank$(p(B))$ is even. In a real implementation, one would choose a polynomial p such that each perfect shift is accompanied by its reciprocal, since the eigenvalues of a symplectic matrix always appear in reciprocal pairs. If $\mu \in \mathbf{R}$ is a perfect shift, then we choose μ^{-1} as a shift as well. If $\mu \in \mathbf{C}$ is a perfect shift, then we choose $\mu, \mu^{-1}, \bar{\mu}$ and $\bar{\mu}^{-1}$ as shifts. Because of this, rank$(p(B))$ will always be even.

THEOREM 4.2 *Let* $B \in \mathbf{R}^{2n \times 2n}$ *be an unreduced symplectic butterfly matrix. Let* $p(B)$ *be a polynomial with* $p(B) \in \mathbf{R}^{2n \times 2n}$ *and* rank$(p(B)) = 2n - \nu =:$ $2k.$ *If* $p(B) = SR$ *exists, then* $\tilde{B} = S^{-1}BS$ *is a symplectic matrix of the form*

where

- $\begin{bmatrix} \tilde{B}_{11} & \tilde{B}_{13} \\ \tilde{B}_{31} & \tilde{B}_{33} \end{bmatrix}$ *is a symplectic butterfly matrix and*

- *the eigenvalues of* $\begin{bmatrix} \tilde{B}_{22} & \tilde{B}_{24} \\ \tilde{B}_{42} & \tilde{B}_{44} \end{bmatrix}$ *are just the* ν *shifts that are eigenvalues of* B.

In order to simplify the notation for the proof of this theorem and the subsequent derivations, we use in the following permuted versions of B, R, and S. Let

$$B_P = PBP^T, \quad R_P = PRP^T, \quad S_P = PSP^T, \quad J_P = PJP^T,$$

with P as in (2.1.2).
From $S^T JS = J$ we obtain

while the permuted butterfly matrix B_P is given by

$$\left[\begin{array}{cc|cc|cc}
b_1 & b_1c_1 - a_1^{-1} & 0 & b_1d_2 & & \\
a_1 & a_1c_1 & 0 & a_1d_2 & & \\
\hline
0 & b_2d_2 & \ddots & & \ddots & \\
0 & a_2d_2 & \ddots & & \ddots & \\
\hline
& & \ddots & & \ddots & \begin{array}{cc} 0 & b_{n-1}d_n \\ 0 & a_{n-1}d_n \end{array} \\
& & & 0 & b_nd_n & \begin{array}{c|cc} b_n & b_nc_n - a_n^{-1} \end{array} \\
& & & 0 & a_nd_n & \begin{array}{c|cc} a_n & a_nc_n \end{array}
\end{array}\right]. \quad (4.1.2)$$

Recall the definition of a generalized Krylov matrix (Definition 2.3)

$$L(A, v, j) = [v, A^{-1}v, A^{-2}v, \ldots, A^{-(j-1)}v, Av, A^2v, \ldots, A^jv]$$

for $A \in \mathbf{R}^{2n \times 2n}$, $v \in \mathbf{R}^{2n}$.

PROOF of Theorem 4.2: B_P is an upper triangular matrix with two additional subdiagonals, where the second additional subdiagonal has a nonzero entry only in every other position (see (4.1.2)). Since R is a J–triangular matrix, R_P is an upper triangular matrix. In the following, we denote by $Z^{2n,2k}$ the first $2k$ columns of a $2n \times 2n$ matrix Z, while Z^{rest} denotes its last $2n - 2k$ columns. $Z^{2k,2k}$ denotes the leading $2k \times 2k$ principal submatrix of a $2n \times 2n$ matrix Z.

Now partition the permuted matrices B_P, S_P, J_P, and R_P as

$$B_P = [B_P^{2n,2k} \mid B_P^{rest}], \qquad S_P = [S_P^{2n,2k} \mid S_P^{rest}],$$

$$J_P = [J_P^{2n,2k} \mid J_P^{rest}], \qquad R_P = \left[\begin{array}{c|c} R_P^{2k,2k} & X \\ \hline 0 & Y \end{array}\right],$$

where the matrix blocks are defined as before: $X \in \mathbf{R}^{2k \times 2(n-k)}$, $Y \in \mathbf{R}^{2(n-k) \times 2(n-k)}$.

First we will show that the first $2k$ columns and rows of $\widetilde{B}_P$ are in the desired form. We will need the following observations. The first $2k$ columns of $p(B_P)$ are linear independent, since B is unreduced. To see this, consider the following identity:

$$p(B)L(B, e_1, n) =$$
$$= [p(B)e_1, p(B)B^{-1}e_1, \ldots, p(B)B^{-(n-1)}e_1, p(B)Be_1, \ldots, p(B)B^ne_1]$$
$$= [p(B)e_1, B^{-1}p(B)e_1, \ldots, B^{-(n-1)}p(B)e_1, Bp(B)e_1, \ldots, B^np(B)e_1]$$
$$= L(B, p(B)e_1, n),$$

where we have used the property $p(B)B^r = B^rp(B)$ for $r = \pm 1, \ldots, \pm n$. From Theorem 3.7 b) we know that, since B is unreduced, $L(B, e_1, n)$ is a nonsingular upper J–triangular matrix. As $\text{rank}(p(B)) = 2k$, $L(B, p(B)e_1, n)$ has rank $2k$. If a matrix of the form

$$L(X, v, n) = [v, X^{-1}v, \ldots, X^{-(n-1)}v, Xv, \ldots, X^n v]$$

has rank $2k$, then the columns

$$v, X^{-1}v, \ldots, X^{-(k-1)}v, Xv, \ldots, X^k v$$

are linear independent. Further we obtain

$$p(B) = L(B, p(B)e_1, n)(L(B, e_1, n))^{-1} =: [p_1, p_2, \ldots, p_{2n}].$$

Due to the special form of $L(B, e_1, n)$ (J–triangular!) and the fact that the columns 1 to k and $n+1$ to $n+k$ of $L(B, p(B)e_1, n)$ are linear independent, the columns

$$p_1, p_2, \ldots, p_k, p_{n+1}, p_{n+2}, \ldots, p_{n+k}$$

of $p(B)$ are linear independent. Hence the first $2k$ columns of

$$p(B_P) = P p(B) P^T$$

are linear independent.

The columns of $S_P^{2n,2k}$ are linear independent, since S_P is nonsingular. It follows that the matrix $R_P^{2k,2k}$ is nonsingular, since

$$p(B_P)I^{2n,2k} = S_P^{2n,2k} R_P^{2k,2k}.$$

Therefore, we have

$$S_P^{2n,2k} = p(B_P)I^{2n,2k}(R_P^{2k,2k})^{-1}. \tag{4.1.3}$$

Moreover, since $\text{rank}(p(B)) = 2k$, we have that $\text{rank}(R_P) = 2k$. Since $\text{rank}(R_P^{2k,2k}) = 2k$, we obtain $\text{rank}(Y) = 0$ and therefore $Y = 0$. From this we see

$$R_P = \left[\begin{array}{c|c} R_P^{2k,2k} & X \\ \hline 0 & 0 \end{array} \right]. \tag{4.1.4}$$

Further we need the following identities

$$\begin{align}
B_P p(B_P) &= p(B_P) B_P, & (4.1.5) \\
B_P^{-1} p(B_P) &= p(B_P) B_P^{-1}, & (4.1.6) \\
B_P^T J_P^{-1} &= J_P^{-1} B_P^{-1}, & (4.1.7) \\
B_P^{-1} &= J_P^{-1} B_P^T J_P, & (4.1.8) \\
S_P^T J_P^{-1} &= J_P^{-1} S_P^{-1}, & (4.1.9) \\
S_P^{-1} &= J_P^{-1} S_P^T J_P, & (4.1.10) \\
S_P J_P^{2n,2k} &= S_P^{2n,2k} J_P^{2k,2k}. & (4.1.11)
\end{align}$$

Equations (4.1.7) – (4.1.11) follow from the fact that B and S are symplectic while (4.1.5) – (4.1.6) result from the fact that Z and $p(Z)$ commute for any matrix Z and any polynomial p.

The first $2k$ columns of $\widetilde{B}_P$ are given by the expression

$$
\begin{aligned}
\widetilde{B}_P^{2n,2k} &= \widetilde{B}_P I^{2n,2k} \\
&= S_P^{-1} B_P S_P I^{2n,2k} \\
&= S_P^{-1} B_P S_P^{2n,2k} \\
&= S_P^{-1} B_P p(B_P) I^{2n,2k} (R_P^{2k,2k})^{-1} \qquad \text{by} \quad (4.1.3) \\
&= S_P^{-1} p(B_P) B_P^{2n,2k} (R_P^{2k,2k})^{-1} \qquad \text{by} \quad (4.1.5) \\
&= R_P B_P^{2n,2k} (R_P^{2k,2k})^{-1}.
\end{aligned}
$$

Using (4.1.4), that $(R_P^{2k,2k})^{-1}$ is an $2k \times 2k$ upper triangular matrix, and that B_P is of the form given in (4.1.2), we obtain that

$$
\widetilde{B}_P^{2n,2k} =
\begin{bmatrix}
x & x & x & x & \cdots & \cdots & x & x \\
x & x & x & x & \cdots & \cdots & x & x \\
0 & x & x & x & & & \vdots & \vdots \\
0 & x & x & x & & & \vdots & \vdots \\
& & \ddots & & \ddots & & x & x \\
& & & \ddots & & \ddots & x & x \\
& & & & 0 & x & x & x \\
& & & & 0 & x & x & x \\
& & & & & & 0 & 0 \\
& & & & & & 0 & 0 \\
0 & & & \cdots & & & & 0 \\
\vdots & & & & & & & \vdots \\
0 & & & \cdots & & & & 0
\end{bmatrix}.
$$

Hence

$$
\widetilde{B}_P =
\begin{bmatrix}
x & x & x & x & \cdots & \cdots & x & x & x & x & \cdots & \cdots & x & x \\
x & x & x & x & \cdots & \cdots & x & x & x & x & \cdots & \cdots & x & x \\
0 & x & x & x & & & \vdots & \vdots & x & x & \cdots & \cdots & x & x \\
0 & x & x & x & & & \vdots & \vdots & x & x & \cdots & \cdots & x & x \\
& & \ddots & & \ddots & & x & x & x & x & \cdots & \cdots & x & x \\
& & & \ddots & & \ddots & x & x & x & x & \cdots & \cdots & x & x \\
& & & & 0 & x & x & x & x & x & \cdots & \cdots & x & x \\
& & & & 0 & x & x & x & x & x & \cdots & \cdots & x & x \\
& & & & & & 0 & 0 & x & x & \cdots & \cdots & x & x \\
& & & & & & 0 & 0 & x & x & \cdots & \cdots & x & x \\
& & & & & & & & x & x & \cdots & \cdots & x & x \\
& & & & & & & & \vdots & \vdots & & & \vdots & \vdots \\
& & & & & & & & x & x & \cdots & \cdots & x & x
\end{bmatrix}.
$$

and thus,

$$
\tilde{B} =
\begin{bmatrix}
\begin{array}{cc|cc}
\diagdown & \square & \diagdown & \square \\
0 & \square & 0 & \square \\
\hline
\diagdown & \square & \diagdown & \square \\
0 & \square & 0 & \square
\end{array}
\end{bmatrix}
\begin{array}{l}
\}k \\
\}n-k \\
\}k \\
\}n-k
\end{array}
\qquad (4.1.12)
$$

$$
\underbrace{\quad}_{k}\ \underbrace{\quad}_{n-k}\ \underbrace{\quad}_{k}\ \underbrace{\quad}_{n-k}
$$

The first $2k$ columns of $(\tilde{B}_P)^T$ are given by the expression

$$
\begin{aligned}
(\tilde{B}_P)^T I^{2n,2k} &= S_P^T B_P^T S_P^{-T} I^{2n,2k} \\
&= S_P^T B_P^T J_P^{-1} S_P J_P I^{2n,2k} && \text{by } (4.1.10) \\
&= S_P^T B_P^T J_P^{-1} S_P J_P^{2n,2k} \\
&= S_P^T B_P^T J_P^{-1} S_P^{2n,2k} J_P^{2k,2k} && \text{by } (4.1.3) \\
&= S_P^T B_P^T J_P^{-1} p(B_P) \begin{bmatrix} (R_P^{2k,2k})^{-1} \\ \hline 0 \end{bmatrix} J_P^{2k,2k} && \text{by } (4.1.11) \\
&= S_P^T J_P^{-1} B_P^{-1} p(B_P) \begin{bmatrix} (R_P^{2k,2k})^{-1} \\ \hline 0 \end{bmatrix} J_P^{2k,2k} && \text{by } (4.1.7) \\
&= S_P^T J_P^{-1} p(B_P) B_P^{-1} \begin{bmatrix} (R_P^{2k,2k})^{-1} \\ \hline 0 \end{bmatrix} J_P^{2k,2k} && \text{by } (4.1.6) \\
&= J_P^{-1} S_P^{-1} p(B_P) B_P^{-1} \begin{bmatrix} (R_P^{2k,2k})^{-1} \\ \hline 0 \end{bmatrix} J_P^{2k,2k} && \text{by } (4.1.9) \\
&= J_P^{-1} R_P B_P^{-1} \begin{bmatrix} (R_P^{2k,2k})^{-1} \\ \hline 0 \end{bmatrix} J_P^{2k,2k} \\
&= (J_P^{-1} R_P J_P^{-1}) B_P^T (J_P \begin{bmatrix} (R_P^{2k,2k})^{-1} \\ \hline 0 \end{bmatrix} J_P^{2k,2k}) && \text{by } (4.1.8).
\end{aligned}
$$

Therefore

$$(\widetilde{B}_P)^T I^{2n,2k} = \begin{bmatrix} x & x & x & x & \cdots & \cdots & x & x \\ x & x & x & x & \cdots & \cdots & x & x \\ 0 & 0 & x & x & & & \vdots & \vdots \\ x & x & x & x & & & \vdots & \vdots \\ & & \ddots & & \ddots & & x & x \\ & & & \ddots & & \ddots & x & x \\ & & & & 0 & 0 & x & x \\ & & & & x & x & x & x \\ & & & & & & 0 & 0 \\ & & & & & & 0 & 0 \\ 0 & & & \cdots & & & & 0 \\ \vdots & & & & & & & \vdots \\ 0 & & & \cdots & & & & 0 \end{bmatrix}.$$

For the last equation we used (4.1.4), that $(R_P^{2k,2k})^{-1}$ a $2k \times 2k$ is an upper triangular matrix, and that B_P is of the form (4.1.2). Hence, we can conclude that

$$\widetilde{B}^T = \left[\begin{array}{cc|cc} \boxminus & \square & \boxminus & \square \\ 0 & \square & 0 & \square \\ \hline \boxminus & \square & \boxminus & \square \\ 0 & \square & 0 & \square \end{array}\right], \tag{4.1.13}$$

where the blocks have the same size as before. Comparing (4.1.12) and (4.1.13) we obtain

$$\widetilde{B} = \left[\begin{array}{cc|cc} \boxminus & 0 & \boxminus & 0 \\ 0 & \square & 0 & \square \\ \hline \boxminus & 0 & \boxminus & 0 \\ 0 & \square & 0 & \square \end{array}\right].$$

This proves the first part of the theorem. The result about the eigenvalues now follows with the arguments as in the proof of Theorem 4.5 in [141]. There, a similar statement for a generic chasing algorithm is proved. $\checkmark$

Assuming its existence, the SR decomposition and the SR step based on unreduced symplectic butterfly matrices possess many of the desirable properties of the QR step. An SR algorithm can thus be formulated similarly to the QR algorithm as already discuss in Section 2.2.1.

let $B_0 = B$ be a symplectic butterfly matrix
for $k = 1, 2, \ldots$

choose a shift polynomial p_k such that $p_k(B_{k-1}) \in \mathbf{R}^{2n \times 2n}$.
compute the SR decomposition $p_k(B_{k-1}) = S_k R_k$
compute $B_k = S_k^{-1} B_{k-1} S_k$

An algorithm for computing S and R explicitly is presented in Section 2.2.1. As with explicit QR steps, the expense of explicit SR steps comes from the fact that $p(B)$ has to be computed explicitly. A preferred alternative is the implicit SR step, an analogue to the Francis QR step [54, 58, 79]. The first implicit transformation S_1 is selected so that the first columns of the implicit and the explicit S are equivalent. That is, a symplectic matrix S_1 is determined such that

$$S_1^{-1} p(B) e_1 = \alpha e_1, \qquad \alpha \in \mathbf{R}.$$

Applying this first transformation to the butterfly matrix yields a symplectic matrix $S_1^{-1} B S_1$ with almost butterfly form having a small bulge. The remaining implicit transformations perform a bulge-chasing sweep down the subdiagonals to restore the butterfly form. That is, a symplectic matrix S_2 is determined such that $S_2^{-1} S_1^{-1} B S_1 S_2$ is of butterfly form again. As the implicit SR step is analogous to the implicit QR step, this technique will not be discussed here in great detail. The algorithm for reducing a symplectic matrix to butterfly form as given in Section 3.2 can be used as a building block for the implicit SR step. An efficient implementation of the SR step for symplectic butterfly matrices involves $\mathcal{O}(n)$ arithmetic operations. Hence a gain in efficiency is obtained compared to the (non-parameterized) SR algorithm on symplectic J–Hessenberg matrices proposed by Flaschka, Mehrmann and Zywietz in [53]. There each SR step involves $\mathcal{O}(n^2)$ arithmetic operations.

4.1.1   LAURENT POLYNOMIALS AS SHIFT POLYNOMIALS

Shift polynomials p for the SR step have to be chosen such that $p(B) \in \mathbf{R}^{2n \times 2n}$. For the standard Hessenberg QR step any polynomial of the form $p(\lambda) = \prod_{i=1}^{k} (\lambda - \mu_i)$ can be chosen. Usually, single or double shift steps are performed in a way to keep all iterates real, although the QR factorization of a complex matrix does exist. For the SR step it is necessary to keep all iterates real, as the set of matrices $A \in \mathbb{C}^{2n \times 2n}$ which have an SR decomposition $A = SR$, where $S^H J S = J$ or $S^H J S = -J$, is not dense in $\mathbb{C}^{2n \times 2n}$. A natural way to choose the spectral transformation function p_j in the butterfly SR algorithm is

- single shift: $p_1(B) = B - \mu I$ for $\mu \in \mathbf{R}$;

- double shift: $p_2(B) = (B - \mu I)(B - \bar{\mu} I)$ for $\mu \in \mathbb{C}$, or
$p_2(B) = (B - \mu I)(B - \frac{1}{\mu} I)$ for $\mu \in \mathbf{R}$;

- quadruple shift: $p_4(B) = (B - \mu I)(B - \bar{\mu} I)(B - \frac{1}{\mu} I)(B - \frac{1}{\bar{\mu}} I)$, for $\mu \in \mathbb{C}$.

In particular the double shift for $\mu \in \mathbb{R}$, or $\mu \in i\mathbb{R}$, and the quadruple shift for $\mu \in \mathbb{C}$ make use of the symmetries of the spectrum of symplectic matrices. But, as explained in the following, a better choice is a Laurent polynomial

$$q_2(\lambda) = p_2(\lambda)\lambda^{-1} \qquad \text{or} \qquad q_4(\lambda) = p_4(\lambda)\lambda^{-2}.$$

Each of these is a function of $\lambda + \lambda^{-1}$. For example,

$$\begin{aligned}
q_4(\lambda) &= (\lambda + \lambda^{-1})^2 - (\mu + \mu^{-1} + \bar{\mu} + \overline{\mu^{-1}})(\lambda + \lambda^{-1}) \\
&\quad + (\mu + \mu^{-1})(\bar{\mu} + \overline{\mu^{-1}}) - 2.
\end{aligned}$$

At first it would appear not to matter whether p_4 or q_4 is used to drive the SR step; the outcome should be essentially the same: An SR iteration driven by p_4 has the form

$$\hat{B} = S^{-1}BS,$$

where S comes from an SR decomposition:

$$p_4(B) = SR.$$

On the other hand,

$$q_4(B) = B^{-2}p_4(B) = (B^{-2}S)R,$$

which is an SR decomposition of $q_4(B)$. Thus an SR iteration driven by q_4 gives

$$(B^{-2}S)^{-1}B(B^{-2}S) = S^{-1}BS = \hat{B},$$

the same as for p_4. This equation ignores the fact that the SR decomposition is not uniquely defined. S is at best unique up to right multiplication by a trivial matrix (3.3.17). Consequently $\hat{B}$ is only unique up to a trivial similarity transformation. The $\hat{B}$ that is obtained in practice will depend upon whether p_4 or q_4 is used to drive the step. In principle any undesirable discrepancy that arises can be corrected by application of a similarity transformation by a trivial matrix. Note, however, that a trivial matrix can be arbitrarily ill conditioned. Thus one transformation could be much better conditioned than the other.

The convergence theory of GR algorithms [142] suggests that Laurent polynomials will be more satisfactory than ordinary polynomials from this standpoint. If symplectic structure is to be preserved throughout the computation, eigenvalues must be deflated in pairs: when λ is removed, λ^{-1} must also be removed. Thus we want eigenvalues λ and λ^{-1} to converge at the same rate. The convergence of GR algorithms is driven by convergence of iterations on a nested sequence of subspaces of dimensions $1, 2, \ldots, 2n - 1$, [115, 142].

If iterations are driven by a function f, the rate of convergence of the subspaces of dimension k is determined by the ratio $|f(\lambda_{k+1})|/|f(\lambda_k)|$, where the eigenvalues of $f(B)$ are numbered in the order

$$|f(\lambda_1)| \geq |f(\lambda_2)| \geq \cdots \geq |f(\lambda_{2n})|. \tag{4.1.14}$$

If f is a Laurent polynomial q of the type we have proposed to use, then $q(\lambda) = q(\lambda^{-1})$ for every λ, so each eigenvalue appears side-by-side with its inverse in the ordering (4.1.14). The odd ratios $|f(\lambda_{2j})|/|f(\lambda_{2j-1})|$ are all equal to one; only the even-dimensional subspaces converge. Reciprocal pairs of eigenvalues converge at the same rate and are deflated at the same time.

In contrast, if f is a regular polynomial $p(\lambda) = \lambda^k q(\lambda)$, then for any eigenvalue λ satisfying $|\lambda| > 1$, we will have $p(\lambda) = \lambda^{2k} p(\lambda^{-1})$, whence $|p(\lambda)| > |p(\lambda^{-1})|$. Thus the underlying subspace iterations will tend to force λ and λ^{-1} to converge at different rates. Suppose, for example, B has a single real eigenvalue λ_1 such that $p(\lambda_1)$ dominates all the other eigenvalues. Then the odd ratio $|p(\lambda_2)|/|p(\lambda_1)|$ is less than one, and the sequence of one-dimensional subspaces will converge. This tends to force a_1, the first entry of B_{21}, toward zero. If, after some iterations, a_1 becomes effectively zero, then b_1, the first entry of B_1, will have converged to λ_1. As we already noted in Remark 3.14, the symplectic structure then forces the $(1,1)$ entry of B_{22} to be λ_1^{-1} and allows a deflation. According to the GR convergence theory, the eigenvalue that should emerge in the $(1,1)$ position of B_{22} is λ_2, where $p(\lambda_2)$ is the second largest eigenvalue of $p(B)$. If, as may happen, $\lambda_2 \neq \lambda_1^{-1}$, we have a conflict between the symplectic structure and the convergence theory. This apparent contradiction is resolved as follows. The convergence of the matrix iterates depends not only on the underlying subspace iterations, but also on the condition numbers of the transforming matrices S [142]. Convergence of the subspace iterations may fail to result in convergence of the matrix iterations if *and only if* the transforming matrices are ill conditioned. The tension between the symplectic structure and the convergence of the subspace iterations is inevitably resolved in favor of the symplectic structure through the production of ill-conditioned transforming matrices. This is clearly something we wish to avoid. Example 4.14 in Section 4.4 demonstrates that situations like this do arise and can have undesirable consequences.

Apart from these considerations, the Laurent polynomial is superior because it allows a more economical implementation than the standard polynomial does. For symplectic butterfly matrices B,

$$p_4(B)e_1 = \sum_{j=1}^{4} (\beta_j e_j + \gamma_j e_{n+j})$$

has (generically) eight nonzero entries, whereas

$$q_4(B)e_1 = \alpha_1 e_1 + \alpha_2 e_2 + \alpha_3 e_3$$

has only three nonzero entries.

For the implicit SR step driven by p_4 or q_4, a symplectic matrix S ($\widetilde{S}$) has to be determined such that $Sp_4(B)e_1 = \alpha e_1$ ($\widetilde{S}q_4(B)e_1 = \alpha e_1$). In order to compute S we eliminate the entries $n+1$ to $n+4$ of $p_4(B)e_1$ by symplectic Givens transformations and the entries 2 to 4 by a symplectic Householder transformation. Hence SBS^{-1} is of the form

```
⎡ x  +  +  +  │  x  x  +  +  +              ⎤
⎢ +  x  +  +  │  x  x  x  +  +              ⎥
⎢ +  +  x  +  │  +  x  x  x  +              ⎥
⎢ +  +  +  x  │  +  +  x  x  x              ⎥
⎢ +  +  +  +  x│  +  +  +  x  x  x          ⎥
⎢            x│           x  x  x          ⎥
⎢              │              ⋱    ⋱        ⎥
⎢ ─────────────┼───────────────────────── ⎥
⎢ x  +  +  +  │  x  x  +  +  +              ⎥
⎢ +  x  +  +  │  x  x  x  +  +              ⎥
⎢ +  +  x  +  │  +  x  x  x  +              ⎥
⎢ +  +  +  x  │  +  +  x  x  x              ⎥
⎢ +  +  +  +  x│  +  +  +  x  x  x          ⎥
⎣            x│           x  x  x     ⋱  ⋱ ⎦
```

where the undesired elements are denoted by $+$. In order to compute $\widetilde{S}$ we eliminate the entries 3 and 2 of $q_4(B)e_1$ by a symplectic Householder transformation. Applying $\widetilde{S}$ to B yields

```
⎡ x  +  +     │  x  x  +  +                 ⎤
⎢ +  x  +     │  x  x  x  +                 ⎥
⎢ +  +  x     │  +  x  x  x                 ⎥
⎢          x  │  +  +  x  x  x              ⎥
⎢            x│           x  x  x           ⎥
⎢             x            x  x  x          ⎥
⎢              │              ⋱   ⋱   ⋱     ⎥
⎢ ─────────────┼───────────────────────── ⎥
⎢ x  +  +     │  x  x  +  +                 ⎥
⎢ +  x  +     │  x  x  x  +                 ⎥
⎢ +  +  x     │  +  x  x  x                 ⎥
⎢          x  │  +  +  x  x  x              ⎥
⎢            x│           x  x  x           ⎥
⎣             x            x  x  x    ⋱  ⋱ ⎦
```

If we use p_4 instead of q_4, the bulge is one column and at least one row larger in each quadrant of the matrix. We prefer q_4 over p_4, because it is cheaper to chase the smaller bulge. Similarly, we use q_2 instead of p_2. To be precise, each step of chasing the bulge one row and one block further down in the matrix requires 8 symplectic Givens transformations, 1 symplectic Gauss transformation, and 2 symplectic Householder transformations (each of the Householder transformations are used to zero 3 matrix elements) when p_4 is used. Using q_4, only 4 symplectic Givens transformations, 1 symplectic Gauss transformation, and 2 symplectic Householder transformations are required. One of the symplectic Householder transformations is used to zero 2 matrix elements, the other one is used to eliminate only 1 matrix element. Moreover, as the bulge created when using q_4 is smaller than when using p_4, applying the transformations is cheaper when q_4 is used to drive the SR step.

As mentioned before, the algorithm for reducing a symplectic matrix to butterfly form as given in Table 3.2 can be used to chase the bulge. For the quadruple shift case the algorithm can be simplified significantly. First the 4 undesired elements in the first column of the above given matrix are annihilated, yielding

$$
\left[
\begin{array}{ccccc|ccccc}
x & + & + &   &   & x & x & + & + &   \\
0 & x & + &   &   & x & x & x & + &   \\
0 & + & x &   &   & + & x & x & x &   \\
  & + & + & x &   & + & + & x & x & x \\
  &   &   &   & x &   &   & x & x & x \\
  &   &   &   &   &   &   &   & \ddots & \ddots & \ddots \\
\hline
x & + & + &   &   & x & x & + & + &   \\
0 & x & + &   &   & x & x & x & + &   \\
0 & + & x &   &   & + & x & x & x &   \\
  & + & + & x &   & + & + & x & x & x \\
  &   &   &   & x &   &   & x & x & x \\
  &   &   &   &   &   &   &   & \ddots & \ddots
\end{array}
\right].
$$
(4.1.15)

Next we obtain

$$
\left[
\begin{array}{ccccc|ccccc}
x & 0 & 0 &   &   & x & x & 0 & 0 &   \\
  & x & + & + &   & x & x & x & + & + \\
  & + & x & + &   & + & x & x & x & + \\
  & + & + & x &   & + & + & x & x & x \\
  &   &   &   & x &   &   & + & + & x & x & x \\
  &   &   &   &   &   &   &   &   & x & x & x \\
\hline
x & 0 & 0 &   &   & x & x & 0 & 0 &   \\
  & x & + & + &   & x & x & x & + & + \\
  & + & x & + &   & + & x & x & x & + \\
  & + & + & x &   & + & + & x & x & x \\
  &   &   &   & x &   &   & + & + & x & x & x \\
  &   &   &   &   & x &   &   &   & x & x & x
\end{array}
\right]
$$

by eliminating the 4 undesired elements in the $(n + 1)$st row (or in the 1st row, depending on the pivoting choice). Bringing the second column into the desired form results in

$$
\left[
\begin{array}{cccc|ccccc}
x & & & & x & & & & \\
 & x & + & + & x & x & x & + & + \\
 & 0 & x & + & 0 & x & x & x & + \\
 & 0 & + & x & 0 & + & x & x & x \\
 & & + & + & x & + & + & x & x & x \\
 & & & & x & & & & x & x & x \\
 & & & & \ddots & & & & \ddots & \ddots & \ddots \\
\hline
x & & & & x & & & & \\
 & x & + & + & x & x & x & + & + \\
 & 0 & x & + & 0 & x & x & x & + \\
 & 0 & + & x & 0 & + & x & x & x \\
 & & + & + & x & + & + & x & x & x \\
 & & & & x & & & & x & x & x \\
 & & & & \ddots & & & & \ddots & \ddots & \ddots
\end{array}
\right].
$$

Now we have the same situation as after the construction of (4.1.15), solely in each block the undesired elements are found one row and column further down. Therefore the bulge can be chased down the diagonal analogous to the last 2 steps.

REMARK 4.3 *A careful flop count shows that this implicit butterfly SR step requires about $1012n - 1276$ flops. If the transformation matrix S is accumulated, an additional $128n^2 - 168n$ flops have to be added. This flop count is based on the fact that $4n - 4$ symplectic Givens transformations, $n - 1$ symplectic Gauss transformation, $n - 2$ symplectic Householder transformations with $v \in \mathbf{R}^3$, and $n - 1$ symplectic Householder transformations with $v \in \mathbf{R}^2$ are used. Moreover, the special structure of the problem is taken into account.*

An implicit butterfly SR step as described above is therefore an order of magnitude cheaper than a Francis QR step, which requires about $40n^2$ flops, see [58]. Of course, the accumulation of the transformation matrix is in both cases an $\mathcal{O}(n^2)$ process.

4.1.2 CHOICE OF SHIFTS AND CONVERGENCE

If the chosen shift is a good approximate eigenvalue, we expect deflation at the end of the SR step as indicated in Theorem 4.2. We propose a shift strategy similar to that used in the standard QR algorithm. For example, for a double shift, we choose the eigenvalues $\mu, \mu^{-1} \in \mathbf{R}$ or $\mu, \overline{\mu} \in \mathbf{C}, |\mu| = 1$ of the 2×2 symplectic submatrix

$$
G = \left[\begin{array}{cc} B_{n,n} & B_{n,2n} \\ B_{2n,n} & B_{2n,2n} \end{array} \right] = \left[\begin{array}{cc} b_n & b_n c_n - a_n^{-1} \\ a_n & a_n c_n \end{array} \right]
$$

(denoting the entries of B by $B_{i,j}$). There is no need to actually compute the eigenvalues of G. Simple algebraic manipulations show that

$$
p_2(B) = B^2 - \beta B + I
$$

where $\beta = \mu + \mu^{-1}$ if $\mu \in \mathbf{R}$ or $\beta = \mu + \bar{\mu}$ otherwise. As the sum of the eigenvalues of a matrix is equal to the trace of that matrix, we have

$$\beta = \text{trace}(G) = B_{n,n} + B_{2n,2n} = b_n + a_n c_n.$$

Hence $q_2(B) = p_2(B)B^{-1} = (B + B^{-1}) - \beta I$ and the first column of $q_2(B)$ is given by

$$\begin{aligned}
q_2(B)e_1 &= (B_{1,1} + B_{n+1,n+1} - \beta)e_1 + B_{n+1,n+2}e_2 \\
&= (b_1 + a_1 c_1 - b_n - a_n c_n)e_1 + a_1 d_2 e_2.
\end{aligned}$$

Similar, for a quadruple shift, we choose the eigenvalues of the 4×4 symplectic submatrix

$$\begin{aligned}
G &= \left[\begin{array}{cc|cc}
B_{n-1,n-1} & & B_{n-1,2n-1} & B_{n-1,2n} \\
& B_{n,n} & B_{n,2n-1} & B_{n,2n} \\
\hline
B_{2n-1,n-1} & & B_{2n-1,2n-1} & B_{2n-1,2n} \\
& B_{2n,n} & B_{2n,2n-1} & B_{2n,2n}
\end{array} \right] \\
&= \left[\begin{array}{cc|cc}
b_{n-1} & & b_{n-1}c_{n-1} - a_{n-1}^{-1} & b_{n-1}d_n \\
& b_n & b_n d_n & b_n c_n - a_n^{-1} \\
\hline
a_{n-1} & & a_{n-1}c_{n-1} & a_{n-1}d_n \\
& a_n & a_n d_n & a_n c_n
\end{array} \right].
\end{aligned}$$

Due to its symplecticity the matrix G has either eigenvalues

a) $\lambda, \bar{\lambda}, \lambda^{-1}, \overline{\lambda^{-1}} \in \mathbf{C}, |\lambda| \neq 1$, or

b) $\lambda_1, \bar{\lambda_1}, \lambda_2, \bar{\lambda_2} \in \mathbf{C}, |\lambda_1| = |\lambda_2| = 1$, or

c) $\lambda_1, \bar{\lambda_1} \in \mathbf{C}, |\lambda_1| = 1, \lambda_2, \lambda_2^{-1} \in \mathbf{R}$, or

d) $\lambda_1, \lambda_1^{-1}, \lambda_2, \lambda_2^{-1} \in \mathbf{R}$.

For ease of notation, we will denote the eigenvalues of G by $\mu, \bar{\mu}, \mu^{-1}, \overline{\mu^{-1}}$ in the following discussion. In case b) this has to be interpreted as

$$\mu = \lambda_1, \ \bar{\mu} = \bar{\lambda_1}, \ \mu^{-1} = \lambda_2, \ \overline{\mu^{-1}} = \bar{\lambda_2},$$

while in case c) we mean

$$\mu = \lambda_1, \ \mu^{-1} = \bar{\lambda_1}, \ \bar{\mu} = \lambda_2, \ \overline{\mu^{-1}} = \lambda_2^{-1}.$$

Similar, in case d), this stands for

$$\mu = \lambda_1, \ \mu^{-1} = \lambda_1^{-1}, \ \bar{\mu} = \lambda_2, \ \overline{\mu^{-1}} = \lambda_2^{-1}.$$

As before, there is no need to compute the eigenvalues of G. Comparing

$$p_4(B) = B^4 - (\mu + \bar{\mu} + \mu^{-1} + \overline{\mu^{-1}})(B^3 + B)$$
$$+ (2 + (\mu + \mu^{-1})(\bar{\mu} + \overline{\mu^{-1}}))B^2 + I$$

with the characteristic polynomial of G we obtain

$$p_4(B) = B^4 - \beta(B^3 + B) + \gamma B^2 + I, \qquad (4.1.16)$$

where

$$\begin{aligned}
\beta &= \text{trace}(G) \\
&= b_{n-1} + b_n + a_{n-1}c_{n-1} + a_n c_n, \\
\gamma &= (B_{n-1,n-1} + B_{2n-1,2n-1})(B_{n,n} + B_{2n,2n}) + 2 - B_{2n-1,2n}B_{2n,2n-1} \\
&= (b_{n-1} + a_{n-1}c_{n-1})(b_n + a_n c_n) + 2 - a_{n-1}a_n d_n^2.
\end{aligned}$$

Hence, $q_4(B) = p_4(B)B^{-2} = (B + B^{-1})^2 - \beta(B + B^{-1}) + (\gamma - 2)I$ and the first column of $q_4(B)$ is given by

$$\begin{aligned}
q_4(B)e_1 &= [(b_1 + a_1c_1)^2 + a_1a_2d_2^2 - \beta(b_1 + a_1c_1) + \gamma - 2]e_1 \\
&+ [a_1d_2(b_2 + a_2c_2 + b_1 + a_1c_1 - \beta)]e_2 \qquad (4.1.17) \\
&+ a_1a_2d_2d_3e_3.
\end{aligned}$$

This is exactly the generalized Rayleigh-quotient strategy for choosing shifts proposed by Watkins and Elsner in [142]. Hence, the convergence theorems proven in [142] can be applied here (for a summary of the results, see Theorem 2.17 and 2.19 in Section 2.2.1). But, in order not to destroy the symplectic structure of the problem, the positive integer k used in these theorems has to be an even number here. In particular, the butterfly SR algorithm is typically cubically convergent. In order to see this, we make use of the fact that any unreduced symplectic butterfly matrix is similar to an unreduced butterfly matrix with $b_i = 1$ and $|a_i| = 1$ for $i = 1, \ldots, n$ and $\text{sign}(a_i) = \text{sign}(d_i)$ for $i = 2, \ldots, n$ (Remark 3.9). Clearly, we can modify the butterfly SR algorithm such that each iterate satisfies these constraints. Consider for each iterate B_i

$$PB_iP^T = \begin{bmatrix} X_{11}^{(i)} & X_{12}^{(i)} \\ X_{21}^{(i)} & X_{22}^{(i)} \end{bmatrix}, \qquad X_{22}^{(i)} \in \mathbf{R}^{2\ell \times 2\ell},$$

where 2ℓ is the degree of the shift polynomial $p_i(\lambda) = \lambda^\ell q_i(\lambda)$ (due to the symmetry of the spectrum of a symplectic matrix, this is a reasonable choice).

As for $j = n - \ell + 1$

$$
X_{21}^{(i)} = \begin{bmatrix} 0 & \cdots & 0 & b_j d_j \\ 0 & \cdots & 0 & a_j d_j \\ 0 & \cdots & 0 & 0 \\ \vdots & & \vdots & \vdots \\ 0 & \cdots & 0 & 0 \end{bmatrix}, \quad
X_{12}^{(i)} = \begin{bmatrix} 0 & 0 & 0 & \cdots & 0 \\ \vdots & \vdots & \vdots & & \vdots \\ 0 & 0 & 0 & \cdots & 0 \\ 0 & b_{j-1} d_j & 0 & \cdots & 0 \\ 0 & a_{j-1} d_j & 0 & \cdots & 0 \end{bmatrix},
$$

and $b_{j-1} = b_j = 1, |a_j| = 1$, we have

$$
\|X_{12}^{(i)}\|_F = \sqrt{2 d_j^2} = \|X_{21}^{(i)}\|_F.
$$

Hence from Theorem 2.19 we obtain that the convergence rate of the butterfly SR algorithm is typically cubic.

Furthermore, the hypothesis $\mathrm{span}\{e_1, \ldots, e_k\} \cap \mathcal{U} = \{0\}$ of Theorem 2.17 holds for unreduced symplectic butterfly matrices (note, as above, that k has to be an even number here). Suppose $y \in \mathrm{span}\{e_1, \ldots, e_k\}$ is nonzero. Let its last nonzero component be $y_r, r \leq k$. If B has unreduced butterfly form, then PBP^T has the form

$$
\begin{bmatrix}
x & x & 0 & x & & & & & \\
x & x & 0 & x & & & & & \\
0 & x & x & x & \ddots & & & & \\
0 & x & x & x & & \ddots & & & \\
& & 0 & x & \ddots & & 0 & x & \\
& & 0 & x & & \ddots & 0 & x & \\
& & & & \ddots & & x & x & 0 & x \\
& & & & & \ddots & x & x & 0 & x \\
& & & & & & 0 & x & x & x \\
& & & & & & 0 & x & x & x
\end{bmatrix}
$$

Consider vectors of the form $PB^j P^T y$, $j = \ldots, -2, -1, 0, 1, 2 \ldots$. Let us first assume that r is odd. Then the last nonzero component (lnc) of these vectors are

j	0	1	-1	2	-2	3	$\ldots$	$-\frac{2n-r-1}{2}$	$\frac{2n-r+1}{2}$
lnc	r	$r+1$	$r+2$	$r+3$	$r+4$	$r+5$	$\ldots$	$2n-1$	$2n$

Hence the vectors

$$
y, PBP^T y, PB^{-1} P^T y, PB^2 P^T y, PB^{-2} P^T y, \ldots, PB^m P^T y
$$

are linearly independent where $m = 2n - k + 1$. Therefore the smallest invariant subspace of PBP^T that contains y has dimension at least $2n - k + 1$. Since $\mathcal{U}$ is invariant under PBP^T and has dimension $2n - k$, it follows that $y \notin \mathcal{U}$. Thus span$\{e_1, \ldots, e_k\} \cap \mathcal{U} = \{0\}$.

A similar argument holds for even r. Then the last nonzero component of the vectors $PB^j P^T y$ are

j	0	-1	1	-2	2	-3	$\ldots$	$-\frac{2n-r}{2}$	$\frac{2n-r}{2}$
lnc	r	$r+1$	$r+2$	$r+3$	$r+4$	$r+5$	$\ldots$	$2n-1$	$2n$

Hence the vectors

$$y, PBP^T y, PB^{-1} P^T y, PB^2 P^T y, PB^{-2} P^T y, \ldots, PB^m P^T y$$

are linearly independent where $m = 2n - k + 1$. Therefore the smallest invariant subspace of PBP^T that contains y has dimension at least $2n - k + 1$. Since $\mathcal{U}$ is invariant under PBP^T and has dimension $2n - k$, it follows that $y \notin \mathcal{U}$. Thus span$\{e_1, \ldots, e_k\} \cap \mathcal{U} = \{0\}$.

4.1.3 THE OVERALL PROCESS

By applying a sequence of double or quadruple SR steps to a symplectic butterfly matrix B, it is possible to reduce the tridiagonal blocks in B to quasi-diagonal form with 1×1 and 2×2 blocks on the diagonal. The eigenproblem decouples into a number of simple symplectic 2×2 or 4×4 eigenproblems. In doing so, it is necessary to monitor the off-diagonal elements in the tridiagonal blocks of B in order to bring about decoupling whenever possible. Decoupling occurs if $d_j = 0$ for some j as

$$B = \begin{bmatrix} B_{11} & B_{12} \\ B_{21} & B_{22} \end{bmatrix}$$

$$= \left[\begin{array}{cccc|cccc} b_1 & & & & b_1 c_1 - a_1^{-1} & b_1 d_2 & & \\ & \ddots & & & & b_2 d_2 & \ddots & \ddots \\ & & \ddots & & & & \ddots & \ddots & b_{n-1} d_n \\ & & & b_n & & & b_n d_n & b_n c_n - a_n^{-1} \\ \hline a_1 & & & & a_1 c_1 & a_1 d_2 & & \\ & \ddots & & & & a_2 d_2 & \ddots & \ddots \\ & & \ddots & & & & \ddots & \ddots & a_{n-1} d_n \\ & & & a_n & & & a_n d_n & a_n c_n \end{array}\right].$$

Or, equivalently, if $(B_{12})_{j,j-1}, (B_{12})_{j-1,j}, (B_{22})_{j,j-1}$, and $(B_{22})_{j-1,j}$ are simultaneously zero.

When dealing with upper Hessenberg matrices, as in the QR setting, decoupling occurs whenever a subdiagonal element becomes zero. In practise, decoupling is said to occur whenever a subdiagonal element in the Hessenberg matrix H is suitably small. For example, in LAPACK [10], basically, if

$$|h_{p+1,p}| \leq cu(|h_{p,p}| + |h_{p+1,p+1}|)$$

for some small constant c and the unit roundoff u, then $h_{p+1,p}$ is declared to be zero (please note that the actual implementation is somewhat more refined). This is justified since rounding errors of order $u\|H\|$ are already present throughout the matrix.

Taking the same approach here, we could check whether

$$\begin{aligned}
max\{|(B_{12})_{i,i-1}|, |(B_{12})_{i,i+1}|\} &\leq \epsilon(|(B_{12})_{i-1,i-1}| + |(B_{12})_{ii}|) \quad (4.1.18) \\
max\{|(B_{22})_{i,i-1}|, |(B_{22})_{i,i+1}|\} &\leq \epsilon(|(B_{22})_{i-1,i-1}| + |(B_{22})_{ii}|)
\end{aligned}$$

are simultaneously satisfied, in this case we will have deflation. Here ϵ denotes a properly chosen small constant c times the unit roundoff u. But, of course, in practice, the conditions $d_j = 0$ and $(B_{12})_{j,j-1} = (B_{12})_{j-1,j} = (B_{22})_{j,j-1} = (B_{22})_{j-1,j} = 0$ are not equivalent. A more refined deflation criterion can involve the explicit computation of the d_j's and a check which d_j can be declared to be zero. Making use of the equalities

$$\begin{aligned}
(B_{22})_{j-1,j} &= a_{j-1}d_j, & (B_{22})_{j,j-1} &= a_j d_j, \\
(B_{12})_{j-1,j} &= b_{j-1}d_j, & (B_{12})_{j,j-1} &= b_j d_j
\end{aligned}$$

and

$$(B_{11})_{jj} = b_j, \quad (B_{21})_{jj} = a_j,$$

we can compute d_j as

$$d_j = \left(\frac{(B_{12})_{j,j-1}}{(B_{11})_{j-1,j-1}} + \frac{(B_{12})_{j-1,j}}{(B_{11})_{j,j}} + \frac{(B_{22})_{j,j-1}}{(B_{21})_{j-1,j-1}} + \frac{(B_{22})_{j-1,j}}{(B_{21})_{j,j}} \right) / 4$$

(assuming for simplicity that $(B_{11})_{kk} \neq 0, (B_{21})_{kk} \neq 0, k = j - 1, j$) and test for decoupling by either simply

$$d_j \leq c\epsilon$$

for some small constant c or, using the above approach

$$d_j \leq \epsilon \min \left\{ \frac{(|(B_{12})_{i-1,i-1}| + |(B_{12})_{ii}|)}{(B_{21})_{jj}}, \frac{(|(B_{12})_{i-1,i-1}| + |(B_{12})_{ii}|)}{(B_{21})_{j-1,j-1}}, \right.$$
$$\left. \frac{(|(B_{22})_{i-1,i-1}| + |(B_{22})_{ii}|)}{(B_{11})_{jj}}, \frac{(|(B_{22})_{i-1,i-1}| + |(B_{22})_{ii}|)}{(B_{11})_{j-1,j-1}} \right\}.$$

Numerical examples have shown that (4.1.18) already yields satisfactorily results.

We proceed with the process of applying double or quadruple SR steps to a symplectic butterfly matrix B until the problem has completely split into subproblems of dimension 2 or 4. The complete process is given in Table 4.1. In a final step we then have to solve these small subproblems in order to compute a real Schur-like form from which eigenvalues and invariant subspaces can be read off. That is, in the 2×2 and 4×4 subproblems we will zero the $(2, 1)$ block (if possible) using a symplectic transformation. In case the 4×4 subproblem has real eigenvalues we will further reduce the $(1, 1)$ and $(2, 2)$ blocks. Moreover, we can sort the eigenvalues such that the eigenvalues inside the unit circle will appear in the $(1, 1)$ block. This can be done similar to the approach taken in [38] in the Hamiltonian SR case.

Let us consider the 2×2 case first. In case the $(2, 1)$ entry is zero, the problem is already in the desired form; but we might have to reorder the eigenvalues on the diagonal such the smaller one is in the $(1, 1)$ position. See below for details. Otherwise, these subproblems are of the form

$$M = \left[\begin{array}{cc} b_j & b_j c_j - a_j^{-1} \\ a_j & a_j c_j \end{array} \right].$$

The eigenvalues are given by $\lambda_{\pm} = (a_i c_j + b_j)/2 \pm \sqrt{(a_i c_j + b_j)^2/4 - 1}$. If these eigenvalues are real, choose the one that is inside the unit circle and denote it by λ. The corresponding eigenvector is given by

$$\left[\begin{array}{c} \lambda - a_j c_j \\ a_j \end{array} \right].$$

Then the orthogonal symplectic matrix

$$Q = \frac{1}{\sqrt{(\lambda - a_j c_j)^2 + a_j^2}} \left[\begin{array}{cc} \lambda - a_j c_j & -a_j \\ a_j & \lambda - a_j c_j \end{array} \right]$$

transforms M into upper triangular form

$$Q^T M Q = \left[\begin{array}{cc} \lambda & \star \\ 0 & \lambda^{-1} \end{array} \right].$$

In case $|\lambda_{\pm}| = 1$, we leave M as it is. Embedding Q into a $2n \times 2n$ symplectic Givens transformation in the usual way, we can update the $2n \times 2n$ problem. The above described process computes the real Schur form of M using a (symplectic) Givens transformation. In our implementation we use the MATLAB routine 'schur' for this purpose instead of the above, explicit approach. In this case we might have to order the eigenvalues on the diagonal as there is no guarantee that 'schur' puts the eigenvalue inside the unit circle into the $(1, 1)$

Algorithm: SR Algorithm for Butterfly Matrices

Given a symplectic butterfly matrix B, the following algorithm computes a symplectic matrix S such that $S^{-1}BS$ is a symplectic matrix in which the $(1,1)$, $(1,2)$, $(2,1)$, and $(2,2)$ blocks are each block-diagonal where all blocks are either 1×1 or 2×2. Moreover, the block structure for all four blocks of $\widehat{B} = S^{-1}BS$ is the same. Thus the eigenproblem for $\widehat{B}$ decouples into 2×2 and 4×4 symplectic eigenproblems. B is overwritten by $S^{-1}BS$.

> choose $\epsilon = cu$ suitably
> **repeat until** $q = n$
>> set all sub- and superdiagonal entries of the $(1,2)$ and $(2,1)$ block of B to zero
>> that satisfy
>>
>> $$|(B_{12})_{i,i-1}| \leq \epsilon(|(B_{12})_{i-1,i-1}| + |(B_{12})_{ii}|)$$
>> $$|(B_{22})_{i,i-1}| \leq \epsilon(|(B_{22})_{i-1,i-1}| + |(B_{22})_{ii}|)$$
>> $$|(B_{12})_{i-1,i}| \leq \epsilon(|(B_{12})_{i+1,i+1}| + |(B_{12})_{ii}|)$$
>> $$|(B_{22})_{i-1,i}| \leq \epsilon(|(B_{22})_{i+1,i+1}| + |(B_{22})_{ii}|)$$
>>
>> find the largest nonnegative q and the smallest nonnegative p such that
>>
>> $$B = \left[\begin{array}{ccc|ccc} X_{11} & & & Z_{11} & & \\ & X_{22} & & & Z_{22} & \\ & & X_{33} & & & Z_{33} \\ \hline Y_{11} & & & V_{11} & & \\ & Y_{22} & & & V_{22} & \\ & & Y_{33} & & & V_{33} \end{array}\right],$$
>>
>> where
>>
>> $$X_{11}, Y_{11}, Z_{11}, V_{11} \in \mathbf{R}^{p \times p},$$
>> $$X_{22}, Y_{22}, Z_{22}, V_{22} \in \mathbf{R}^{(n-p-q) \times (n-p-q)},$$
>> $$X_{33}, Y_{33}, Z_{33}, V_{33} \in \mathbf{R}^{q \times q},$$
>>
>> then $X_{33}, Y_{33}, Z_{33}, V_{33}$ are block diagonal with 1×1 and 2×2 blocks and $\left[\begin{smallmatrix} X_{22} & Z_{22} \\ Y_{22} & V_{22} \end{smallmatrix}\right]$ is an unreduced butterfly matrix.
>> **if** $q < n$
>>> perform a double or quadruple shift SR step using the algorithm given in Table 3.2 on $\left[\begin{smallmatrix} X_{22} & Z_{22} \\ Y_{22} & V_{22} \end{smallmatrix}\right]$
>>> update B and S accordingly
>> **end**
> **end**
> solve the 2×2 and 4×4 subproblems as described in the text and accumulate the transformations

Table 4.1. SR Algorithm for Butterfly Matrices

position. Assume, using 'schur', we have achieved

$$Q_s^T M Q_s = \begin{bmatrix} \alpha & t \\ 0 & \alpha^{-1} \end{bmatrix}.$$

If $|\alpha| \geq 1$, the reordering can be done as described in [58, Section 7.6.2] using a Givens rotation Q_D such that the second component of

$$Q_D^T \begin{bmatrix} t \\ \alpha - \alpha^{-1} \end{bmatrix}$$

is zero. If $Q = Q_S Q_D$, then

$$Q^T M Q = \begin{bmatrix} \alpha^{-1} & \pm t \\ 0 & \alpha \end{bmatrix}$$

where $|\alpha^{-1}| < 1$.

In the 4×4 case, the subproblem can be a general 4×4 symplectic matrix M with no additional structure. One way to solve these problems is an approach analogous to the one described in [38] for the 4×4 Hamiltonian case: Transform M to butterfly form. Assume that M is an unreduced butterfly matrix and choose two eigenvalues λ and μ of M such that $\widetilde{M} = (M - \lambda I)(M - \mu I) \in \mathbf{R}^{4 \times 4}$, rank$(\widetilde{M}) \leq 2$. Compute an orthogonal symplectic matrix Q that transforms $\widetilde{M}$ into upper triangular form such that the $(1,1)$ and $(2,2)$ entries are nonzero. Then, as rank$(\widetilde{M}) \leq 2$

$$Q^T \widetilde{M} Q = \begin{bmatrix} x & x & x & x \\ 0 & x & x & x \\ 0 & 0 & 0 & 0 \\ 0 & 0 & 0 & 0 \end{bmatrix}.$$

The last two rows of Q^T span the left invariant subspace of M corresponding to the eigenvalues λ and μ, that is

$$Q^T M Q = \begin{bmatrix} \Delta & \star \\ 0 & \Delta^{-T} \end{bmatrix},$$

where $\Delta \in \mathbf{R}^{2 \times 2}, \sigma(\Delta) = \{\lambda^{-1}, \mu^{-1}\}$. If M is not unreduced, the problem decouples as described in Remark 3.14. For a thorough discussion of this approach, see [13]. The reduction to butterfly form involves two symplectic Givens, one symplectic Householder and one symplectic Gauss transformations, while the computation of Q involves three symplectic Givens and one symplectic Householder transformations. The procedure described next solves the 4×4 problem using less transformations. In particular, depending on the eigenvalue structure of M, we might be able to solve the problem without employing a symplectic Gauss transformation.

We will work directly on M without first reducing it to any specific form. Let λ and μ be eigenvalues of M such that $\widetilde{M} = (M - \lambda I)(M - \mu I) \in \mathbf{R}^{4 \times 4}$, $\operatorname{rank}(\widetilde{M}) \leq 2$. If we eliminate the entries of $\widetilde{M}$ in positions $(4, 1), (3, 1), (2, 1)$, and $(4, 2)$ using orthogonal symplectic transformations as indicated in the following scheme (where we use the same notation as before)

$$
\begin{bmatrix}
\star & \star & \star & \star \\
3, H^{pre} & \star & \star & \star \\
2, G^{pre} & \star & \star & \star \\
1, G^{pre} & 4, G^{pre} & 0_4 & 0_4
\end{bmatrix}
$$

then the entries $(1, 1)$ and $(2, 2)$ are nonzero and the additional zeros in the last row follow as the resulting matrix is still of rank ≤ 2. In case the element $(3, 2)$ is nonzero, we can eliminate the element $(2, 2)$ using a symplectic Gauss transformation yielding

$$
\begin{bmatrix}
\star & \star & \star & \star \\
0 & 0 & 0 & 0 \\
0 & \star & \star & \star \\
0 & 0 & 0 & 0
\end{bmatrix}
\tag{4.1.19}
$$

as the resulting matrix is still of rank ≤ 2. In case the element $(3, 2)$ is zero, we have

$$
\begin{bmatrix}
\star & \star & \star & \star \\
0 & \star & \star & \star \\
0 & 0 & 0 & 0 \\
0 & 0 & 0 & 0
\end{bmatrix}
\tag{4.1.20}
$$

as the matrix is of rank ≤ 2. Accumulate all necessary transformations into the symplectic matrix S such that $S^{-1}\widetilde{M}$ is either of form (4.1.19) or of form (4.1.20).

Let us consider the case (4.1.20) first. The last two rows of S^{-1} span the left invariant subspace of M corresponding to λ^{-1}, μ^{-1}. Therefore, we get

$$
S^{-1}MS = \begin{bmatrix}
\Delta & \star \\
0 & \Delta^{-T}
\end{bmatrix}
\tag{4.1.21}
$$

where $\Delta \in \mathbf{R}^{2 \times 2}$ has the eigenvalues λ^{-1} and μ^{-1}. If λ and μ are real, Δ can be further reduced to upper triangular form.

In the case (4.1.19), the second and fourth row of S^{-1} span the left invariant subspace of M corresponding to λ^{-1}, μ^{-1}. Therefore, we obtain

$$S^{-1}MS = \left[\begin{array}{cc|cc} x_1 & 0 & x_2 & 0 \\ 0 & y_1 & 0 & y_2 \\ \hline x_3 & 0 & x_4 & 0 \\ 0 & y_3 & 0 & y_4 \end{array}\right] \tag{4.1.22}$$

where $\left[\begin{smallmatrix} y_1 & y_2 \\ y_3 & y_4 \end{smallmatrix}\right]$ has the eigenvalues λ and μ. The problem decouples into two symplectic 2×2 problems which can be solved as described before. Moreover, as each of the two subproblems is symplectic again, each 2×2 problem can have either complex eigenvalues $\lambda, \overline{\lambda}$ with $|\lambda| = 1$ or real eigenvalues λ and λ^{-1}.

Then structure of the eigenvalues of M determines which form we can obtain:

- If M has 4 complex eigenvalues $\lambda, \overline{\lambda}, \lambda^{-1}, \overline{\lambda^{-1}}$ with $|\lambda| > 1$, then we have to choose $\mu = \overline{\lambda}$. We can not achieve the form (4.1.22) as that would imply that $|\lambda| = 1$, but $|\lambda| > 1$. Hence, by the above described process M will be put into the form (4.1.21).

- If M has 4 complex eigenvalues $\lambda, \overline{\lambda}$ and $\beta, \overline{\beta}$ with $|\lambda| = |\beta| = 1, \lambda \neq \overline{\beta}$, then we have to choose $\mu = \overline{\lambda}$. We can not achieve the form (4.1.21) as that would imply that $\lambda, \overline{\lambda}$ and $\lambda^{-1}, \overline{\lambda^{-1}}$ are eigenvalues of M, but $\beta \neq \lambda^{-1}$. Hence, by the above described process M will be put into the form (4.1.22).

- If M has 4 complex eigenvalues $\lambda, \overline{\lambda}$ and $\beta, \overline{\beta}$ with $|\lambda| = |\beta| = 1, \lambda = \beta$, then we have to choose $\mu = \overline{\lambda}$. This results in a matrix $\widehat{M}$ of rank 0 or of rank 2 depending on the Jordan structure of the matrix. In the first case, nothing can be achieved. In the latter case, a reduction is possible.

- If M has 4 real eigenvalues λ, λ^{-1} and β, β^{-1} with $|\lambda| \geq 1, |\beta| \geq 1, \lambda \neq \beta$, then we choose $\mu = \beta$. In this case we can not reach the form (4.1.22) as this would imply $\beta = \lambda^{-1}$, but $|\beta| \geq 1, |\lambda| \geq 1, \beta \neq \lambda$. Hence, by the above described process M will be put into the form (4.1.21).

- If M has 4 real eigenvalues λ, λ^{-1} and β, β^{-1} with $|\lambda| > 1, |\beta| > 1, \lambda = \beta$, then we choose $\mu = \beta$. In this case we can not reach the form (4.1.22) as this would imply $\beta = \lambda^{-1}$, but $\lambda = \beta$ and $|\lambda| > 1$. Hence, by the above described process M will be put into the form (4.1.21).

- If M has 4 real eigenvalues of absolute value 1, we can reduce M to desired form. E.g., if all eigenvalues are equal to one, then we have to choose $\lambda = \mu = 1$. Depending on the Jordan structure of the matrix M, either $\widehat{M} = M - \lambda I$ or $\widehat{M}$ will be of rank 2. If $\widehat{M}$ is of rank 2, we will work with

$\widehat{M}$, otherwise with $\widetilde{M}$. In any case, the above process will put M either into the form (4.1.22) or (4.1.21). In case, we reach the form (4.1.22), each symplectic 2×2 subproblem can be put into real Schur form. On the other hand, if we reach the form (4.1.21), each 2×2 diagonal block can be put into triangular form.

- If M has 2 real eigenvalues λ, λ^{-1} and two complex eigenvalues $\beta, \bar{\beta}$ with $|\lambda| \geq 1, |\beta| = 1$, then we have to choose $\mu = \lambda^{-1}$. We can not achieve the form (4.1.21) as that would imply that $\lambda, \lambda^{-1}, \bar{\lambda}$ and $\overline{\lambda^{-1}}$ are eigenvalues of M, but $\beta \neq \bar{\lambda}$. Hence, by the above described process M will be put into the form (4.1.22).

REMARK 4.4 *The SR algorithm for butterfly matrices as given in Table 4.1 requires about $675n^2$ flops. If the transformation matrix S is accumulated, an additional $85n^3$ flops have to be added. This estimate is based on the observation that $\frac{2}{3}$ SR iterations per eigenvalue are necessary. Hence, the SR algorithm for a $2n \times 2n$ symplectic matrix requires about $25n^3 + 675n^2$ flops for the reduction to butterfly form and the SR iteration, plus $113n^3$ flops for the accumulation of the transformation matrix.*

This a considerably cheaper than applying the QR algorithm to a $2n \times 2n$ matrix. That would require about $80n^3$ flops for the reduction to Hessenberg form and the QR iteration, plus $120n^3$ flops for the accumulation of the transformation matrix. Recall, that an iteration step of the implicit butterfly SR algorithm is an $\mathcal{O}(n)$ process, while the Francis QR step is an $\mathcal{O}(n^2)$ process.

This almost completes the discussion of the butterfly SR algorithm. It remains to consider the potential loss of J–orthogonality due to round-off errors. This was an important issue for the structure-preserving method for the symplectic eigenproblem presented by Flaschka, Mehrmann, and Zywietz in [53] (see Section 3.1). That method first reduces the symplectic matrix M to symplectic J–Hessenberg form, that is to a matrix of the form

where the $(1,1), (2,1)$ and $(2,2)$ blocks are upper triangular and the $(1,2)$ block is upper Hessenberg. The SR iteration preserves this form at each step and is supposed to converge to a form from which the eigenvalues can be read off. The authors report the loss of the symplectic structure due to roundoff errors after only a few SR steps. As a symplectic J–Hessenberg matrix looks like a general J–Hessenberg matrix, it is not easy to check and to guarantee that the structure is kept invariant in the presence of roundoff errors. In [53], two examples are given demonstrating the loss of the symplectic structure.

The symplectic butterfly SR algorithm discussed here also destroys the symplectic structure of the butterfly matrix due to roundoff errors. However, the very compact butterfly form allows one to restore the symplectic structure of the iterates easily and cheaply whenever necessary. This can be done using either one of the two decompositions (3.2.15), (3.2.7) of a symplectic butterfly matrix discussed in Section 3.2. Whichever decomposition is used, one assumes that the two diagonal blocks of the butterfly matrix are exact. That is, one assumes that the parameters $a_1, \ldots, a_n, b_1, \ldots, b_n$, which can be read off of the butterfly matrix directly, are correct. Then one uses them to compute the other $2n - 1$ parameters. Using, e.g., the decomposition (3.2.7) one obtains different formulae for the other parameters:

$$
\begin{aligned}
c_k &= B_{k+n,k+n}/a_k \\
&= (B_{k,k+n} + a_k^{-1})/b_k, \\
d_k &= B_{k,k+n-1}/b_k \\
&= B_{k-1,k+n}/b_{k-1} \\
&= B_{k+n,k+n-1}/a_k \\
&= B_{k+n-1,k+n}/a_{k-1}.
\end{aligned}
$$

Adding the terms on the right hand sides and averaging, corrected values for the parameters c_k and d_k are obtained (in actual computations one should use only those terms for which the numerical computations are save, e.g. in case b_k is zero or very small, the equations with this term are not used). Using the so obtained parameters, one computes new entries for the $(1, 2)$ and $(2, 2)$ block of the butterfly matrix. Using this procedure to force the symplectic structure whenever necessary, the SR algorithm based on the butterfly form has no problems in solving the two abovementioned examples given by Flaschka, Mehrmann, and Zywietz in [53], even cubic convergence can be observed; see Section 4.4.

4.2   THE PARAMETERIZED SR ALGORITHM FOR SYMPLECTIC BUTTERFLY MATRICES

The introduction of the symplectic butterfly form in Section 3 was motivated by the Schur parameterization of unitary Hessenberg matrices. Using such a Schur parameterization, one step of the shifted QR algorithm for unitary Hessenberg matrices (*UHQR algorithm*) can be carried out in $\mathcal{O}(n)$ arithmetic operations. Coupled with the shift strategy of Eberlein and Huang [46], this will permit computation of the spectrum of a unitary Hessenberg matrix, to machine precision, in $\mathcal{O}(n^2)$ operations. The key observation that lead to the development of the UHQR algorithm is the following [61]: One step of the QR algorithm with shift $\tau \in \mathbb{C}$, applied to a matrix $H \in \mathbb{C}^{n \times n}$ may be described as follows. Factor $H - \tau I = QR$ with Q unitary and R upper triangular. Then put $\tilde{H} = Q^H H Q$. If H is a unitary Hessenberg matrix, then so are Q and $\tilde{H}$. Any unitary Hessenberg matrix has a unique Schur factorization of the form $G_1(\gamma_1)G_2(\gamma_2) \cdots G_n(\gamma_n)$, where

$$G_k = \text{diag}(I^{k-1,k-1}, \begin{bmatrix} -\gamma_k & \sigma_k \\ \sigma_k & \overline{\gamma_k} \end{bmatrix}, I^{n-k-2,n-k-2})$$

for $k = 1, \ldots, n - 1$ ($|\gamma_k|^2 + \sigma_k^2 = 1, \sigma_k \in \mathbf{R}$) and

$$G_n = \text{diag}(I^{n-1,n-1}, -\gamma_n), |\gamma_n| = 1.$$

Hence $H, \tilde{H}$ and Q may be represented by their Schur factorizations

$$
\begin{aligned}
H &= H(\gamma_1, \ldots, \gamma_n) = G_1(\gamma_1)G_2(\gamma_2) \cdots G_n(\gamma_n), \\
\tilde{H} &= \tilde{H}(\tilde{\gamma}_1, \ldots, \tilde{\gamma}_n) = G_1(\tilde{\gamma}_1)G_2(\tilde{\gamma}_2) \cdots G_n(\tilde{\gamma}_n), \\
Q &= Q(\alpha_1, \ldots, \alpha_n) = G_1(\alpha_1)G_2(\alpha_2) \cdots G_n(\alpha_n).
\end{aligned}
$$

Taking a close look at the implicit implementation of the QR step, it turns out that $G_1(\alpha_1)$ is uniquely determined by H and the shift τ. In the first step, H is transformed by a similarity transformation with $G_1(\alpha_1)$, introducing a small bulge in the upper Hessenberg form. The remaining factors of Q are then used to chase this bulge down the subdiagonal. Taking a closer look at what happens, one sees that the sequence of transformations can be interpreted in a very nice way. Let us start with $H_1 = G_1^H(\alpha_1)H$. This matrix is an upper Hessenberg matrix just like H. Next we have to complete the similarity transformation with G_1, this will introduce a small bulge: $H_1 G_1(\alpha_1)$ is upper Hessenberg, apart from a positive element in position $(3, 1)$. Premultiplication of $H_1 G_1(\alpha_1)$ by $G_2(\alpha_2)$ to form H_2 must create a zero in position $(3, 1)$ (otherwise, the later (unitary) transformations of H_2 would lead to a matrix $\hat{H}$ with a nonnull vector below position $(2, 1)$). Hence, $G_2(\alpha_2)^H H_1 G_1(\alpha_1) = H_2$ is an upper Hessenberg matrix. Now define

$$H_k = G_k(\alpha_k)^H H_{k-1} G_{k-1}(\alpha_{k-1}), \quad H_0 = H, \quad G_0(\alpha_0) = I.$$

The matrix H_{k-1} is a unitary upper Hessenberg matrix. Then

$$\tilde{H}_k = H_{k-1} G_{k-1}(\alpha_{k-1})$$

completes the similarity transformation with $G_{k-1}(\alpha_{k-1})$. Moreover, $\tilde{H}_k$ is upper Hessenberg, apart from a positive element in position $(k + 1, k - 1)$. Premultiplication of $H_{k-1} G_{k-1}(\alpha_{k-1})$ by $G_k(\alpha_k)$ to form H_k must create a zero in position $(k + 1, k - 1)$ (otherwise, the later (unitary) transformations of H_k would lead to a matrix $\hat{H}$ with a nonnull vector below position $(k, k - 1)$).

Continuing with this argument all matrices H_k are unitary upper Hessenberg. In particular, each H_k can be written as

$$H_k = H(\tilde{\gamma}_1, \ldots, \tilde{\gamma}_{k-1}, -\omega\tilde{\alpha}_k, -\omega\gamma_{k+1}, \ldots, -\omega\gamma_n) \cdot D,$$

where

$$D = \text{diag}(I^{k-1,k-1}, \overline{\omega}, I^{n-k,n-k}),$$

$|\tilde{\alpha}_k|^2 + \tilde{\beta}_k^2 = 1$ and $\omega = \text{sign}(\tau)$. Hence,

$$H_n = H(\tilde{\gamma}_1, \ldots, \tilde{\gamma}_{n-1}, -\tilde{\alpha}_k\gamma_n).$$

Since $|\alpha_n| = 1$,

$$\tilde{H} = Q^H H Q = H_n G_n(\alpha_n) = H(\tilde{\gamma}_1, \ldots, \tilde{\gamma}_{n-1}, \gamma_n).$$

That is, after k steps in the implicit bulge chasing process, the first $k - 1$ Schur parameters of the resulting matrix $\tilde{H}$ can already be read off of the intermediate matrix H_k. In each step, one new Schur parameter is determined. At any point in the implicit QR step only·a certain, limited number of rows and columns is involved in the current computation; the leading and the trailing rows and columns are not affected. A careful implementation of the algorithm works only with the Schur parameters, computing one new Schur parameter at a time.

In this section we will develop a parameterized SR algorithm for computing the eigeninformation of a parameterized symplectic butterfly matrix. The algorithm will work only on the parameters similar to the approach described above for the development of the UHQR algorithm. One step of the SR algorithm with shift polynomial q applied to a matrix $B \in \mathbf{R}^{2n \times 2n}$ may be described as follows: Factor $q(B) = SR$ with S symplectic and R J–triangular. Then put $\tilde{B} = S^{-1}BS$. If B is an unreduced symplectic butterfly matrix, then so is $\tilde{B}$. Hence, B and $\tilde{B}$ can be given in parameterized form as in (3.2.15) or (3.2.7). Unfortunately, unlike in the single shift unitary QR case discussed above, the transformation matrix S in the symplectic butterfly SR algorithm does not have the same structure as the matrix being transformed. S is symplectic, but not of butterfly form. Therefore, S can not be given in parameterized form, making the following derivations slightly more complicated than in the unitary QR case.

4.2.1   THE BASIC IDEA

The key to the development of such an algorithm working only on the parameters is the observation that at any point in the implicit SR step only a certain, limited number of rows and columns of the symplectic butterfly matrix is worked on. In the leading part of the intermediate matrices the butterfly form is already retained and is not changed any longer, while the trailing part has not been changed yet. Hence, from the leading part the first parameters of the resulting butterfly matrix can be read off, while from the trailing part the last parameters of the original butterfly matrix can still be read off. Recall the implicit SR step as described in Section 4.1. The first implicit transformation S_1 is selected in order to introduce a bulge into the symplectic butterfly matrix B. That is, a symplectic matrix S_1 is determined such that

$$S_1^{-1}q(B)e_1 = \alpha e_1, \qquad \alpha \in \mathbf{R},$$

where $q(B)$ is an appropriately chosen Laurent polynomial. Applying this first transformation to the butterfly matrix yields a symplectic matrix $S_1^{-1}BS_1$ with almost butterfly form having a small bulge. The remaining implicit transformations perform a bulge-chasing sweep down the subdiagonals to restore the butterfly form. That is, a symplectic matrix S_2 is determined such that

$$\widetilde{B} = S_2^{-1}S_1^{-1}BS_1S_2$$

is of butterfly form again. If B is an unreduced butterfly matrix and $q(B)$ is of full rank, that is $\mathrm{rank}(q(B)) = 2n$, then $\widetilde{B}$ is also an unreduced butterfly matrix. Hence, there will be parameters $\tilde{a}_1, \dots, \tilde{a}_n, \tilde{b}_1, \dots, \tilde{b}_n, \tilde{c}_1, \dots, \tilde{c}_n,$ $\tilde{d}_2, \dots, \tilde{d}_n$ which determine $\widetilde{B}$. During the bulge-chasing sweep the bulge is successively moved down the subdiagonals, one row and one column at a time. Consider for simplicity a double shift implicit SR step. The bulge is introduced by a transformation of the form

$$S_1 = \left[\begin{array}{cc|cc} \alpha & \beta & & \\ -\beta & \alpha & & \\ & I^{n-2,n-2} & & \\ \hline & & \alpha & \beta \\ & & -\beta & \alpha \\ & & & I^{n-2,n-2} \end{array}\right]. \tag{4.2.23}$$

In a slight abuse of notation, we will call matrices of the form (4.2.23) symplectic Householder transformations in the following, although they are the direct sum of two Givens transformations. Whenever a transformation of the form (4.2.23) is used in the following, one can just as well use a symplectic Householder transformation as defined in Section 2.1.2.

Applying a transformation of the form (4.2.23) to B to introduce a bulge, results in a matrix of the form

$$
S_1^{-1} B S_1 = \left[
\begin{array}{cccc|cccc}
x & + &   &   & x & x & + &   \\
+ & x &   &   & x & x & x &   \\
  &   & x &   & + & x & x' & x \\
  &   &   & x &   &   & x & x \\
  &   &   &   &   &   &   & \ddots \\
\hline
x & + &   &   & x & x & + &   \\
+ & x &   &   & x & x & x &   \\
  &   & x &   & + & x & x & x \\
  &   &   & x &   &   & x & x \\
  &   &   &   &   &   &   & \ddots
\end{array}
\right].
$$

Now a symplectic Givens transformation to eliminate the $(n+2,1)$ element and a symplectic Gauss transformation to eliminate the $(2,1)$ element are applied, resulting in

$$
\left[
\begin{array}{cccc|cccc}
x & + &   &   & x & x & + &   \\
  & x &   &   & x & x & x &   \\
  & + & x &   & + & x & x & x \\
  &   &   & x &   &   & x & x \\
  &   &   &   &   &   &   & \ddots \\
\hline
x & + &   &   & x & x & + &   \\
  & x &   &   & x & x & x &   \\
  & + & x &   & + & x & x & x \\
  &   &   & x &   &   & x & x \\
  &   &   &   &   &   &   & \ddots
\end{array}
\right].
$$

This bulge is chased down the subdiagonals one row and one column at a time. The $(1,1)$ and the $(n+1,1)$ element are not altered in any subsequent transformation. Hence, at this point we can already read off $\tilde{a}_1$ and $\tilde{b}_1$. The bulge-chase is done using the algorithm for reducing a symplectic matrix to butterfly form as given in Table 3.2 (without pivoting). In a first step, a sequence of symplectic Givens, Householder, and Gauss transformations is applied resulting in

$$
\left[
\begin{array}{cccc|cccc}
x &   &   &   & x & x &   &   \\
  & x & + &   & x & x & x & + \\
  &   & x &   &   & x & x & x \\
  &   & + & x & + & x & x &   \\
  &   &   &   &   &   &   & \ddots \\
\hline
x &   &   &   & x & x &   &   \\
  & x & + &   & x & x & x & + \\
  &   & x &   &   & x & x & x \\
  &   & + & x & + & x & x &   \\
  &   &   &   &   &   &   & \ddots
\end{array}
\right].
$$

Next the same sequence of symplectic Givens, Householder, and Gauss transformations (of course, operating in different rows and columns as before) is applied in order to achieve

$$
\begin{bmatrix}
x & & & & & & x & x & & & \\
& x & & & & & x & x & x & & \\
& & x & + & & & x & x & x & + & \\
& & & x & & & & x & x & x & \\
& & & + & x & & & + & x & x & \\
& & & & & \ddots & & & & & \ddots \\
\hline
x & & & & & & x & x & & & \\
& x & & & & & x & x & x & & \\
& & x & + & & & x & x & x & + & \\
& & & x & & & & x & x & x & \\
& & & + & x & & & + & x & x & \\
& & & & & \ddots & & & & & \ddots
\end{bmatrix}.
$$

During this step, rows 2 and $n + 1$ and columns 1 and $n + 1$ are not changed anymore. The parameters $\tilde{a}_2, \tilde{b}_2, \tilde{c}_1$, and $\tilde{d}_2$ of the resulting matrix $\tilde{B}$ can be read off. In general, once the bulge is chased down j rows and columns, the leading j rows and columns of each block are not changed anymore. The parameters $\tilde{a}_1, \ldots, \tilde{a}_j, \tilde{b}_1, \ldots, \tilde{b}_j, \tilde{c}_1, \ldots, \tilde{c}_{j-1}, \tilde{d}_2, \ldots, \tilde{d}_j$ of the resulting matrix $\tilde{B}$ can be read off.

In the following we will derive an algorithm that computes the parameters

$$
\tilde{a}_1, \ldots, \tilde{a}_n, \tilde{b}_1, \ldots, \tilde{b}_n, \tilde{c}_1, \ldots, \tilde{c}_n, \tilde{d}_2, \ldots, \tilde{d}_n
$$

of $\tilde{B}$ one set (that is, $a_{j+1}, b_{j+1}, c_j, d_{j+1}$) at a time given the parameters

$$
a_1, \ldots, a_n, b_1, \ldots, b_n, c_1, \ldots, c_n, d_2, \ldots, d_n
$$

of B. The matrices B and $\tilde{B}$ are never formed explicitly. In order to derive such a method, we will work with the factorization $B = K^{-1}N$ (3.2.15) or (3.2.7), as the parameters of B can be read off of K and N directly. Fortunately, K and N can be expressed as products of even simpler matrices.

Let us start with the factorization $B = B_u = K_u^{-1}N_u$ as in (3.2.7),

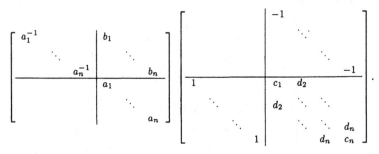

K_u^{-1} can be decomposed into a product of even simpler symplectic matrices

$$
X_1 X_2 \cdots X_n = K_u^{-1}
$$

where

$$
X_k = \left[
\begin{array}{cc|cc}
I^{k-1,k-1} & & & \\
 & \begin{matrix} a_k^{-1} \\ I^{n-k,n-k} \end{matrix} & b_k & \\
\hline
 & & I^{k-1,k-1} & \\
 & & a_k & \\
 & & & I^{n-k,n-k}
\end{array}
\right].
$$

Similarly, N_u can be decomposed

$$
Y_n Y_{n-1} \cdots Y_1 J^T = N_u
$$

where

$$
Y_k = \left[
\begin{array}{cccc|ccc}
I_a & & & & & & \\
 & 1 & & & & & \\
 & & 1 & & & & \\
 & & & I_b & & & \\
\hline
 & & & & I_a & & \\
 & -c_k & -d_{k+1} & & & 1 & \\
 & -d_{k+1} & & & & & 1 \\
 & & & & & & I_b
\end{array}
\right],
$$

$$
Y_n = \left[
\begin{array}{cc|cc}
I^{n-1,n-1} & & & \\
 & 1 & & \\
\hline
 & & I^{n-1,n-1} & \\
 & -c_n & & 1
\end{array}
\right].
$$

Here, we use the notation $I_a = I^{k-1,k-1}$ and $I_b = I^{n-k-1,n-k-1}$.

Because of their special structure, most of the X_k, Y_k, the symplectic Givens transformations G_j, the symplectic Householder transformations H_j, and the symplectic Gauss transformations L_j as defined in Section 2.1.2 commute:

$$
\begin{aligned}
X_j X_k &= X_k X_j && \text{for all } j, k, \\
Y_j Y_k &= Y_k Y_j && \text{for all } j, k, \\
X_j Y_k &= Y_k X_j && \text{for } j \neq k, j \neq k - 1, \\
G_j X_k &= X_k G_j && \text{for } j \neq k, \\
H_j X_k &= X_k H_j && \text{for } j \neq k, j \neq k + 1, \\
L_j X_k &= X_k L_j && \text{for } j \neq k, j \neq k - 1, \\
G_j Y_k &= Y_k G_j && \text{for } j \neq k, j \neq k - 1, \\
H_j Y_k &= Y_k H_j && \text{for } j \neq k, j \neq k - 1, j \neq k + 1, \\
L_j Y_k &= Y_k L_j && \text{for } j \neq k, j \neq k - 1, j \neq k + 1.
\end{aligned}
$$

Here we assume that

$$H_k = \text{diag}(I^{k-1,k-1}, P, I^{n-k-1,n-k-1}, I^{k-1,k-1}, P, I^{n-k-1,n-k-1})$$

where $P \in \mathbf{R}^{2 \times 2}$ is a Givens transformation, as all H_k considered in this section are of this special form. Hence, we can write

$$B_u = X_n Y_n X_{n-1} Y_{n-1} \cdots X_2 Y_2 X_1 Y_1 J^T. \qquad (4.2.24)$$

The factorization $B = B_s = K_s^{-1} N_s$ as in (3.2.15) is given by

$$
\left[\begin{array}{ccc|ccc}
b_1^{-1} & & & & & \\
& \ddots & & & & \\
& & b_n^{-1} & & & \\
\hline
a_1 & & & b_1 & & \\
& \ddots & & & \ddots & \\
& & a_n & & & b_n
\end{array}\right]
\left[\begin{array}{ccc|ccc}
1 & & & u_1 & v_2 & \\
& \ddots & & v_2 & \ddots & \ddots \\
& & & & \ddots & \ddots & v_n \\
& & 1 & & v_n & u_n \\
\hline
& & & 1 & & \\
& & & & \ddots & \\
& & & & & 1
\end{array}\right].
$$

Recall that the tridiagonal blocks in N_s and N_u are related by

$$
\left[\begin{array}{cccc}
u_1 & v_2 & & \\
v_2 & \ddots & \ddots & \\
& \ddots & \ddots & v_n \\
& & v_n & u_n
\end{array}\right]
=
\left[\begin{array}{cccc}
c_1 - a_1^{-1} b_1^{-1} & d_2 & & \\
d_2 & \ddots & \ddots & \\
& \ddots & \ddots & d_n \\
& & d_n & c_n - a_n^{-1} b_n^{-1}
\end{array}\right]
$$

in case all a_j and b_j are nonzero (see Remark 3.14).

Similar to K_u and N_u, K_s and N_s can be decomposed into products of simpler symplectic matrices:

$$U_1 U_2 \cdots U_n = K_s^{-1},$$

where

$$
U_k =
\left[\begin{array}{ccc|ccc}
I^{k-1,k-1} & & & & & \\
& b_k^{-1} & & & & \\
& & I_{n-k,n-k} & & & \\
\hline
& & & I^{k-1,k-1} & & \\
& a_k & & & b_k & \\
& & & & & I^{n-k,n-k}
\end{array}\right],
$$

and

$$V_n V_{n-1} \cdots V_1 = N_s,$$

where

$$V_k = \left[\begin{array}{cccc|cccc} I_a & & & & & & & \\ & 1 & & & & u_k & v_k & \\ & & 1 & & & v_k & & \\ & & & I_b & & & & \\ \hline & & & & I_a & & & \\ & & & & & 1 & & \\ & & & & & & 1 & \\ & & & & & & & I_b \end{array}\right],$$

$$V_n = \left[\begin{array}{cc|cc} I^{n-1,n-1} & & & \\ & 1 & & u_n \\ \hline & & I^{n-1,n-1} & \\ & & & 1 \end{array}\right].$$

Here, we use the notation $I_a = I^{k-1,k-1}$ and $I_b = I^{n-k-1,n-k-1}$.

As before, because of their special structure most of the U_k, V_k, the symplectic Givens transformations G_j, the symplectic Householder transformations H_j, and the symplectic Gauss transformations L_j as defined in Section 2.1.2 commute:

$$\begin{array}{lll} U_j U_k & = & U_k U_j \qquad \text{for all } j, k, \\ V_j V_k & = & V_k V_j \qquad \text{for all } j, k, \\ U_j V_k & = & V_k U_j \qquad \text{for } j \neq k, j \neq k - 1, \\ G_j U_k & = & U_k G_j \qquad \text{for } j \neq k, \\ H_j U_k & = & U_k H_j \qquad \text{for } j \neq k, j \neq k + 1, \\ L_j U_k & = & U_k L_j \qquad \text{for } j \neq k, j \neq k - 1, \\ G_j V_k & = & V_k G_j \qquad \text{for } j \neq k, j \neq k - 1, \\ H_j V_k & = & V_k H_j \qquad \text{for } j \neq k, j \neq k - 1, j \neq k + 1, \\ L_j V_k & = & V_k L_j \qquad \text{for } j \neq k, j \neq k - 1, j \neq k + 1. \end{array}$$

Hence, we can write

$$B_s = U_n V_n U_{n-1} V_{n-1} \cdots U_2 V_2 U_1 V_1. \qquad (4.2.25)$$

Now let us take a closer look at a double shift bulge chase. We will start with an unreduced symplectic butterfly matrix B either decomposed as in (4.2.24) or as in (4.2.25). The resulting matrix $\tilde{B}$ will have a decomposition of the same form as B,

$$\begin{array}{lll} \tilde{B}_u & = & \tilde{X}_n \tilde{Y}_n \tilde{X}_{n-1} \tilde{Y}_{n-1} \cdots \tilde{X}_2 \tilde{Y}_2 \tilde{X}_1 \tilde{Y}_1 J^T, \\ \tilde{B}_s & = & \tilde{U}_n \tilde{V}_n \tilde{U}_{n-1} \tilde{V}_{n-1} \cdots \tilde{U}_2 \tilde{V}_2 \tilde{U}_1 \tilde{V}_1. \end{array}$$

As noted before, the bulge is introduced by the transformation $S_1^{-1}BS_1$ with a matrix S_1 of the form (4.2.23). This leads to a matrix of the form

$$
S_1^{-1}BS_1 =
\left[
\begin{array}{cccccc|cccccc}
\otimes & \oplus & & & & & \otimes & \otimes & \oplus & & & \\
\oplus & \otimes & & & & & \otimes & \otimes & \otimes & & & \\
 & & x & & & & \oplus & \otimes & \otimes & x & & \\
 & & & x & & & & & x & x & & \\
 & & & & \ddots & & & & & & \ddots & \\
\hline
\otimes & \oplus & & & & & \otimes & \otimes & \oplus & & & \\
\oplus & \otimes & & & & & \otimes & \otimes & \otimes & & & \\
 & & x & & & & \oplus & \otimes & x & & \\
 & & & x & & & & & x & x & & \\
 & & & & \ddots & & & & & & \ddots & \\
\end{array}
\right]
$$

where x denotes desired entries in the butterfly form, $+$ undesired entries, and $\otimes$ and $\oplus$ desired and undesired elements that are changed by the current transformation. As S_1 is a symplectic Householder transformation, S_1 and most of the factors of B_u and B_s commute:

$$
\begin{aligned}
S_1^{-1}B_uS_1 &= X_nY_n\cdots X_3Y_3S_1^{-1}X_2Y_2X_1Y_1J^TS_1, \\
S_1^{-1}B_sS_1 &= U_nV_n\cdots U_3V_3S_1^{-1}U_2V_2U_1V_1S_1.
\end{aligned}
$$

Since S_1 is unitary and symplectic, we have $S_1^{-1} = S_1^T$ and $J^TS_1 = S_1J^T$. Hence,

$$
\begin{aligned}
S_1^TB_uS_1 &= X_nY_n\cdots X_3Y_3S_1^TX_2Y_2X_1Y_1S_1J^T, \\
S_1^TB_sS_1 &= U_nV_n\cdots U_3V_3S_1^TU_2V_2U_1V_1S_1.
\end{aligned}
$$

Next a symplectic Givens transformation G_2 is applied to zero the $(n+2,1)$ element:

$$
G_2S_1^TBS_1G_2^T =
\left[
\begin{array}{cccccc|cccccc}
x & \oplus & & & & & x & \otimes & + & & & \\
\oplus & \otimes & & & & & \otimes & \otimes & \otimes & & & \\
 & \oplus & x & & & & + & \otimes & x & x & & \\
 & & & x & & & & & x & x & & \\
 & & & & \ddots & & & & & & \ddots & \\
\hline
x & \oplus & & & & & x & \otimes & + & & & \\
0 & \otimes & & & & & \otimes & \otimes & \otimes & & & \\
 & \oplus & x & & & & + & \otimes & x & x & & \\
 & & & x & & & & & x & x & & \\
 & & & & \ddots & & & & & & \ddots & \\
\end{array}
\right].
$$

As G_2 and most of the factors of B_u and B_s commute and as G_2 is unitary and symplectic (hence, $J^T G_2^T = G_2^T J^T$) we obtain

$$G_2 S_1^T B_u S_1 G_2^T = X_n Y_n \cdots X_3 Y_3 G_2 S_1^T X_2 Y_2 X_1 Y_1 S_1 G_2^T J^T,$$
$$G_2 S_1^T B_s S_1 G_2^T = U_n V_n \cdots U_3 V_3 G_2 S_1^T U_2 V_2 U_1 V_1 S_1 G_2^T.$$

Now a symplectic Gauss transformation L_2 is chosen to eliminate the $(2,1)$ element such that

$$B^{(1)} := L_2 G_2 S_1^T B S_1 G_2^T L_2^{-1}$$

$$=
\begin{bmatrix}
\begin{array}{cccc|cccc}
\otimes & \oplus & & & \otimes & \otimes & \oplus & \\
0 & \otimes & & & \otimes & \otimes & \otimes & \\
& \oplus & x & & \oplus & \otimes & x & x \\
& & & x & & & x & x \\
& & & & \ddots & & & & \ddots \\
\hline
\otimes & \oplus & & & \otimes & \otimes & \oplus & \\
& \otimes & & & \otimes & \otimes & \otimes & \\
& \oplus & x & & \oplus & \otimes & x & x \\
& & & x & & & x & x \\
& & & & \ddots & & & & \ddots \\
\end{array}
\end{bmatrix}.$$

At this point the actual bulge, which is chased down the subdiagonal, is formed. That is, now a sequence of symplectic Givens, Householder and Gauss transformations is applied to successively chase the bulge of the above form down the subdiagonal.

L_2 is symplectic, but not unitary. Hence, $J^T L_2^{-1} = L_2^T J^T$. Moreover, as L_2 and most of the factors of B_u and B_s commute, we have

$$B_u^{(1)} := L_2 G_2 S_1^T B_u S_1 G_2^T L_2^{-1}$$
$$= X_n Y_n \cdots X_3 Y_3 L_2 G_2 S_1^T X_2 Y_2 X_1 Y_1 S_1 G_2^T L_2^T J^T,$$
$$B_s^{(1)} := L_2 G_2 S_1^T B_s S_1 G_2^T L_2^{-1}$$
$$= U_n V_n \cdots U_3 V_3 L_2 G_2 S_1^T U_2 V_2 U_1 V_1 S_1 G_2^T L_2^{-1}.$$

Recall that for symplectic matrices S the inverse S^{-1} is given by $S^{-1} = -J S^T J$ such that in the above formula for $B_s^{(1)}$ we do not have to invert L_2^{-1} explicitly.

The $(1,1)$ and the $(n+1,1)$ elements of $B^{(1)}$ are not altered by any subsequent transformation. Therefore, at this point we can read off $\widetilde{a}_1$ and $\widetilde{b}_1$ of the final $\widetilde{B}$. In other words, we can rewrite

$$L_2 G_2 S_1^T X_2 Y_2 X_1 Y_1 S_1 G_2^T L_2^T J^T$$

in terms of $\widetilde{X}_1$ times an appropriate symplectic matrix Z_1 times J^T, and

$$L_2 G_2 S_1^T U_2 V_2 U_1 V_1 S_1 G_2^T L_2^{-1}$$

in terms of $\widetilde{U}_1$ times an appropriate symplectic matrix W_1. That is,

$$
\begin{aligned}
L_2 G_2 S_1^T X_2 Y_2 X_1 Y_1 S_1 G_2^T L_2^T J^T &= \widetilde{X}_1 Z_1 J^T, \qquad (4.2.26)\\
L_2 G_2 S_1^T U_2 V_2 U_1 V_1 S_1 G_2^T L_2^{-1} &= \widetilde{U}_1 W_1,
\end{aligned}
$$

where Z_1 and W_1 are symplectic. Moreover, as $\widetilde{X}_1$ commutes with

$$X_n, \ldots, X_3, Y_n, \ldots, Y_3$$

and $\widetilde{U}_1$ commutes with

$$U_n, \ldots, U_3, V_n, \ldots, V_3$$

we obtain

$$
\begin{aligned}
B_u^{(1)} &= \widetilde{X}_1 X_n Y_n \cdots X_3 Y_3 Z_1 J^T,\\
B_s^{(1)} &= \widetilde{U}_1 U_n V_n \cdots U_3 V_3 W_1.
\end{aligned}
$$

Now the bulge is chased down the subdiagonals one row and one column at a time. This is done using the algorithm for reducing a symplectic matrix to butterfly form as given in Table 3.2. First a symplectic Givens transformation is applied to eliminate the $(n + 1, 2)$ element. This yields

$$
G_2^T B^{(1)} G_2 =
\left[
\begin{array}{cccc|cccc}
x & 0 & & & x & \otimes & + & \\
 & \otimes & & & \otimes & \otimes & \otimes & \\
 & \oplus & x & & + & \otimes & x & x \\
 & & & x & & & x & x \\
 & & & & \ddots & & & \ddots \\
\hline
x & 0 & & & x & \otimes & + & \\
 & \otimes & & & \otimes & \otimes & \otimes & \\
 & \oplus & x & & + & \otimes & x & x \\
 & & & x & & & x & x \\
 & & & & \ddots & & & \ddots
\end{array}
\right],
$$

or in terms of $B_u^{(1)}$ and $B_s^{(1)}$

$$
\begin{aligned}
G_2^T B_u^{(1)} G_2 &= \widetilde{X}_1 X_n Y_n \cdots X_3 Y_3 G_2^T Z_1 G_2 J^T,\\
G_2^T B_s^{(1)} G_2 &= \widetilde{U}_1 U_n V_n \cdots U_3 V_3 G_2^T W_1 G_2.
\end{aligned}
$$

Then a symplectic Householder transformation H_2 is used to zero the $(n + 1, n + 3)$ element:

$$H_2^T G_2^T B^{(1)} G_2 H_2 = \begin{bmatrix} x & & & & & x & \otimes & 0 & & \\ & \otimes & \oplus & & & \otimes & \otimes & \otimes & \oplus & \\ & \oplus & \otimes & & & \oplus & \otimes & \otimes & \otimes & \\ & & & x & & & \oplus & \otimes & x & \\ & & & & \ddots & & & & & \ddots \\ \hline x & & & & & x & \otimes & 0 & & \\ & \otimes & \oplus & & & \otimes & \otimes & \otimes & \oplus & \\ & \oplus & \otimes & & & \oplus & \otimes & \otimes & \otimes & \\ & & & x & & & \oplus & \otimes & x & \\ & & & & \ddots & & & & & \ddots \end{bmatrix}.$$

Using again the commuting properties and the fact that H_2 is unitary and symplectic, we obtain

$$H_2^T G_2^T B_u^{(1)} G_2 H_2 = \tilde{X}_1 X_n Y_n \cdots X_4 Y_4 H_2^T X_3 Y_3 G_2^T Z_1 G_2 H_2 J^T,$$
$$H_2^T G_2^T B_s^{(1)} G_2 H_2 = \tilde{U}_1 U_n V_n \cdots U_4 V_4 H_2^T U_3 V_3 G_2^T W_1 G_2 H_2.$$

A symplectic Givens transformation G_3 annihilates the $(n+3, 2)$ element. This yields

$$G_3 H_2^T G_2^T B^{(1)} G_2 H_2 G_3^T = \begin{bmatrix} x & & & & & x & x & & & \\ & x & \oplus & & & x & x & \otimes & + & \\ & \oplus & \otimes & & & \oplus & \otimes & \otimes & \otimes & \\ & & \oplus & x & & & + & \otimes & x & \\ & & & & \ddots & & & & & \ddots \\ \hline x & & & & & x & x & & & \\ & x & \oplus & & & x & x & \otimes & + & \\ & 0 & \otimes & & & 0 & \otimes & \otimes & \otimes & \\ & & \oplus & x & & & + & \otimes & x & \\ & & & & \ddots & & & & & \ddots \end{bmatrix},$$

and

$$G_3 H_2^T G_2^T B_u^{(1)} G_2 H_2 G_3^T$$
$$= \tilde{X}_1 X_n Y_n \cdots X_4 Y_4 G_3 H_2^T X_3 Y_3 G_2^T Z_1 G_2 H_2 G_3^T J^T,$$
$$G_3 H_2^T G_2^T B_s^{(1)} G_2 H_2 G_3^T$$
$$= \tilde{U}_1 U_n V_n \cdots U_4 V_4 G_3 H_2^T U_3 V_3 G_2^T W_1 G_2 H_2 G_3^T.$$

Finally, a symplectic Gauss transformation L_3 to eliminate the $(3,2)$ element completes the bulge chase:

$$B^{(2)} := L_3 G_3 H_2^T G_2^T B^{(1)} G_2 H_2 G_3^T L_3^{-1}$$

$$= \begin{bmatrix}
\begin{array}{cccc|cccc}
x & & & & x & \otimes & & \\
 & \otimes & \oplus & & \otimes & \otimes & \otimes & \oplus \\
 & 0 & \otimes & & 0 & \otimes & \otimes & \otimes \\
 & & \oplus & x & & \oplus & \otimes & x \\
 & & & & \ddots & & & \ddots \\ \hline
x & & & & x & \otimes & & \\
 & \otimes & \oplus & & \otimes & \otimes & \otimes & \oplus \\
 & & \otimes & & & \otimes & \otimes & \otimes \\
 & & \oplus & x & & \oplus & \otimes & x \\
 & & & & \ddots & & & \ddots
\end{array}
\end{bmatrix}.$$

The bulge has been chased exactly one row and one column down the subdiagonal in each block. The form of $B^{(2)}$ is the same as the form of $B^{(1)}$, just the bulge can be found one row and one column further down in each block. The same sequence of symplectic Givens, Householder and Gauss transformation as in the last four steps can be used to chase the bulge one more row and column down in each block.

Furthermore, due to the commuting properties and the symplecticity of L_3 we have

$$
\begin{aligned}
B_u^{(2)} &:= L_3 G_3 H_2^T G_2^T B_u^{(1)} G_2 H_2 G_3^T L_3^{-1} \\
&= \tilde{X}_1 X_n Y_n \cdots X_4 Y_4 L_3 G_3 H_2^T X_3 Y_3 G_2^T Z_1 G_2 H_2 G_3^T L_3^T J^T, \\
B_s^{(2)} &:= L_3 G_3 H_2^T G_2^T B_s^{(1)} G_2 H_2 G_3^T L_3^{-1} \\
&= \tilde{U}_1 U_n V_n \cdots U_4 V_4 L_3 G_3 H_2^T U_3 V_3 G_2^T W_1 G_2 H_2 G_3^T L_3^{-1}.
\end{aligned}
$$

In subsequent transformations the elements of $B^{(2)}$ in the positions $(2,2)$, $(n+2,2)$, $(1,n+1)$, $(1,n+2)$, $(2,n+1)$, $(n+1,n+1)$, $(n+1,n+2)$ and $(n+2,n+1)$ are not altered. Hence, at this point we can read off $\tilde{a}_2$, $\tilde{b}_2$, $\tilde{c}_1$, and $\tilde{d}_2$ of the final $\tilde{B}$. Note that $\tilde{X}_2$ and $\tilde{Y}_1$, resp. $\tilde{U}_2$ and $\tilde{V}_1$, do not commute. In other words, we can rewrite

$$L_3 G_3 H_2^T X_3 Y_3 G_2^T Z_1 G_2 H_2 G_3^T L_3^T J^T$$

in terms of $\tilde{X}_2 \tilde{Y}_1$ times an appropriate symplectic matrix Z_2 times J^T, and

$$L_3 G_3 H_2^T U_3 V_3 G_2^T W_1 G_2 H_2 G_3^T L_3^{-1}$$

in terms of $\widetilde{U}_2\widetilde{V}_1$ times an appropriate symplectic matrix W_2. That is,

$$
\begin{aligned}
L_3 G_3 H_2^T X_3 Y_3 G_2^T Z_1 G_2 H_2 G_3^T L_3^T J^T &= \widetilde{X}_2 \widetilde{Y}_1 Z_2 J^T, \\
L_3 G_3 H_2^T U_3 V_3 G_2^T W_1 G_2 H_2 G_3^T L_3^{-1} &= \widetilde{U}_2 \widetilde{V}_1 W_2.
\end{aligned}
$$

As $\widetilde{X}_2$ and $\widetilde{Y}_1$ commute with most of the factors of $B_u^{(2)}$ we obtain

$$
B_u^{(2)} = \widetilde{X}_1 \widetilde{X}_2 \widetilde{Y}_1 X_n Y_n \cdots X_4 Y_4 Z_2 J^T,
$$

and analogously,

$$
B_s^{(2)} = \widetilde{U}_1 \widetilde{U}_2 \widetilde{V}_1 U_n V_n \cdots U_4 V_4 W_2.
$$

Continuing in this fashion, we obtain for $j = 2, \ldots, n-1$

$$
B^{(j)} := L_{j+1} G_{j+1} H_j^T G_j^T B^{(j-1)} G_j H_j G_{j+1}^T L_{j+1}^{-1},
$$

and

$$
\begin{aligned}
B_u^{(j)} :=&\ L_{j+1} G_{j+1} H_j^T G_j^T B_u^{(j-1)} G_j H_j G_{j+1}^T L_{j+1}^{-1} \\
=&\ \widetilde{X}_1 \cdots \widetilde{X}_{j-1} \widetilde{Y}_1 \cdots \widetilde{Y}_{j-2} X_n Y_n \cdots X_{j+2} Y_{j+2} \cdot \\
&\qquad \cdot L_{j+1} G_{j+1} H_j^T X_{j+1} Y_{j+1} G_j^T Z_{j-1} G_j H_j G_{j+1}^T L_{j+1}^T J^T \\
=&\ \widetilde{X}_1 \cdots \widetilde{X}_{j-1} \widetilde{X}_j \widetilde{Y}_1 \cdots \widetilde{Y}_{j-2} \widetilde{Y}_{j-1} X_n Y_n \cdots X_{j+2} Y_{j+2} Z_j J^T, \\
B_s^{(j)} :=&\ L_{j+1} G_{j+1} H_j^T G_j^T B_s^{(2)} G_j H_j G_{j+1}^T L_{j+1}^{-1} \\
=&\ \widetilde{U}_1 \cdots \widetilde{U}_{j-1} \widetilde{V}_1 \cdots \widetilde{V}_{j-2} U_n V_n \cdots U_{j+2} V_{j+2} \cdot \\
&\qquad \cdot L_{j+1} G_{j+1} H_j^T U_{j+1} V_{j+1} G_j^T W_{j-1} G_j H_j G_{j+1}^T L_{j+1}^{-1} \\
=&\ \widetilde{U}_1 \cdots \widetilde{U}_{j-1} \widetilde{U}_j \widetilde{V}_1 \cdots \widetilde{V}_{j-2} \widetilde{V}_{j-1} U_n V_n \cdots U_{j+2} V_{j+2} W_j,
\end{aligned}
$$

where $X_{n+1} = Y_{n+1} = U_{n+1} = V_{n+1} = I$. Thus,

$$
B^{(n-1)} := L_n G_n H_{n-1}^T G_{n-1}^T B^{(n-2)} G_{n-1} H_{n-1} G_n^T L_n^{-1},
$$

and

$$
\begin{aligned}
B_u^{(n-1)} &= \widetilde{X}_1 \cdots \widetilde{X}_{n-1} \widetilde{Y}_1 \cdots \widetilde{Y}_{n-2} Z_{n-1} J^T, \\
B_s^{(n-1)} &= \widetilde{U}_1 \cdots \widetilde{U}_{n-1} \widetilde{V}_1 \cdots \widetilde{V}_{n-2} W_{n-1}.
\end{aligned}
$$

One last symplectic Givens transformation has to be applied to $B^{(n-1)}$ to obtain the new butterfly matrix $\widetilde{B}$

$$
G_n^T B^{(n-1)} G_n = \widetilde{B}.
$$

Hence,

$$
\begin{aligned}
G_n^T Z_{n-1} G_n &= \tilde{X}_n \tilde{Y}_{n-1} \tilde{Y}_n, \\
G_n^T W_{n-1} G_n &= \tilde{U}_n \tilde{V}_{n-1} \tilde{V}_n,
\end{aligned}
$$

and

$$
\begin{aligned}
\tilde{B}_u &= \tilde{X}_1 \cdots \tilde{X}_n \tilde{Y}_1 \cdots \tilde{Y}_n J^T, \\
\tilde{B}_s &= \tilde{U}_1 \cdots \tilde{U}_n \tilde{V}_1 \cdots \tilde{V}_n.
\end{aligned}
$$

4.2.2   THE DETAILS

How can the above observations be used to derive an algorithm which works solely on the parameters that determine B without forming B, $\tilde{B}$ or any of the intermediate matrices? We will consider this question here, concentrating on the decomposition $B_u = K_u^{-1} N_u$ of B. The same derivations can be done for the decomposition $B_s = K_s^{-1} N_s$. Let us start with (4.2.26),

$$
L_2 G_2 S_1^T X_2 Y_2 X_1 Y_1 S_1 G_2^T L_2^T = \tilde{X}_1 Z_1.
$$

X_1, X_2, Y_1, and Y_2 are known. S_1 is determined by the choice of the Laurent polynomial which drives the current SR step. As discussed in Section 4.1.2 for a double shift the shift polynomial $q_2(B) = (B + B^{-1}) - \beta I$ should be chosen where $\beta = \mu + \mu^{-1}$ if $\mu \in \mathbf{R}$ or $\beta = \mu + \bar{\mu}$ for $\mu \in \mathbf{C}, |\mu| = 1$. Here the shift μ is chosen corresponding to the generalized Rayleigh-quotient strategy. This implies

$$
q_2(B)e_1 = (b_1 + a_1 c_1 - b_n - a_n c_n)e_1 + a_1 d_2 e_2.
$$

Hence, for S_1 as in (4.2.23), α and β have to be determined such that

$$
\begin{bmatrix} \alpha & -\beta \\ \beta & \alpha \end{bmatrix}
\begin{bmatrix} b_1 + a_1 c_1 - b_n - a_n c_n \\ a_1 d_2 \end{bmatrix}
= \begin{bmatrix} \star \\ 0 \end{bmatrix}.
$$

Next a symplectic Givens transformation G_2 has to be determined such that

$$
(G_2 S_1^T X_2 Y_2 X_1 Y_1 S_1 G_2^T)_{(n+2,n+1)} = 0.
$$

This implies that $G_2 = G(2, \alpha_2, \beta_2)$ has to be chosen such that

$$
\begin{bmatrix} \alpha_2 & \beta_2 \\ -\beta_2 & \alpha_2 \end{bmatrix}
\begin{bmatrix} (S_1^T X_2 Y_2 X_1 Y_1 S_1)_{2,n+1} \\ (S_1^T X_2 Y_2 X_1 Y_1 S_1)_{n+2,n+1} \end{bmatrix}
= \begin{bmatrix} \star \\ 0 \end{bmatrix}
$$

where

$$
\begin{aligned}
(S_1^T X_2 Y_2 X_1 Y_1 S_1)_{2,n+1} &= \beta \alpha (b_1 - b_2), \\
(S_1^T X_2 Y_2 X_1 Y_1 S_1)_{n+2,n+1} &= \beta \alpha (a_1 - a_2).
\end{aligned}
$$

Now a symplectic Gauss transformation $L_2 = L_2(\tau_1, \psi_1)$ is used such that

$$(L_2 G_2 S_1^T X_2 Y_2 X_1 Y_1 S_1 G_2^T L_2^T)_{2,n+1} = 0.$$

Hence, we have to compute τ_1 and ψ_1 such that

$$\left[\begin{array}{c|c} \begin{array}{cc} \tau_1 & \\ & \tau_1 \end{array} & \begin{array}{cc} & \psi_1 \\ \psi_1 & \end{array} \\ \hline & \begin{array}{cc} \tau_1^{-1} & \\ & \tau_1^{-1} \end{array} \end{array}\right] \left[\begin{array}{c} (G_2 S_1^T X_2 Y_2 X_1 Y_1 S_1 G_2^T)_{1,n+1} \\ (G_2 S_1^T X_2 Y_2 X_1 Y_1 S_1 G_2^T)_{2,n+1} \\ (G_2 S_1^T X_2 Y_2 X_1 Y_1 S_1 G_2^T)_{n+1,n+1} \\ 0 \end{array}\right] = \left[\begin{array}{c} \star \\ 0 \\ \star \\ 0 \end{array}\right],$$

where

$$
\begin{array}{rcl}
(G_2 S_1^T X_2 Y_2 X_1 Y_1 S_1 G_2^T)_{1,n+1} &=& \alpha^2 b_1 + \beta^2 b_2, \\
(G_2 S_1^T X_2 Y_2 X_1 Y_1 S_1 G_2^T)_{2,n+1} &=& \alpha_2 \beta \alpha (b_1 - b_2) + \beta_2 \beta \alpha (a_1 - a_2), \\
(G_2 S_1^T X_2 Y_2 X_1 Y_1 S_1 G_2^T)_{n+1,n+1} &=& \alpha^2 a_1 + \beta^2 a_2.
\end{array}
$$

Now we can read off $\widetilde{a}_1$ and $\widetilde{b}_1$

$$
\begin{array}{rcl}
\widetilde{a}_1 &=& (\alpha^2 a_1 + \beta^2 a_2)/\tau_1^2, \\
\widetilde{b}_1 &=& \alpha^2 b_1 + \beta^2 b_2.
\end{array}
$$

Moreover, $L_2 G_2 S_1^T X_2 Y_2 X_1 Y_1 S_1 G_2^T L_2^T$ is a matrix of the form

$$\left[\begin{array}{ccccc|ccc} x & x & x & & & x & x & \\ x & x & x & & & & x & \\ & & x & & & & & \\ & & & \ddots & & & & \\ \hline x & x & x & & & x & x & \\ x & x & x & & & & x & \\ x & x & & & & x & x & \\ & & & & & & & \ddots \end{array}\right].$$

Now we form

$$\widetilde{X}_1 = \left[\begin{array}{cc|cc} \begin{array}{cc} \widetilde{a}_1^{-1} & \\ & \ddots \end{array} & & \begin{array}{cc} \widetilde{b}_1 & \\ & \end{array} & \\ & I^{n-1,n-1} & & \\ \hline & & \begin{array}{cc} \widetilde{a}_1 & \\ & \end{array} & \\ & & & I^{n-1,n-1} \end{array}\right],$$

and build $Z_1 = \tilde{X}_1^{-1} L_2 G_2 S_1^T X_2 Y_2 X_1 Y_1 S_1 G_2^T L_2^T$. This is a matrix of the form

$$
\begin{bmatrix}
1 & & & & & & & \\
\delta_{21} & \delta_{22} & \delta_{23} & & & \varepsilon_{22} & & \\
& & 1 & & & & & \\
& & & \ddots & & & & \\
\hline
\mu_{11} & \mu_{12} & \mu_{13} & & 1 & \zeta_{12} & & \\
\mu_{21} & \mu_{22} & \mu_{23} & & & \zeta_{22} & & \\
\mu_{31} & \mu_{32} & & & & \zeta_{32} & 1 & \\
& & & & & & & \ddots
\end{bmatrix},
$$

where the entries that will be used in the subsequent transformations are given by

$$
\begin{aligned}
\delta_{22} &= \tau_1^2(\alpha_2 h_1 + \beta_2 h_2) + \tau_1 \psi_1 h_5, \\
\delta_{23} &= -\tau_1 \alpha_2 \alpha b_2 d_3 - \tau_1 \beta_2 \alpha d_3 a_2, \\
\varepsilon_{22} &= -\beta_2 h_1 + \alpha_2 h_2 + \psi_1 h_6/\tau_1, \\
\mu_{11} &= (\tau_1^2 g_1 + \psi_1 \tau_1 h_6)/(\alpha^2 a_1 + \beta^2 b_1), \\
\mu_{12} &= \psi_1 \tau_1 + \tau_1^2 h_5/(\alpha^2 a_1 + \beta^2 b_1), \\
\mu_{13} &= -\tau_1 \beta d_3 a_2/(\alpha^2 a_1 + \beta^2 b_1), \\
\mu_{22} &= \alpha_2 h_3 + \beta_2 h_4, \\
\mu_{23} &= (\beta_2 \alpha b_2 d_3 - \alpha_2 \alpha d_3 a_2)/\tau_1, \\
\mu_{32} &= -\tau_1 \alpha_2 \alpha d_3, \\
\zeta_{12} &= h_6/(\alpha^2 a_1 + \beta^2 b_1), \\
\zeta_{22} &= (-\beta_2 h_3 + \alpha_2 h_4)/\tau_1^2, \\
\zeta_{32} &= \beta_2 \alpha d_3/\tau_1,
\end{aligned}
$$

where

$$
\begin{aligned}
h_1 &= \alpha_2 g_2 + \beta_2 g_3, \\
h_2 &= \alpha_2(\beta^2 b_1 + \alpha^2 b_2) + \beta_2(\beta^2 a_1 + \alpha^2 a_2), \\
h_3 &= \alpha_2 g_3 - \beta_2 g_2, \\
h_4 &= \alpha_2(\beta^2 a_1 + \alpha^2 a_2) - \beta_2(\beta^2 b_1 + \alpha^2 b_2), \\
h_5 &= \alpha_2 g_4 + \beta_2 \alpha \beta(a_1 - a_2), \\
h_6 &= \alpha_2 \alpha \beta(a_1 - a_2) - \beta_2 g_4,
\end{aligned}
$$

and

$$
\begin{aligned}
g_1 &= -\alpha^2 a_1 c_1 + \alpha\beta d_2(a_1 - a_2) - \beta^2 c_2 a_2, \\
g_2 &= \beta^2(a_1^{-1} - b_1 c_1) - \alpha\beta d_2(b_1 + b_2) + \alpha^2(a_2^{-1} - b_2 c_2), \\
g_3 &= -\beta^2 a_1 c_1 - \alpha\beta d_2(a_1 + a_2) - \alpha^2 a_2 c_2, \\
g_4 &= -\alpha^2 a_1 d_2 + \alpha\beta(c_2 a_2 - a_1 c_1) + \beta^2 a_2 d_2.
\end{aligned}
$$

Next we have to consider

$$
L_3 G_3 H_2^T X_3 Y_3 G_2^T Z_1 G_2 H_2 G_3^T L_3^T J^T.
$$

First a symplectic Givens transformation G_2 eliminates the $(n+1, n+2)$ element of Z_1. This implies that $G_2 = G_2(\alpha_3, \beta_3)$ has to be chosen such that

$$
[\mu_{12} \ \ \zeta_{12}]
\begin{bmatrix}
\alpha_3 & \beta_3 \\
-\beta_3 & \alpha_3
\end{bmatrix}
= [\ast \ \ 0]. \tag{4.2.27}
$$

The resulting transformed matrix is given by

$$
G_2^T Z_1 G_2 =
\left[
\begin{array}{cccc|ccc}
1 & & & & & & \\
\delta_{21}^{(1)} & \delta_{22}^{(1)} & \delta_{23}^{(1)} & & \varepsilon_{22}^{(1)} & & \\
& & 1 & & & & \\
& & & \ddots & & & \\
\hline
\mu_{11}^{(1)} & \mu_{12}^{(1)} & \mu_{13}^{(1)} & & 1 & & \\
\mu_{21}^{(1)} & \mu_{22}^{(1)} & \mu_{23}^{(1)} & & \zeta_{22}^{(1)} & & \\
\mu_{31}^{(1)} & \mu_{32}^{(1)} & & & \zeta_{32}^{(1)} & 1 & \\
& & & & & & \ddots
\end{array}
\right],
$$

where the relevant entries are

$$
\begin{aligned}
\mu_{12}^{(1)} &= \alpha_3 \mu_{12} - \beta_3 \zeta_{12}, \\
\mu_{22}^{(1)} &= \alpha_3^2 \mu_{22} + \beta_3 \alpha_3(\delta_{22} - \zeta_{22}) - \beta_3^2 \varepsilon_{22}, \\
\mu_{23}^{(1)} &= \beta_3 \delta_{23} + \alpha_3 \mu_{23}, \\
\mu_{32}^{(1)} &= \alpha_3 \mu_{32} - \beta_3 \zeta_{32}, \\
\delta_{22}^{(1)} &= \alpha_3^2 \delta_{22} - \alpha_3 \beta_3(\varepsilon_{22} + \mu_{22}) + \beta_3^2 \zeta_{22}, \qquad (4.2.28)\\
\delta_{23}^{(1)} &= \alpha_3 \delta_{23} - \beta_3 \mu_{23}, \\
\varepsilon_{22}^{(1)} &= \alpha_3^2 \varepsilon_{22} + \beta_3 \alpha_3(\delta_{22} - \zeta_{22}) - \beta_3^2 \mu_{22}, \\
\zeta_{22}^{(1)} &= \alpha_3^2 \zeta_{22} + \alpha_3 \beta_3(\mu_{22} + \varepsilon_{22}) + \beta_3^2 \delta_{22}, \\
\zeta_{32}^{(1)} &= \beta_3 \mu_{32} + \alpha_3 \zeta_{32}.
\end{aligned}
$$

The $(1,1)$ and the $(1,3)$ entry are not altered by this transformation: $\mu_{11}^{(1)} = \mu_{11}$ and $\mu_{13}^{(1)} = \mu_{13}$.

Next a symplectic Householder transformation H_2 is used to zero the $(n+1,3)$ element of $G_2^T Z_1 G_2$. H_2 is a matrix of the form (4.2.23); we denote its entries by α_4 and β_4. The scalars α_4 and β_4 have to be chosen such that

$$[\mu_{12}^{(1)} \quad \mu_{13}^{(1)}] \begin{bmatrix} \alpha_4 & \beta_4 \\ -\beta_4 & \alpha_4 \end{bmatrix} = [\star \quad 0]. \tag{4.2.29}$$

This results in $H_2^T X_3 Y_3 G_2^T Z_1 G_2 H_2$

$$\begin{bmatrix}
1 & & & & & & & & \\
\delta_{21}^{(2)} & \delta_{22}^{(2)} & \delta_{23}^{(2)} & \delta_{24}^{(2)} & & & \varepsilon_{22}^{(2)} & \varepsilon_{23}^{(2)} & \\
\delta_{31}^{(2)} & \delta_{32}^{(2)} & \delta_{33}^{(2)} & \delta_{34}^{(2)} & & & \varepsilon_{32}^{(2)} & \varepsilon_{33}^{(2)} & \\
& & & 1 & & & & & \\
& & & & \ddots & & & & \\
\hline
\mu_{11}^{(2)} & \mu_{12}^{(2)} & & & & 1 & & & \\
\mu_{21}^{(2)} & \mu_{22}^{(2)} & \mu_{23}^{(2)} & \mu_{24}^{(2)} & & & \zeta_{22}^{(2)} & \zeta_{23}^{(2)} & \\
\mu_{31}^{(2)} & \mu_{32}^{(2)} & \mu_{33}^{(2)} & \mu_{34}^{(2)} & & & \zeta_{32}^{(2)} & \zeta_{33}^{(2)} & \\
& \mu_{42}^{(2)} & \mu_{43}^{(2)} & & & & & & 1 \\
& & & & & & & & \ddots
\end{bmatrix},$$

where the relevant entries are given by

$$\begin{aligned}
\delta_{33}^{(2)} &= \alpha_4^2(a_3^{-1} - b_3 c_3) + \alpha_4 \beta_4(\mu_{32}^{(1)} b_3 + \delta_{23}^{(1)}) + \beta_4^2 \delta_{22}^{(1)}, \\
\delta_{34}^{(2)} &= -\alpha_4 b_3 d_4, \\
\varepsilon_{22}^{(2)} &= \alpha_4^2 \varepsilon_{22}^{(1)} - \alpha_4 \beta_4 \zeta_{32}^{(1)} b_3 + \beta_4^2 b_3, \\
\varepsilon_{32}^{(2)} &= \alpha_4^2 \zeta_{32}^{(1)} b_3 + \alpha_4 \beta_4(\varepsilon_{22}^{(1)} - b_3), \\
\varepsilon_{33}^{(2)} &= \alpha_4^2 b_3 + \alpha_4 \beta_4 \zeta_{32}^{(1)} b_3 + \beta_4^2 \varepsilon_{22}^{(1)}, \\
\mu_{12}^{(2)} &= \alpha_4 \mu_{12}^{(1)} - \beta_4 \mu_{13}^{(1)}, \\
\mu_{22}^{(2)} &= \alpha_4^2 \mu_{22}^{(1)} - \alpha_4 \beta_4(\mu_{23}^{(1)} + \mu_{32}^{(1)} a_3) - \beta_4^2 a_3 c_3, \\
\mu_{23}^{(2)} &= \alpha_4^2 \mu_{23}^{(1)} + \alpha_4 \beta_4(a_3 c_3 + \mu_{22}^{(1)}) - \beta_4^2 \mu_{32}^{(1)} a_3, \\
\mu_{24}^{(2)} &= \beta_4 a_3 d_4, \\
\mu_{33}^{(2)} &= -\alpha_4^2 a_3 c_3 + \alpha_4 \beta_4(\mu_{23}^{(1)} + \mu_{32}^{(1)} a_3) + \beta_4^2 \mu_{22}^{(1)}, \\
\mu_{34}^{(2)} &= -\alpha_4 a_3 d_4, \\
\mu_{43}^{(2)} &= -\alpha_4 d_4,
\end{aligned} \tag{4.2.30}$$

and

$$\zeta_{22}^{(2)} = \alpha_4^2 \zeta_{22}^{(1)} - \alpha_4 \beta_4 \zeta_{32}^{(1)} a_3 + \beta_4^2,$$

$$\zeta_{23}^{(2)} = \alpha_4 \beta_4 (\zeta_{22}^{(1)} - a_3) - \beta_4^2 \zeta_{32}^{(1)} a_3, \qquad (4.2.31)$$

$$\zeta_{32}^{(2)} = \alpha_4^2 \zeta_{32}^{(1)} a_3 + \alpha_4 \beta_4 (\zeta_{22}^{(1)} - 1),$$

$$\zeta_{33}^{(2)} = \alpha_4^2 a_3 + \alpha_4 \beta_4 \zeta_{32}^{(1)} a_3 + \beta_4^2 \zeta_{22}^{(1)}.$$

The $(1,1)$ entry is not altered by this transformation: $\mu_{11}^{(2)} = \mu_{11}^{(1)} = \mu_{11}$.

A symplectic Givens transformation G_3 is employed to zero the $(n+3, n+2)$ element in $H_2^T X_3 Y_3 G_2^T Z_1 G_2 H_2$. This implies that $G_3 = G_3(\alpha_5, \beta_5)$ has to be chosen such that

$$\begin{bmatrix} \alpha_5 & \beta_5 \\ -\beta_5 & \alpha_5 \end{bmatrix} \begin{bmatrix} \varepsilon_{32}^{(2)} \\ \zeta_{32}^{(2)} \end{bmatrix} = \begin{bmatrix} \star \\ 0 \end{bmatrix}. \qquad (4.2.32)$$

The resulting matrix $G_3 H_2^T X_3 Y_3 G_2^T Z_1 G_2 H_2 G_3^T$ is given by

$$\begin{bmatrix}
1 & & & & & & & \\
\delta_{21}^{(3)} & \delta_{22}^{(3)} & \delta_{23}^{(3)} & \delta_{24}^{(3)} & & \varepsilon_{22}^{(3)} & \varepsilon_{23}^{(3)} & \\
\delta_{31}^{(3)} & \delta_{32}^{(3)} & \delta_{33}^{(3)} & \delta_{34}^{(3)} & & \varepsilon_{32}^{(3)} & \varepsilon_{33}^{(3)} & \\
& & & 1 & & & & \\
& & & & \ddots & & & \\
\hline
\mu_{11}^{(3)} & \mu_{12}^{(3)} & & & 1 & & & \\
\mu_{21}^{(3)} & \mu_{22}^{(3)} & \mu_{23}^{(3)} & \mu_{24}^{(3)} & & \zeta_{22}^{(3)} & \zeta_{23}^{(3)} & \\
& \mu_{32}^{(3)} & \mu_{33}^{(3)} & \mu_{34}^{(3)} & & & \zeta_{33}^{(3)} & \\
& \mu_{42}^{(3)} & \mu_{43}^{(3)} & & & & \zeta_{43}^{(3)} & 1 \\
& & & & & & & \ddots
\end{bmatrix},$$

where the relevant entries are given by

$$\delta_{33}^{(3)} = \alpha_5^2 \delta_{33}^{(2)} + \alpha_5 \beta_5 (\mu_{33}^{(2)} + \varepsilon_{33}^{(2)}) + \beta_5^2 \zeta_{33}^{(2)},$$

$$\varepsilon_{33}^{(3)} = \alpha_5^2 \varepsilon_{33}^{(2)} + \alpha_5 \beta_5 (\zeta_{33}^{(2)} - \delta_{33}^{(2)}) - \beta_5^2 \mu_{33}^{(2)},$$

$$\mu_{33}^{(3)} = \alpha_5^2 \mu_{33}^{(2)} + \alpha_5 \beta_5 (\zeta_{33}^{(2)} - \delta_{33}^{(2)}) - \beta_5^2 \varepsilon_{33}^{(2)}, \qquad (4.2.33)$$

$$\zeta_{33}^{(3)} = \alpha_5^2 \zeta_{33}^{(2)} - \alpha_5 \beta_5 (\mu_{33}^{(2)} + \varepsilon_{33}^{(2)}) + \beta_5^2 \delta_{33}^{(2)},$$

and

$$\mu_{43}^{(3)} = \alpha_5 \mu_{43}^{(2)}, \qquad \qquad \zeta_{43}^{(3)} = -\beta_5 \mu_{43}^{(2)},$$

$$\delta_{34}^{(3)} = \alpha_5 \delta_{34}^{(2)} + \beta_5 \mu_{34}^{(2)}, \qquad \varepsilon_{32}^{(3)} = \alpha_5 \varepsilon_{32}^{(2)} + \beta_5 \zeta_{32}^{(2)},$$

$$\mu_{23}^{(3)} = \alpha_5 \mu_{23}^{(2)} + \beta_5 \zeta_{23}^{(2)}, \qquad \mu_{34}^{(3)} = \alpha_5 \mu_{34}^{(2)} - \beta_5 \delta_{34}^{(2)}, \qquad (4.2.34)$$

$$\zeta_{23}^{(3)} = \alpha_5 \zeta_{23}^{(2)} - \beta_5 \mu_{23}^{(2)}.$$

Some of the relevant entries do not change:

$$\mu_{11}^{(3)} = \mu_{11}^{(2)} = \mu_{11}^{(1)} = \mu_{11},$$

and

$$
\begin{array}{llll}
\varepsilon_{22}^{(3)} & = & \varepsilon_{22}^{(2)}, & \quad \mu_{12}^{(3)} = \mu_{12}^{(2)}, & \quad \mu_{22}^{(3)} = \mu_{22}^{(2)}, \\
\mu_{24}^{(3)} & = & \mu_{24}^{(2)}, & \quad \zeta_{22}^{(3)} = \zeta_{22}^{(2)}.
\end{array}
$$

Finally the $(3, n+2)$ element of $G_3 H_2^T X_3 Y_3 G_2^T Z_1 G_2 H_2 G_3^T$ is annihilated using a symplectic Gauss transformation L_3. Hence, we have to compute τ_2 and ψ_2 such that

$$
\left[
\begin{array}{c|c}
\tau_2 & \psi_2 \\
\hline
\tau_2 & \psi_2 \\
\tau_2^{-1} & \\
& \tau_2^{-1}
\end{array}
\right]
\left[
\begin{array}{c}
\varepsilon_{22}^{(3)} \\
\varepsilon_{32}^{(3)} \\
\zeta_{22}^{(3)} \\
0
\end{array}
\right]
=
\left[
\begin{array}{c}
\star \\
0 \\
\star \\
0
\end{array}
\right].
\tag{4.2.35}
$$

We obtain $L_3 G_3 H_2^T X_3 Y_3 G_2^T Z_1 G_2 H_2 G_3^T L_3^T$

$$
\left[
\begin{array}{ccccc|cc}
1 & & & & & & \\
\delta_{21}^{(4)} & \delta_{22}^{(4)} & \delta_{23}^{(4)} & \delta_{24}^{(4)} & & \varepsilon_{22}^{(4)} & \varepsilon_{23}^{(4)} \\
& \delta_{32}^{(4)} & \delta_{33}^{(4)} & \delta_{34}^{(4)} & & & \varepsilon_{33}^{(4)} \\
& & & 1 & & & \\
& & & & \ddots & & \\
\hline
\mu_{11}^{(4)} & \mu_{12}^{(4)} & & & & 1 & \\
\mu_{21}^{(4)} & \mu_{22}^{(4)} & \mu_{23}^{(4)} & \mu_{24}^{(4)} & & \zeta_{22}^{(4)} & \zeta_{23}^{(4)} \\
& \mu_{32}^{(4)} & \mu_{33}^{(4)} & \mu_{34}^{(4)} & & & \zeta_{33}^{(4)} \\
& \mu_{42}^{(4)} & \mu_{43}^{(4)} & & & & \zeta_{43}^{(4)} \quad 1 \\
& & & & & & \ddots
\end{array}
\right],
$$

where the relevant entries are given by

$$
\begin{array}{rcl}
\delta_{33}^{(4)} & = & \tau_2^2 \delta_{33}^{(3)} + \psi_2 \tau_2 \mu_{23}^{(3)}, \\
\delta_{34}^{(4)} & = & \tau_2 \delta_{34}^{(3)} + \psi_2 \mu_{24}^{(3)}, \\
\mu_{22}^{(4)} & = & \mu_{22}^{(3)} + \psi_2 \zeta_{23}^{(3)} / \tau_2, \\
\mu_{23}^{(4)} & = & \mu_{23}^{(3)} + \psi_2 \zeta_{22}^{(3)} / \tau_2, \\
\varepsilon_{33}^{(4)} & = & \varepsilon_{33}^{(3)} + \psi_2 \zeta_{23}^{(3)} / \tau_2.
\end{array}
\tag{4.2.36}
$$

and

$$
\begin{array}{llll}
\mu_{12}^{(4)} &=& \mu_{12}^{(3)}\tau_2, & \qquad \mu_{24}^{(4)} &=& \mu_{24}^{(3)}/\tau_2, \\
\mu_{34}^{(4)} &=& \mu_{34}^{(3)}/\tau_2, & \qquad \mu_{43}^{(4)} &=& \mu_{43}^{(3)}\tau_2, \\
\zeta_{22}^{(4)} &=& \zeta_{22}^{(3)}/\tau_2^2, & \qquad \zeta_{23}^{(4)} &=& \zeta_{23}^{(3)}/\tau_2^2, \\
\zeta_{33}^{(4)} &=& \zeta_{33}^{(3)}/\tau_2^2, & \qquad \zeta_{43}^{(4)} &=& \zeta_{43}^{(3)}/\tau_2,
\end{array}
\tag{4.2.37}
$$

Again, some of the relevant entries are not altered:

$$
\varepsilon_{22}^{(4)} = \varepsilon_{22}^{(3)} = \varepsilon_{22}^{(2)},
$$
$$
\mu_{11}^{(4)} = \mu_{11}^{(3)} = \mu_{11}^{(2)} = \mu_{11}^{(1)} = \mu_{11},
$$
$$
\mu_{33}^{(4)} = \mu_{33}^{(3)}.
$$

Now the parameters $\widetilde{a}_2, \widetilde{b}_2, \widetilde{c}_1,$ and $\widetilde{d}_2$ can be read off:

$$
\begin{array}{rcl}
\widetilde{a}_2 &=& \zeta_{22}^{(4)}, \\
\widetilde{b}_2 &=& \varepsilon_{22}^{(4)} = \varepsilon_{22}^{(3)}, \\
\widetilde{c}_1 &=& -\mu_{11}^{(4)} = -\mu_{11}, \\
\widetilde{d}_2 &=& -\mu_{12}^{(4)}.
\end{array}
$$

Forming $\widetilde{X}_2\widetilde{Y}_1$ we see that $Z_2 = \widetilde{Y}_1^{-1}\widetilde{X}_2^{-1}L_3G_3H_2^T X_3 Y_3 G_2^T Z_1 G_2 H_2 G_3^T L_3^T$ is given by

$$
Z_2 = \left[
\begin{array}{ccccc|ccccc}
1 \\
& 1 \\
& \delta_{32}^{(5)} & \delta_{33}^{(5)} & \delta_{34}^{(5)} & & & \varepsilon_{33}^{(5)} \\
& & & 1 \\
& & & & \ddots \\
\hline
& & & & & 1 \\
& \mu_{22}^{(5)} & \mu_{23}^{(5)} & \mu_{24}^{(5)} & & 1 & \zeta_{23}^{(5)} \\
& \mu_{32}^{(5)} & \mu_{33}^{(5)} & \mu_{34}^{(5)} & & & \zeta_{33}^{(5)} \\
& \mu_{42}^{(5)} & \mu_{43}^{(5)} & & & & \zeta_{43}^{(5)} & 1 \\
& & & & & & & & \ddots
\end{array}
\right],
$$

where only the elements

$$
\begin{array}{rcll}
\mu_{2k}^{(5)} &=& \mu_{2k}^{(4)}/\zeta_{22}^{(4)}, & \qquad k = 2,3,4 \\
\zeta_{23}^{(5)} &=& \zeta_{23}^{(4)}/\zeta_{22}^{(4)}.
\end{array}
\tag{4.2.38}
$$

changed.

Comparing Z_1 and Z_2, the bulge has been chased down exactly one row and column in each block. The same sequence of symplectic Givens, Householder and Gauss transformations as in the last four steps can be used to chase the bulge one more row and column down in each block. Therefore, renaming

$$
\begin{aligned}
\delta_{22} &= \delta_{33}^{(5)}, & \delta_{23} &= \delta_{34}^{(5)}, \\
\varepsilon_{22} &= \varepsilon_{33}^{(5)}, & \mu_{11} &= \mu_{22}^{(5)}, \\
\mu_{12} &= \mu_{23}^{(5)}, & \mu_{13} &= \mu_{24}^{(5)}, \\
\mu_{22} &= \mu_{33}^{(5)}, & \mu_{23} &= \mu_{34}^{(5)}, \\
\mu_{32} &= \mu_{43}^{(5)}, & \zeta_{12} &= \zeta_{23}^{(5)}, \\
\zeta_{22} &= \zeta_{33}^{(5)}, & \zeta_{32} &= \zeta_{43}^{(5)},
\end{aligned}
$$

and repeating the computations (4.2.27) – (4.2.38) we obtain

$$
\begin{aligned}
\tilde{a}_3 &= \zeta_{22}^{(4)}, \\
\tilde{b}_3 &= \varepsilon_{22}^{(4)}, \\
\tilde{c}_2 &= -\mu_{11}^{(4)}, \\
\tilde{d}_3 &= -\mu_{12}^{(4)}.
\end{aligned}
$$

Iterating like this, the parameters $\tilde{a}_1, \ldots, \tilde{a}_{n-1}, \tilde{b}_1, \ldots, \tilde{b}_{n-1}, \tilde{c}_1, \ldots, \tilde{c}_{n-2}$, and $\tilde{d}_2, \ldots, \tilde{d}_{n-1}$ can be computed.

For the final step of the algorithm, let us consider the matrix Z_{n-1}. It has the form

$$
\left[
\begin{array}{ccc|ccc}
I^{n-2,n-2} & & & & & \\
& 1 & & & & \\
& \delta & \delta & & & \varepsilon \\
\hline
& & & I^{n-2,n-2} & & \\
& \mu & \mu & & 1 & \zeta \\
& \mu & \mu & & & \zeta
\end{array}
\right].
$$

A symplectic Givens transformation G_n has to be applied to zero the $(2n-1, 2n)$ entry of Z_{n-1}. The transformation $G_n^T Z_{n-1} G_n$ does not cause any fill-in. Hence, the remaining parameters $\tilde{a}_n, \tilde{b}_n, \tilde{c}_{n-1}, \tilde{c}_n$, and $\tilde{d}_n$ can be read off, as

$$
\tilde{X}_n \tilde{Y}_{n-1} \tilde{Y}_n =
\left[
\begin{array}{ccc|ccc}
I^{n-2,n-2} & & & & & \\
& 1 & & & & \\
& -\tilde{b}_n \tilde{d}_n & \tilde{a}_n^{-1} - \tilde{b}_n \tilde{c}_n & & & \tilde{b}_n \\
\hline
& & & I^{n-2,n-2} & & \\
& -\tilde{c}_{n-1} & -\tilde{d}_n & & 1 & \\
& -\tilde{a}_n \tilde{d}_n & -\tilde{a}_n \tilde{c}_n & & & \tilde{a}_n
\end{array}
\right],
$$

and $G_n^T Z_{n-1} G_n$ are symplectic butterfly matrices of the same form.

Using the same renaming convention as above, this implies that for the Givens transformation G_n, the scalars α_6 and β_6 have to be determined such that

$$[\mu_{12} \ \ \zeta_{12}] \begin{bmatrix} \alpha_6 & \beta_6 \\ -\beta_6 & \alpha_6 \end{bmatrix} = [* \ \ 0].$$

Applying the transformation, the following matrix entries change:

$$
\begin{aligned}
\mu_{12} &= \alpha_6 \mu_{12} - \beta_6 \zeta_{12}, \\
\mu_{22} &= \beta_6(\alpha_6 \delta_{22} - \beta_6 \varepsilon_{22}) + \alpha_6(\alpha_6 \mu_{22} - \beta_6 \zeta_{22}), \\
\zeta_{22} &= \alpha_6(\beta_6 \mu_{22} + \alpha_6 \zeta_{22}) + \beta_6(\beta_6 \delta_{22} + \alpha_6 \varepsilon_{22}), \\
\varepsilon_{22} &= \alpha_6(\beta_6 \delta_{22} + \alpha_6 \varepsilon_{22}) - \beta_6(\beta_6 \mu_{22} + \alpha_6 \zeta_{22}), \\
\delta_{22} &= \alpha_6^2 \delta_{22} - \alpha_6 \beta_6(\varepsilon_{22} + \mu_{22}) + \beta_6^2 \zeta_{22}.
\end{aligned}
$$

The parameters $\tilde{a}_n, \tilde{b}_n, \tilde{c}_{n-1}, \tilde{c}_n$, and $\tilde{d}_n$ are given by

$$
\begin{aligned}
\tilde{a}_n &= \zeta_{22}, & \tilde{b}_n &= \varepsilon_{22}, \\
\tilde{c}_{n-1} &= -\mu_{11}, & \tilde{c}_n &= -\mu_{22}/a_n, \\
\tilde{d}_n &= -\mu_{12}.
\end{aligned}
$$

REMARK 4.5 *a) No 'optimality' is claimed for the form of the algorithm as discussed above either with regard to operation counts or numerical stability. Variants are certainly possible.*

b) A careful flop count shows that one parameterized SR step as described above requires $219n - 233$ flops (assuming that the parameters of a symplectic Givens transformation and of a symplectic Householder transformation are computed using 6 flops, while those of a symplectic Gauss are computed using 7 flops). This can be seen as follows: The initial step requires 166 flops, the final one 39 flops. For the rest of the computation we have that

the computation of	requires
(4.2.28)	47 flops
(4.2.30),(4.2.31)	90 flops
(4.2.33), (4.2.34)	48 flops
(4.2.36), (4.2.37)	32 flops
(4.2.38)	2 flops

These computations have to be repeated $n - 2$ times, resulting in $219(n-2)$ flops. These flops counts assume that quantities like $\alpha_j^2, \beta_j^2, \alpha_j \beta_j$ which are used more than once are computed only once in order to save computational time. If the transformation matrix S is required, then $64n^2 - 128n$ flops have to be added as $2n - 4$ symplectic Givens transformations, $n - 2$ symplectic

Gauss transformations, and $n - 2$ symplectic Householder transformations with $v \in \mathbf{R}^2$ are used.

c) *Similar to the above derivations a double shift SR step based on the factorization $B_s = K_s^{-1} N_s$ can be developed. Although the development of a quadruple shift SR step for the factorizations B_u or B_s is possible.*

d) *The presented parameterized double shift SR algorithm can not be used to mimic a quadruple shift. For a quadruple shift the Laurent polynomial*

$$q_4(\lambda) = (\lambda + \lambda^{-1})^2 - (\mu + \mu^{-1} + \bar{\mu} + \overline{\mu^{-1}})(\lambda + \lambda^{-1})$$
$$+ (\mu + \mu^{-1})(\bar{\mu} + \overline{\mu^{-1}}) - 2$$

is used. The shift μ should be chosen according to the generalized Rayleigh-quotient strategy as explained in Section 4.1.2. That is, for a quadruple shift, the eigenvalues of the 4×4 symplectic matrix

$$G = \left[\begin{array}{cc|cc} b_{n-1} & & b_{n-1}c_{n-1} - a_{n-1}^{-1} & b_{n-1}d_n \\ & b_n & b_n d_n & b_n c_n - a_n^{-1} \\ \hline a_{n-1} & & a_{n-1}c_{n-1} & a_{n-1}d_n \\ & a_n & a_n d_n & a_n c_n \end{array}\right]$$

are chosen. We can not work with a double shift step in the case that the matrix G has eigenvalues $\mu, \bar{\mu}, \mu^{-1}, \overline{\mu^{-1}} \in \mathbb{C}, |\mu| \neq 1$. One might have the idea to first apply a double SR step with the driving polynomial

$$q_2^{(1)} = (B - \mu I)(B - \bar{\mu} I)B^{-1}$$

followed by a double shift SR step with the driving polynomial

$$q_2^{(2)} = (B - \mu^{-1} I)(B - \overline{\mu^{-1}} I)B^{-1}$$

as this is equivalent to applying a quadruple SR step. The vectors $q_2^{(1)} e_1$ and $q_2^{(2)} e_1$ are of the form

$$\xi_1 e_1 + \xi_2 e_2 + \xi_3 e_{n+1}.$$

But the parameterized double shift SR step relies on the fact that for the driving polynomial q_2 we have

$$q_2(B)e_1 = \varsigma_1 e_1 + \varsigma_2 e_2.$$

4.2.3   THE OVERALL PROCESS

By applying a sequence of parameterized double shift SR steps to a symplectic butterfly matrix B, it is possible to reduce the tridiagonal blocks in B to diagonal form if B has only real eigenvalues or eigenvalues on the unit circle. The eigenproblem decouples into simple symplectic 2×2 eigenproblems. Decoupling occurs if $d_j = 0$ for some j. Therefore it is necessary to monitor the parameters d_j in order to bring about decoupling whenever possible. We proceed with the process of applying double shift SR steps until the problem has completely split into subproblems of dimension 2. That is, until all parameters d_j are equal to zero. The complete process is given in Table 4.2. In a final step we then have to solve the small subproblems. This has already been discussed in Section 4.1.3.

REMARK 4.6 *The parameterized double shift SR algorithm for butterfly matrices as given in Table 4.2 requires about $146n^2$ flops. If the transformation matrix S is accumulated, an additional $28n^3$ flops have to be added. This estimate is based on the observation that $\frac{2}{3}$ SR iterations per eigenvalue are necessary.*

Preliminary tests show that the parameterized double shift SR algorithm computes the eigenvalues of randomly generated symplectic matrices with about the same accuracy as the double shift SR algorithm using usually less iterations. For symplectic matrices with clustered eigenvalues, the parameterized double shift SR algorithm seems to perform much better than the usual double shift SR algorithm for butterfly matrices. See Section 4.4 for details.

4.3 THE BUTTERFLY SZ ALGORITHM

In the previous sections we have developed an SR algorithm for symplectic butterfly matrices and a parameterized version of it. The first algorithm works on the butterfly matrix B and transforms it into a butterfly matrix $\widetilde{B}$ which decouples into simple 2×2 or 4×4 symplectic eigenproblems. The latter algorithm works only on the $4n - 1$ parameters that determine a symplectic butterfly matrix B. It computes the parameters which determine the matrix $\widetilde{B}$ without ever forming B or $\widetilde{B}$. Here we will develop another algorithm that works only on the parameters. We have seen that B can be factored into $B = K^{-1}N$ as in (3.2.15) or (3.2.7). The eigenvalue problem $K^{-1}Nx = \lambda x$ is equivalent to $(\lambda K - N)x = 0$ and $(K - \lambda N)x = 0$ because of the symmetry of the spectrum. In the latter equations the $4n - 1$ parameters are given directly. The idea here is that instead of considering the eigenproblem for B, we can just as well consider the generalized eigenproblem $(K - \lambda N)x = 0$. An SZ algorithm will be developed to solve this generalized eigenproblem. The SZ algorithm is the analogue of the SR algorithm for the generalized eigenproblem, just as the QZ algorithm is the analogue of the QR algorithm for the generalized eigenproblem, see Section 2.2.2. Both are instances of the GZ algorithm [143].

Algorithm: Parameterized Double Shift SR Algorithm
for Butterfly Matrices

Given the parameters $a_1, \ldots, a_n, b_1, \ldots, b_n, c_1, \ldots, c_n, d_2, \ldots, d_n$ of a symplectic butterfly matrix B, the following algorithm computes the parameters $\widetilde{a}_1, \ldots, \widetilde{a}_n, \widetilde{b}_1, \ldots, \widetilde{b}_n, \widetilde{c}_1, \ldots, \widetilde{c}_n, \widetilde{d}_2, \ldots, \widetilde{d}_n$ of a symplectic butterfly matrix $\widetilde{B}$ that is similar to B. All $\widetilde{d}_j$ are zero. Thus the eigenproblem for $\widetilde{B}$ decouples into 2×2 symplectic eigenproblems. B is assumed to have only real eigenvalues or eigenvalues on the unit circle.

> let $d_1 = 0$
> let $q = n + 1$
> let $p = 1$
> **repeat until** $q = p$
>> set all d_j to zero that satisfy $d_j \leq \epsilon$
>> find the largest nonnegative q and the smallest nonnegative p
>> such that
>>
>> $$d_1 = \cdots = d_p = 0 \neq d_{p+1}$$
>> $$d_{q-1} \neq d_q = \cdots = d_n = 0$$
>>
>> **if** $q \neq p$
>>> perform a parameterized double shift SR step on
>>>
>>> $$a_{p+1}, \ldots, a_{q-1}, b_{p+1}, \ldots, b_{q-1},$$
>>> $$c_{p+1}, \ldots, c_{q-1}, d_{p+1}, \ldots, d_{q-1}$$
>>>
>> **end**
> **end**
> solve the 2×2 subproblems as described in the text

Table 4.2.   Parameterized Double Shift SR Algorithm for Butterfly Matrices

---

4.3.1   THE BUTTERFLY SZ STEP

Each iteration step of the butterfly SZ algorithm begins with K and N such that the corresponding butterfly matrix $B = K^{-1}N$ is unreduced. Choose a spectral transformation function q and compute a symplectic matrix Z_1 such that

$$Z_1^{-1} q(K^{-1}N)e_1 = \alpha e_1$$

for some scalar α. Then transform the pencil to

$$\widehat{K} - \lambda\widehat{N} = (K - \lambda N)Z_1.$$

This introduces a bulge into the matrices $\widehat{K}$ and $\widehat{N}$. Now transform the pencil to

$$\widetilde{K} - \lambda\widetilde{N} = S^{-1}(\widehat{K} - \lambda\widehat{N})\widehat{Z},$$

where $\widetilde{K}$ and $\widetilde{N}$ are in form (3.4.18) or (3.4.19), depending on the form of K and N, S and $\widehat{Z}$ are symplectic, and $\widehat{Z}e_1 = e_1$. This concludes the iteration step.

Letting $Z = Z_1\widehat{Z}$, we have

$$\widetilde{K} - \lambda\widetilde{N} = S^{-1}(K - \lambda N)Z.$$

The symplectic matrices $\widetilde{K}^{-1}\widetilde{N}$ and $\widetilde{N}\widetilde{K}^{-1}$ are similar to $K^{-1}N$ and NK^{-1}, respectively. Indeed

$$\widetilde{K}^{-1}\widetilde{N} = Z^{-1}(K^{-1}N)Z \quad \text{and} \quad \widetilde{N}\widetilde{K}^{-1} = S^{-1}NK^{-1}S.$$

The following theorem shows that these similarity transformations amount to iterations of the SR algorithm on $K^{-1}N$ and NK^{-1}.

For the proof of the following theorem recall the definition of a generalized Krylov matrix (Definition 2.3)

$$L(A, v, j) = [v, A^{-1}v, A^{-2}v, \ldots, A^{-(j-1)}v, Av, A^2v, \ldots, A^jv]$$

for $A \in \mathbf{R}^{2n \times 2n}, v \in \mathbf{R}^{2n}$.

THEOREM 4.7 *There exist J–triangular matrices R and U such that*

$$q(NK^{-1}) = SR \quad \text{and} \quad q(K^{-1}N) = ZU.$$

PROOF: The transforming matrix Z was constructed so that

$$Ze_1 = Z_1\widehat{Z}e_1 = Z_1e_1 = \alpha^{-1}q(B)e_1,$$

where $B = K^{-1}N$. Now

$$\begin{aligned} q(B)L(B, e_1, n) &= L(B, q(B)e_1, n) \\ &= \alpha L(B, Ze_1, n) \\ &= \alpha Z L(\widetilde{B}, e_1, n), \end{aligned}$$

where $\widetilde{B} = Z^{-1}BZ = \widetilde{K}^{-1}\widetilde{N}$. By Theorem 3.7, $L(B, e_1, n)$ and $L(\widetilde{B}, e_1, n)$ are J–triangular, and $L(B, e_1, n)$ is nonsingular. Hence $q(K^{-1}N) = ZU$, where

$$U = \alpha L(\widetilde{B}, e_1, n)L(B, e_1, n)^{-1}$$

is J–triangular.

The proof that $q(NK^{-1})$ equals SR depends on which of the decompositions (3.4.18) or (3.4.19) is being used. If (3.4.19) is being used, K and $\widetilde{K}$ are J–triangular matrices, so $\widetilde{K}^{-1}e_1 = \beta e_1$ and $Ke_1 = \gamma e_1$ for some $\beta, \gamma \in \mathbf{R}$. Since $\widetilde{K} = S^{-1}KZ$, we have

$$
\begin{aligned}
Se_1 &= KZ\widetilde{K}^{-1}e_1 \\
&= \beta KZe_1 \\
&= \beta\alpha^{-1}Kq(K^{-1}N)e_1 \\
&= \beta\alpha^{-1}q(NK^{-1})Ke_1 \\
&= \beta\alpha^{-1}\gamma q(NK^{-1})e_1.
\end{aligned}
$$

Thus $q(NK^{-1})e_1 = \delta Se_1$ for some nonzero $\delta \in \mathbf{R}$. Since the matrices $C = NK^{-1}$ and $\widetilde{C} = \widetilde{N}\widetilde{K}^{-1}$ are butterfly matrices, and C is unreduced, we can now repeat the argument of the previous paragraph with B replaced by C to get

$$
q(NK^{-1}) = SR,
$$

where

$$
R = \delta L(\widetilde{C}, e_1, n)L(C, e_1, n)^{-1}
$$

is J–triangular.

If the decomposition (3.4.18) is being used, K is not J–upper triangular. However, since $\widetilde{N}^{-1}e_1 = e_1$ and $Ne_1 = e_1$, we can use the equation $\widetilde{N} = S^{-1}NZ$ in the form $S = NZ\widetilde{N}^{-1}$ to prove that $q(NK^{-1})e_1 = \delta Se_1$ for some δ, as above. In this case C and $\widetilde{C}$ are not butterfly matrices, but their inverses are. Thus one can show, as above, that

$$
q(NK^{-1}) = SR,
$$

where $R = \delta L(\widetilde{C}^{-1}, e_1, n)L(C^{-1}, e_1, n)^{-1}$. $\checkmark$

We now consider the details of implementing an SZ iteration for the symplectic pencil $K - \lambda N$ as in (3.4.19). The spectral transformation function q_4 will be chosen as discussed in Section 4.1:

$$
q_4(\lambda) = (\lambda - \mu)(\lambda - \mu^{-1})(\lambda - \overline{\mu})(\lambda - \overline{\mu}^{-1})\lambda^{-2}
$$

for $\mu \in \mathbf{C}$. Here the shift μ is chosen corresponding to the generalized Rayleigh-quotient strategy. This implies (see (4.1.17))

$$
\begin{aligned}
q_4(B)e_1 &= [(b_1 + a_1c_1)^2 + a_1a_2d_2^2 - \beta(b_1 + a_1c_1) + \gamma - 2]e_1 \\
&\quad + [a_1d_2(b_2 + a_2c_2 + b_1 + a_1c_1 - \beta)]e_2 + a_1a_2d_2d_3e_3,
\end{aligned}
$$

where

$$\begin{aligned}
\beta &= b_{n-1} + b_n + a_{n-1}c_{n-1} + a_n c_n, \\
\gamma &= (b_{n-1} + a_{n-1}c_{n-1})(b_n + a_n c_n) + 2 - a_{n-1}a_n d_n^2.
\end{aligned}$$

Hence Z_1 has to be chosen such that $Z_1^{-1}q_4(B)e_1 = \alpha e_1$. Applying Z_1^{-1} to $K - \lambda N$ introduces a bulge. The main part of the iteration is a bulge chasing process that restores KZ_1^{-1} and NZ_1^{-1} to their original forms. This is done using the algorithm given in Table 3.4 for reducing a symplectic matrix pencil $K - \lambda N$, where K and N are symplectic, to a butterfly matrix pencil (3.4.19) as discussed in Section 3.4. The algorithm uses the symplectic Givens transformations G_k, the symplectic Householder transformations H_k, and the symplectic Gauss transformations L_k introduced in Section 2.1.2. In this elimination process zeros in the rows of K and N will be introduced by applying one of the above mentioned transformations from the right, while zeros in the columns will be introduced by applying the transformations from the left.

For ease of reference, the basic idea of the algorithm as given in Section 3.4 is summarized here again.

bring the first column of N into the desired form
for $j = 1$ to n
    bring the jth row of K into the desired form
    bring the jth column of K into the desired form
    bring the $(n + j)$th column of N into the desired form
    bring the jth row of N into the desired form

The remaining rows and columns in K and N that are not explicitly touched during the process will be in the desired form due to the symplectic structure.

In the following we will consider such a bulge chasing step for a bulge that is created using a quadruple shift Laurent polynomial. Choosing the spectral transformation function q_4 results in a symplectic transformation matrix of the form

$$Z_1 = \left[\begin{array}{ccc|ccc}
x & x & x & & & \\
x & x & x & & & \\
x & x & x & & & \\
& & I^{n-3,n-3} & & & \\
\hline
& & & x & x & x \\
& & & x & x & x \\
& & & x & x & x \\
& & & & & I^{n-3,n-3}
\end{array}\right].$$

In the following we assume that $K - \lambda N$ is in the form (3.4.19):

$$K_u - \lambda N_u = \begin{bmatrix} \diagdown & \diagdown \\ 0 & \diagdown \end{bmatrix} - \lambda \begin{bmatrix} 0 & \diagdown \\ \diagdown & \diagdown\diagdown \end{bmatrix}.$$

Applying Z_1 to $K_u - \lambda N_u$ results in the matrix pencil $K_1 - \lambda N_1 = (K_u - \lambda N_u)Z_1^{-1}$, where K_1 and N_1 are of the following forms:

$$
K_1 = \left[\begin{array}{ccccc|ccccc}
x & + & + & & & x & + & + & & \\
+ & x & + & & & + & x & + & & \\
+ & + & x & & & + & + & x & & \\
  &   &   & x & & & & & x & \\
  &   &   & & \ddots & & & & & \ddots \\
\hline
  &   &   &   &   & x & + & + & & \\
  &   &   &   &   & + & x & + & & \\
  &   &   &   &   & + & + & x & & \\
  &   &   &   &   &   &   &   & x & \\
  &   &   &   &   &   &   &   &   & \ddots
\end{array}\right]
$$

and

$$
N_1 = \left[\begin{array}{ccccc|cccccc}
  &   &   &   &   & x & + & + & & & \\
  &   &   &   &   & + & x & + & & & \\
  &   &   &   &   & + & + & x & & & \\
  &   &   &   &   &   &   &   & x & & \\
  &   &   &   &   &   &   &   &   & x & \\
  &   &   &   &   &   &   &   &   &   & \ddots \\
\hline
x & + & + & & & x & x & + & & & \\
+ & x & + & & & x & x & x & & & \\
+ & + & x & & & + & x & x & x & & \\
  &   &   & x & & + & + & x & x & x & \\
  &   &   & & x & & & & x & x & x \\
  &   &   & & & \ddots & & & & & \ddots
\end{array}\right],
$$

where $+$ denotes additional, unwanted entries in the matrices.

Now we can apply the algorithm for reducing a symplectic matrix pencil to a butterfly pencil as given in Table 3.4 to the pencil $K_1 - \lambda N_1$. The pencil will be reduced to the original forms of K_u and N_u, resulting in the pencil $\widetilde{K}_u - \lambda \widetilde{N}_u$. Due to the special structure of K_1 and N_1 the algorithm greatly simplifies. First the unwanted entries in the first column of N_1 will be annihilated. The elements $(n+2, 1)$ and $(n+3, 1)$ of N_1 are zeroed by a symplectic Householder transformation H_1: $N_2 = H_1 N_1 = S_1 N_1$ where

$$
N_2 = \left[\begin{array}{ccccc|cccccc}
  &   &   &   &   & x & + & + & & & \\
  &   &   &   &   & + & x & + & & & \\
  &   &   &   &   & + & + & x & & & \\
  &   &   &   &   &   &   &   & x & & \\
  &   &   &   &   &   &   &   &   & x & \\
  &   &   &   &   &   &   &   &   &   & \ddots \\
\hline
x & + & + & & & x & x & + & + & & \\
0 & x & + & & & x & x & x & + & & \\
0 & + & x & & & + & x & x & x & & \\
  &   &   & x & & + & + & x & x & x & \\
  &   &   & & x & & & & x & x & x \\
  &   &   & & & \ddots & & & & & \ddots
\end{array}\right].
$$

As N_2 is symplectic, there are additional zeros in N_2. Obviously, the $(n+1, 1)$ entry of N_2 is nonzero. From

$$e_1^T (N_2^T J N_2) = e_1^T J = e_{n+1}^T \quad \Longrightarrow \quad [0 \ \ldots \ 0 | \beta_1 \ \beta_2 \ \beta_3 \ 0 \ \ldots \ 0] = e_{n+1}^T,$$

where

$$\beta_j = -(N_2)_{n+1,1}(N_2)_{1,n+j}, \quad j = 1, 2, 3,$$

we conclude that the entries of the first row of N_2 have to be zero, only the $(1, n+1)$ entry of N_2 is nonzero, it is the negative of the inverse of the $(n+1, 1)$ entry of N_2. Moreover, the $(1, n + 1)$ and the $(n + 1, 1)$ element of N_2 are not altered in any subsequent transformation step. Recall that the resulting matrix pencil has to be of the form (3.4.19). Hence, we should have

$$(N_2)_{1,n+1} = -(N_2)_{n+1,1}^{-1} = 1;$$

in case our algorithm has not normalized $(N_2)_{1,n+1}$ to be equal to 1, we have to apply a trivial matrix

$$\begin{bmatrix} C & 0 \\ 0 & C \end{bmatrix}$$

with

$$C = \begin{bmatrix} (N_2)_{1,n+1}^{-1} & \\ & I^{n-1,n-1} \end{bmatrix}$$

from the left to scale $(N_2)_{1,n+1}$ to be 1. For the rest of the discussion it will be assumed that this will be done whenever necessary without explicitly mentioning it again. Therefore,

$$N_2 = \begin{bmatrix}
 & & & & & 1 & 0 & 0 & & & \\
 & & & & & + & x & + & & & \\
 & & & & & + & + & x & & & \\
 & & & & & & & & x & & \\
 & & & & & & & & & x & \\
 & & & & & & & & & & \ddots \\
\hline
-1 & + & + & & & x & x & + & + & & \\
0 & x & + & & & x & x & x & + & & \\
0 & + & x & & & + & x & x & x & & \\
 & & & x & & + & + & x & x & x & \\
 & & & & x & & & & x & x & x \\
 & & & & & \ddots & & & & & \ddots
\end{bmatrix}.$$

To preserve eigenvalues, we have to update K_1 in exactly the same way as N_1: $K_2 = S_1 K_1$. This transformation neither causes additional nonzero entries in K_2 nor does it introduce any zeros in K_2. Next we reduce the first row of K_2 to the desired form while preserving the zeros just created in N_2. First, Givens transformations G_2 and G_3 are determined to zero the $(1, 2)$ and $(1, 3)$ elements in K_2. Next a Householder transformation H_2 is used to zero the $(1, n + 3)$

element. The reduction is completed by a Gauss transformation L_2 to eliminate the $(1, n + 2)$ entry in K_2. This transformation might not exist, if the $(1, 1)$ element of K_2 is zero. Then no reduction to the desired form is possible. If the $(1, 1)$ element of K_2 is almost zero, the resulting Gauss transformation will be ill-conditioned. In both cases we discard the shift polynomial and perform one step using a random shift, as proposed by Bunse-Gerstner and Mehrmann in the context of the Hamiltonian SR algorithm in [38]. For the rest of the discussion here, we will assume that no such breakdown will occur. We have

$$K_3 \; = \; K_2 G_2 G_3 H_2 L_2 = K_2 Z_2$$

$$= \begin{bmatrix}
x & 0 & 0 & & & x & 0 & 0 & & \\
+ & x & + & & & + & x & + & & \\
+ & + & x & & & + & + & x & & \\
& & & x & & & & & x & \\
& & & & \ddots & & & & & \ddots \\
\hline
& + & + & & & x & + & + & & \\
& + & + & & & + & x & + & & \\
& + & + & & & + & + & x & & \\
& & & & & & & & x & \\
& & & & & & & & & \ddots
\end{bmatrix} .$$

As K_3 is symplectic, there are additional zeros in K_3. The $(1, 1)$ entry of K_3 will only be multiplied by some constant $c \neq 0$ during the rest of the algorithm.

$$(K_3^T J K_3) e_1 = J e_1 = e_{n+1}$$

implies that

$$[0 \; x \; x \; x \; 0 \; \ldots \; 0 | \beta_1 \; \beta_2 \; \beta_3 \; 0 \; \ldots \; 0] = e_{n+1},$$

where $\beta_j = (K_3)_{1,1}(K_3)_{n+j,n+1}$ for $j = 1, 2, 3$. Hence, as $(K_3)_{1,1}$ is nonzero, the $(n + 2, n + 1)$ and $(n + 3, n + 1)$ entries of K_3 have to be zero. The $(n + 1, n + 1)$ entry is the inverse of the $(1, 1)$ entry. Therefore,

$$K_3 = \begin{bmatrix}
\alpha & 0 & 0 & & & x & 0 & 0 & & \\
+ & x & + & & & + & x & + & & \\
+ & + & x & & & + & + & x & & \\
& & & x & & & & & x & \\
& & & & \ddots & & & & & \ddots \\
\hline
& + & + & & & \alpha^{-1} & + & + & & \\
& + & + & & & 0 & x & + & & \\
& + & + & & & 0 & + & x & & \\
& & & & & & & & x & \\
& & & & & & & & & \ddots
\end{bmatrix} .$$

Updating N_2 in the same way the zeros in the first row and column of N_2 are preserved, but additional entries are created. We obtain

$$N_3 \;=\; N_2 Z_2$$

$$
= \left[
\begin{array}{cccc|ccccc}
 & & & & 1 & & & & \\
 & + & + & & + & x & + & & \\
 & + & + & & + & + & x & & \\
 & & & & & & & x & \\
 & & & & & & & & x \\
 & & & & & & & & & \ddots \\
\hline
-1 & + & + & & x & x & + & + & \\
 & x & + & & x & x & x & + & \\
 & + & x & & + & x & x & x & \\
 & + & + & x & + & + & x & x & x \\
 & & & x & & & x & x & x \\
 & & & & & & & & & \ddots
\end{array}
\right].
$$

Now we turn our attention to the first column of K_3. The undesired elements can be annihilated by a suitable transformation S_2: $S_2 K_3 e_1 = \alpha e_1$. For this, we first use Givens transformations G_3 and G_2 to zero the elements $(3,1)$ and $(2,1)$ (due to the special structure, this can be done by permuting the 2nd and 3rd row with the $(n+2)$nd and $(n+3)$rd row). This is followed by a Householder transformation H_2 to eliminate the element $(n+3,1)$. The element $(n+2,1)$ is then zeroed using a transposed Gauss transformation L_2^T. This yields

$$K_4 \;=\; L_2^T H_2 G_2 G_3 K_3 = S_2 K_3$$

$$
= \left[
\begin{array}{cccc|ccccc}
\tilde{a}_1 & & & & -\tilde{b}_1 & & & & \\
0 & x & + & & 0 & x & + & & \\
0 & + & x & & 0 & + & x & & \\
 & & & x & & & & x & \\
 & & & & x & & & & x \\
 & & & & & & & & & \ddots \\
\hline
 & 0 & 0 & & \tilde{a}_1^{-1} & 0 & 0 & & \\
0 & + & + & & 0 & x & + & & \\
0 & + & + & & 0 & + & x & & \\
 & & & & & & & x & \\
 & & & & & & & & x \\
 & & & & & & & & & \ddots
\end{array}
\right].
$$

The additional zeros follow again because K_4 is symplectic: The additional zeros in the (n+1)st row follow from

$$e_1^T K_4 J K_4^T = e_{n+1}^T$$

while the zeros in the $(n+1)$st column follow from

$$K_4^T J K_4 e_{n+1} = J e_{n+1} = -e_1.$$

The $(1,1)$, $(n+1, n+1)$ and the $(1, n+1)$ entry of K_4 are not altered in the subsequent transformations. Hence at this point the first parameters of the resulting matrix pencil $\widetilde{K} - \lambda \widetilde{N}$ can be read off.

Updating N_3 in the same way the zeros in the first row of N_4 are preserved; additional nonzero entries are created:

$$
N_4 \;=\; S_2 N_3
$$

$$
= \left[
\begin{array}{cccccc|cccccc}
 & + & + & & & & 1 & + & x & + & + & \\
 & + & + & & & & & + & + & x & + & \\
 & & & & & & & & & & x & \\
 & & & & & & & & & & & x \\
 & & & & & & & & & & & & \ddots \\
\hline
 & -1 & + & + & & & & x & x & + & + & \\
 & & x & + & & & & x & x & x & + & \\
 & & + & x & & & & + & x & x & x & \\
 & & + & + & x & & & + & + & x & x & x \\
 & & & & & x & & & & x & x & x \\
 & & & & & & \ddots & & & & & & \ddots
\end{array}
\right].
$$

Hence the first and $(n+1)$st row and column of K_4 are in the desired form, while only the first row and column of N_4 are in the desired form. Next we attack the $(n+1)$st column of N_4. Givens transformations G_2 and G_3 can be used to eliminate the $(3, n+1)$ and $(2, n+1)$ elements, while a Householder transformation H_2 kills off the $(n+4, n+1)$ and $(n+3, n+1)$ entries:

$$
N_5 \;=\; H_2 G_3 G_2 N_4 = S_3 N_4
$$

$$
= \left[
\begin{array}{cccccc|cccccc}
 & + & + & & & & 1 & 0 & x & + & + & \\
 & + & + & & & & & 0 & + & x & + & \\
 & + & + & & & & & + & + & x & \\
 & & & & & & & & & & x & \\
 & & & & & & & & & & & \ddots \\
\hline
 & -1 & + & + & & & & x & x & + & + & \\
 & & x & + & + & & & x & x & x & + & + \\
 & & + & x & + & & & 0 & x & x & x & + \\
 & & + & + & x & & & 0 & + & x & x & x \\
 & & & & & x & & & & x & x & x \\
 & & & & & & \ddots & & & & & & \ddots
\end{array}
\right].
$$

No additional zeros are introduced in N_5. These transformations preserve the zeros created so far in K_4, but additional nonzero elements are introduced;

$$K_5 = S_3 K_4$$

$$
= \begin{bmatrix}
\tilde{a}_1 & & & & & -\tilde{b}_1 & & & & & \\
& x & + & + & & & x & + & + & & \\
& + & x & + & & & + & x & + & & \\
& + & + & x & & & + & + & x & & \\
& & & & x & & & & & x & \\
& & & & & \ddots & & & & & \ddots \\
\hline
& & & & & \tilde{a}_1^{-1} & & & & & \\
& + & + & & & & x & + & + & & \\
& + & + & & & & + & x & + & & \\
& + & + & & & & + & + & x & & \\
& & & & & & & & & x & \\
& & & & & & & & & & \ddots
\end{bmatrix}.
$$

Next we annihilate the undesired entries in the second row of N_5. The elements $(2,2)$ and $(2,3)$ are eliminated by Givens transformations G_2 and G_3, followed by a Householder transformation H_1 to kill the elements $(2, n+3)$ and $(2, n+4)$:

$$N_6 = N_5 G_2 G_3 H_1 = N_5 Z_3$$

$$
= \begin{bmatrix}
& & & & & 1 & & & & & \\
& 0 & 0 & & & & 1 & 0 & 0 & & \\
& 0 & + & + & & & + & x & + & & \\
& 0 & + & + & & & + & + & x & & \\
& & & & x & & & & & x & \\
& & & & & & & & & & \ddots \\
\hline
-1 & 0 & 0 & & & \tilde{c}_1 & \tilde{d}_2 & 0 & 0 & & \\
& -1 & + & + & & \tilde{d}_2 & x & x & + & + & \\
& 0 & x & + & & & x & x & x & + & \\
& 0 & + & x & & & + & x & x & x & \\
& & & & x & & & + & x & x & x \\
& & & & & \ddots & & & & & \ddots
\end{bmatrix}.
$$

The additional zeros in the second column follow as before from

$$e_2^T N_5 J N_5^T = e_2^T J = e_{n+2}^T.$$

The additional zeros in the $(n+1)$st row follow from

$$N_5^T J N_5 e_{n+1} = e_1.$$

The entries of N_6 in the positions $(n+1, n+1)$, $(n+1, n+2)$, and $(n+2, n+1)$ are not altered by any subsequent transformation.

Again, we have to update K_5 accordingly: $K_6 = K_5 Z_3$. All previously created zeros in K_5 are preserved. New nonzero elements are introduced:

$$K_6 = \begin{bmatrix} \tilde{a}_1 & & & & & -\tilde{b}_1 & & & & \\ & x & + & + & & & x & + & + & \\ & + & x & + & & & + & x & + & \\ & + & + & x & & & + & + & x & \\ & & & & x & & & & & x \\ & & & & & \ddots & & & & & \ddots \\ \hline & & & & & \tilde{a}_1^{-1} & & & & \\ & + & + & + & & & x & + & + & \\ & + & + & + & & & + & x & + & \\ & + & + & + & & & + & + & x & \\ & & & & & & & & & x \\ & & & & & & & & & & \ddots \end{bmatrix} .$$

Now we proceed by reducing the second row of K_6 to the desired form just like reducing the first row of K_2 to the desired form. First Givens transformations G_3 and G_4 are determined to zero the $(2,3)$ and $(2,4)$ element in K_6. Next a Householder transformation H_3 is used to zero the $(2, n+4)$ element. The reduction is completed by a Gauss transformation L_3 to eliminate the $(1, n+3)$ entry in K_6.

$$K_7 = K_6 G_3 G_4 H_3 L_3 = K_6 Z_4$$

$$= \begin{bmatrix} \tilde{a}_1 & & & & & -\tilde{b}_1 & & & & \\ & \gamma & 0 & 0 & & & x & 0 & 0 & \\ & + & x & + & & & + & x & + & \\ & + & + & x & & & + & + & x & \\ & & & & x & & & & & x \\ & & & & & \ddots & & & & & \ddots \\ \hline & & & & & \tilde{a}_1^{-1} & & & & \\ & 0 & + & + & & & \gamma^{-1} & + & + & \\ & + & + & + & & & 0 & x & + & \\ & + & + & + & & & 0 & + & x & \\ & & & & & & & & & x \\ & & & & & & & & & & \ddots \end{bmatrix} .$$

The additional zeros in the positions $(n+3, n+2)$ and $(n+4, n+2)$ follow by an argument just like the one for the additional zeros in K_3. The additional zero in the position $(n+2, 2)$ follows by a more difficult argument. The $(n+2, 2)$ element will not change in the subsequent step (besides being multiplied by some constant $c \neq 0$). Looking at the next iterate K_8, it follows from

$$e_2^T K_8^T J K_8 e_2 = 2(K_8)_{2,2}(K_8)_{n+2,2} = e_2^T J e_2 = 0,$$

that $(K_8)_{n+2,2} = 0$. As $(K_8)_{n+2,2} = c(K_7)_{n+2,2}$, with $c \neq 0$, we have $(K_7)_{n+2,2} = 0$.

Updating N_6 introduces new nonzero elements:

$$N_7 = N_6 Z_4 = \left[\begin{array}{cccc|cccccc}
 & & & & 1 & & & & & \\
 & & & & & 1 & & & & \\
 & + & + & & & + & x & + & & \\
 & + & + & & & + & + & x & & \\
 & & & & & & & & x & \\
 & & & & & & & & & \ddots \\
\hline
-1 & & & & & \tilde{c}_1 & \tilde{d}_2 & & & \\
 & -1 & + & + & & \tilde{d}_2 & x & x & + & + \\
 & & x & + & & & x & x & x & + \\
 & & + & x & & & + & x & x & x \\
 & & + & + & x & & + & + & x & x \\
 & & & & & \ddots & & & & \ddots
\end{array}\right].$$

Now we proceed by reducing the second column of K_7 to the desired form, just like reducing the first column of K_4. For this, we first use Givens transformations G_2 and G_3 to zero the elements $(3,2)$ and $(4,2)$. This is followed by a Householder transformation H_3 to eliminate the element $(n+4, 2)$. The element $(n+3, 2)$ is then zeroed using a transposed Gauss transformation L_3^T,

$$\begin{aligned}
K_8 &= L_3^T H_3 G_3 G_2 K_7 = S_4 K_7 \\[2mm]
&= \left[\begin{array}{cccc|cccc}
\tilde{a}_1 & & & & -\tilde{b}_1 & & & \\
 & \tilde{a}_2 & & & & -\tilde{b}_2 & & \\
 & 0 & x & + & & 0 & x & + \\
 & 0 & + & x & & 0 & + & x \\
 & & & x & & & & x \\
 & & & & \ddots & & & \ddots \\
\hline
 & & & & \tilde{a}_1^{-1} & & & \\
 & 0 & 0 & & & \tilde{a}_2^{-1} & 0 & 0 \\
 & 0 & + & + & & & x & + \\
 & 0 & + & + & & & + & x \\
 & & & & & & & x \\
 & & & & & \ddots & & \ddots
\end{array}\right].
\end{aligned}$$

The additional zeros follow as before. Updating N_7 results in

$$N_8 = S_4 N_7 = \left[\begin{array}{cccc|cccccc}
 & & & & 1 & & & & & \\
 & & & & & 1 & & & & \\
 & + & + & & & + & x & + & + & \\
 & + & + & & & + & + & x & + & \\
 & & & & & & & & x & \\
 & & & & & & & & & \ddots \\
\hline
-1 & & & & & \tilde{c}_1 & \tilde{d}_2 & & & \\
 & -1 & + & + & & \tilde{d}_2 & x & x & + & + \\
 & & x & + & & & x & x & x & + \\
 & & + & x & & & + & x & x & x \\
 & & + & + & x & & + & + & x & x & x \\
 & & & & & \ddots & & & & \ddots
\end{array}\right].$$

Now we have the same situation as after the construction of K_4, N_4, solely in each block the undesired elements are found one row and column further down. Therefore these undesired elements can be chased down the diagonal analogous to the last 4 steps.

REMARK 4.8 *a) A careful implementation of this process will just work with the $4n - 1$ parameters and some additional variables instead of with the matrices K and N.*

b) A careful flop count show that this implicit butterfly SZ step requires about $1530n - 2528$ flops. The accumulation of the the transformation matrix S or (Z) requires additional $128n^2$ flops. This flop count is based on the fact that $4n - 6$ symplectic Givens transformations, $n - 1$ symplectic Gauss transformation, $n - 3$ symplectic Householder transformations with $v \in \mathbf{R}^3$, and $n - 1$ symplectic Householder transformations with $v \in \mathbf{R}^2$ are used for the computation of S (Z). Moreover, the special structure of the problem is taken into account.

c) An implicit butterfly SZ step as described above is an order of a magnitude cheaper than an implicit QZ step, which requires about $88n^2$ flops, see [58]. Of course, the accumulation of the transformation matrix is in both cases an $\mathcal{O}(n^2)$ process.

d) It is possible to incorporate pivoting into the process in order to make it more stable. E.g., in the process as described the jth column of K will be brought into the desired form. Due to symplecticity, the $(n + j)$th column of K will then be of desired form as well. One could just as well attack the $(n + j)$th column of K, the jth column will then be of desired form due to symplecticity.

e) In intermediate steps there are even more zeros than shown in the discussion of the bulge chasing process. E.g., in N_2 the elements $(n+1, 2)$ and $(n+1, 3)$ are zero as well. This follows easily from the construction of N_2:

$$
\begin{aligned}
N_2 &= S_1 N_1 \\
&= S_1 N Z_1 \\
&= S_1 \begin{bmatrix} 0 & I^{n,n} \\ -I^{n,n} & T \end{bmatrix} \operatorname{diag}(P, I^{n-3,n-3}, P, I^{n-3,n-3}).
\end{aligned}
$$

S_1 *is computed such that* $S_1 N_1 e_1 = \alpha e_{n+1}$, *that is* S_1 *is a symplectic Householder transformation of the form* $\operatorname{diag}(\widehat{P}, I^{n-3,n-3}, \widehat{P}, I^{n-3,n-3})$, *where* $\widehat{P}$ *is chosen such that*

$$
\widehat{P} P e_1 = \alpha e_1.
$$

As $\widehat{P}$ and P are orthogonal matrices, we have $\widehat{P}Pe_1 = \pm e_1$ and $e_1^T\widehat{P}P = \pm e_1^T$. Hence, the entries $(n+1,2)$ and $(n+1,3)$ of N_2 are zero. We did not show these intermediate zeros, as the next step of the reduction process will cause fill-in for these entries. A careful implementation working only with the $4n-1$ parameters should make use of these additional zeros.

f) *The use of symplectic transformations throughout the reduction process assures that the factors K and N remain symplectic separately. If the objective is only to preserve the symplectic property of the pencil ($KJK^T = NJN^T$), one has greater latitude in the choice of transformations. Only the right-hand (Z) transformations need to be symplectic; the left (S) transforms can be more general as long as they are regular.*

Banse [13] developed an elimination process to reduce a symplectic matrix pencil with symplectic matrices to the reduced form $K_s - \lambda N_s$ (3.4.18). This algorithm can also be used as a building block for an SZ algorithm based on the reduced form $K_s - \lambda N_s$. It turns out that in that setting, for a double or quadruple shift step, there are slightly more nonzero entries in the matrices K and N than there are in the setting discussed here. This implies that more elementary symplectic transformations have to be used. In particular, additional $n-1$ symplectic Gauss transformations of type II have to be used, which are not needed in the above bulge chasing process.

4.3.2 THE OVERALL PROCESS

By applying a sequence of double or quadruple SZ steps to the symplectic matrix pencil $K - \lambda N$ of the form (3.4.19) it is possible to reduce the symmetric tridiagonal matrix T in the lower right block of N to quasi-diagonal form with 1×1 and 2×2 blocks on the diagonal. The eigenproblem decouples into a number of simple 2×2 or 4×4 eigenproblems. In doing so, it is necessary to monitor T's subdiagonal in order to bring about decoupling whenever possible. The complete process is given in Table 4.3.

REMARK 4.9 a) *The SZ algorithm for butterfly matrix pencils as given in Table 4.3 requires about $1020n^2$ flops. If the transformation matrix S (Z) is accumulated, an additional $85n^3$ flops have to be added. This estimate is based on the observation that $\frac{2}{3}$ SZ iterations per eigenvalue are necessary. Hence, the SZ algorithm for a $2n \times 2n$ symplectic matrix pencil requires about $75n^3 + 1004n^2$ flops for the reduction to butterfly form and the SZ iteration, plus $113n^3$ flops for the accumulation of each transformation matrix.*

This is considerably cheaper than applying the QZ algorithm to a $2n \times 2n$ matrix pencil. That would require about $240n^3$ flops for the reduction to Hessenberg-triangular form and the QZ iteration, plus $128n^3$ flops for the

---

Algorithm: SZ Algorithm for Butterfly Pencils

Given a symplectic matrix pencil $K - \lambda N$ of the form (3.4.19), the following algorithm computes symplectic matrices Z and S such that for $\widehat{K} := S^{-1}KZ$ and $\widehat{N} := S^{-1}NZ$, $\widehat{B} := \widehat{K}^{-1}\widehat{N}$ is a symplectic matrix in which the $(1,1)$, $(1,2)$, $(2,1)$, and $(2,2)$ blocks are each block-diagonal where all blocks are either 1×1 or 2×2. Moreover, the block structure for all four blocks of $\widehat{B}$ is the same. Thus the eigenproblem for $\widehat{B}$ decouples into 2×2 and 4×4 symplectic eigenproblems. K is overwritten by $S^{-1}KZ$ and N by $S^{-1}NZ$.

> **choose** $\epsilon = c\mathbf{u}$ suitably
> **repeat until** $q = n$
>> set all sub- and superdiagonal entries $t_{i,i-1} = t_{i-1,i}$ in T to zero
>> that satisfy
>>
>> $$|t_{i,i-1}| \leq \epsilon(|t_{i-1,i-1}| + |t_{ii}|)$$
>>
>> find the largest nonnegative q and the smallest nonnegative p such
>> that if
>>
>> $$N = \begin{bmatrix} & & & -I^{p,p} & & \\ & & & & -I^{n-p-q,n-p-q} & \\ & & & & & -I^{q,q} \\ \hline I^{p,p} & & & T_{11} & & \\ & I^{n-p-q,n-p-q} & & & T_{22} & \\ & & I^{q,q} & & & T_{33} \end{bmatrix},$$
>>
>> where $T_{11} \in \mathbf{R}^{p \times p}, T_{22} \in \mathbf{R}^{(n-p-q) \times (n-p-q)}$, and $T_{33} \in \mathbf{R}^{q \times q}$,
>> then T_{33} is block diagonal with 1×1 and 2×2 blocks and
>> T_{22} is unreduced symmetric tridiagonal.
>> partition K conformably
>>
>> $$K = \begin{bmatrix} X_{11} & & & Y_{11} & & \\ & X_{22} & & & Y_{22} & \\ & & X_{33} & & & Y_{33} \\ \hline & & & X_{11}^{-1} & & \\ & & & & X_{22}^{-1} & \\ & & & & & X_{33}^{-1} \end{bmatrix}$$
>>
>> where $X_{11}, Y_{11} \in \mathbf{R}^{p \times p}, X_{22}, Y_{22} \in \mathbf{R}^{(n-p-q) \times (n-p-q)}$, and
>> $X_{33}, Y_{33} \in \mathbf{R}^{q \times q}$.
>> **if** $q < n$
>>> perform a double or quadruple shift SZ step using the algorithm.
>>> given in Table 3.4 on
>>>
>>> $$\begin{bmatrix} X_{22} & Y_{22} \\ \hline & X_{22}^{-1} \end{bmatrix} - \lambda \begin{bmatrix} & -I^{n-p-q,n-p-q} \\ \hline I^{n-p-q,n-p-q} & T_{22} \end{bmatrix}$$
>>>
>>> update K and N accordingly
>> **end**
> **end**

Table 4.3.   SZ Algorithm for Butterfly Pencils

accumulation of the left transformation matrix and $160n^3$ *flops for the accumulation of the right transformation matrix. Recall, that an iteration step of the implicit butterfly* SZ *algorithm is an* $\mathcal{O}(n)$ *process, while the implicit* QZ *step is an* $\mathcal{O}(n^2)$ *process.*

b) *Example 4.11 in Section 4.4 indicates that eigenvalues of symplectic butterfly pencils computed by this algorithm are significantly more accurate than those computed by the* SR *algorithm and often competitive to those computed by the* QR *algorithm. Hence if a symplectic matrix/matrix pencil is given in parameterized form as in the context of the symplectic Lanczos algorithm (see Section 5) one should not form the corresponding butterfly matrix, but compute the eigenvalues via the* SZ *algorithm.*

4.4 NUMERICAL EXAMPLES

The SR and SZ algorithms for computing the eigenvalues of symplectic matrices/matrix pencils as discussed in Sections 4.1, 4.2, and 4.3 were implemented in MATLAB Version 5.1. Numerical experiments were performed on a SPARC Ultra 1 creator workstation.

In order to detect deflation, subdiagonal elements were declared to be zero during the iterations when a condition of the form

$$|h_{p+1,p}| \leq 10 \cdot n \cdot eps(|h_{pp}| + |h_{p+1,p+1}|) \qquad (4.4.39)$$

was fulfilled, where the dimension of the problem is $2n \times 2n$ and $eps \approx 2.2204 * 10^{-16}$ is MATLAB's floating point relative accuracy.

The experiments presented here will illustrate the typical behavior of the proposed algorithms. For a general symplectic matrix or a symplectic matrix pencil with both matrices symplectic, our implementation first reduces the matrix/matrix pencil to butterfly form (3.2.10)/a pencil of the form (3.4.19) and then iterates using in general only quadruple shift steps. The parameterized SR algorithm was implemented for double shift steps only. The shifts are chosen according to the generalized Rayleigh-quotient strategy discussed in Section 4.1. Tests were run using

- randomly generated symplectic matrices/matrix pencils;

- randomly generated parameters $a_1, \ldots, a_n, b_1, \ldots, b_n, c_1, \ldots, c_n, d_2, \ldots, d_n \in \mathbf{R}$ from which a butterfly matrix and the corresponding symplectic matrix pencil were constructed;

- examples from the benchmark collection [22];

- the examples discussed in [53].

Our observations have been the following.

- The methods did always converge; not once did we encounter an example where an exceptional SR/SZ step with a random shift was necessary (although, no doubt, such an example can be constructed).

- Cubic convergence can be observed.

- The parameterized SR algorithm converges slightly faster than the double shift SR algorithm. The eigenvalues are computed to about the same accuracy.

- The SZ algorithm is considerably better than the SR algorithm in computing the eigenvalues of a parameterized symplectic matrix/matrix pencil.

- The number of (quadruple-shift) iterations needed for convergence for each eigenvalue is about $2/3$.

EXAMPLE 4.10 *For the first set of tests, 100 symplectic matrices for each of the dimensions $2n \times 2n$ for $n = 5 : 5 : 50$ were generated by computing the SR decomposition of random $2n \times 2n$ matrices:*

$$\texttt{A = rand(2*n);   [M,R] = sr(A);}$$

where M is symplectic and R is J–triangular such that $A = MR$. Some of the results we obtained are summarized in Tables 4.4 and 4.5. In each table, the first column indicates the size of the problem.

As the generated matrices M are only symplectic modulo roundoff errors, symplecticity was tested via $\|M^T JM - J\|$ for all examples. The second column of Table 4.4 reports the maximal norm observed for each dimension. It is obvious that for increasing dimension, symplecticity is more and more lost. Hence, we may expect our algorithm to have some difficulties performing well, as its theoretical foundation is the symplecticity of the matrix/matrix pencil treated. The SR algorithm computes a symplectic matrix S and a symplectic matrix B such that in exact arithmetic, $S^{-1}MS = B$ is of butterfly-like form and B decouples into a number of 2×2 and 4×4 subproblems. In order to see how well the computed S and B obey this relation, $\|S^{-1}MS - B\|_2$ was computed for each example, and the maximal and minimal value of these norms for each dimension is reported in the third column of Table 4.4. In the course of the iterations, symplectic Gauss transformations have to be used. All other involved transformations are orthogonal. These are known to be numerically stable. Hence, the Gauss transformations are the only source for instability. The column 'condmax' of the table displays the maximal condition number of all Gauss transformations applied during all 100 examples of each dimension. The condition number of the Gauss transformations were never too large (i.e., exceeding the tolerance threshold, chosen here as $1/\sqrt{eps}$), hence no exceptional SR step with a random shift was required. The last column *of Table 4.4 gives the average number of iterations needed for convergence of each eigenvalue. This number tends to be around* $2/3$ *iterations per eigenvalue.*

Table 4.5 reports on the accuracy of the computed eigenvalues. For this purpose, the MATLAB *function* `eig` *was called in order to solve the* 2×2 *and* 4×4 *subproblems of B to generate a list of eigenvalues computed via the* SR *algorithm*

$$|\lambda_1^{SR}| \geq |\lambda_2^{SR}| \geq \cdots \geq |\lambda_{2n}^{SR}|.$$

If $|\lambda_j^{SR}| = |\lambda_{j+1}^{SR}|$ *than the eigenvalue with negative imaginary part will be* λ_j^{SR}, *the one with positive real part* λ_{j+1}^{SR}. *These eigenvalues were compared to the eigenvalues* $\lambda_j^{\mathbf{eig}}$ *of M obtained via* `eig`, *sorted as the eigenvalues* λ_j^{SR}. *The eigenvalues* $\lambda_j^{\mathbf{eig}}$ *were considered to be the 'exact' eigenvalues. This assumption is justified for the randomly generated examples using as a criterion* $\sigma_\lambda = \sigma_{\min}(M - \lambda I)$ *(where* $\sigma_{min}(A)$ *denotes the smallest singular value of the matrix A). It is well known that* σ_λ *is the 2–norm of the perturbation matrix E of smallest 2–norm for which* λ *is an exact eigenvalue of* $M + E$. *In other words,*

$$\sigma_\lambda = \min\{\|E\|_2 \mid \lambda \in \sigma(M + E)\}.$$

Thus, σ_λ *is the magnitude of the "backward error" in* λ *as an approximate eigenvalue of M. In particular, the smaller* σ_λ, *the better is* λ *as an approximate eigenvalue of M. Here we have that* σ_λ *is of order eps for eigenvalues computed via* `eig` *while for the eigenvalues computed via the* SR *algorithm, this 'residual' is larger by a factor of order* $\mathcal{O}(10^d)$ *where d is the number of digits lost as indicated by our relative error measure.*

The column max(`relerr`) *reports the maximal relative error*

$$\max_j \frac{|\lambda_j^{\mathbf{eig}} - \lambda_j^{SR}|}{|\lambda_j^{SR}|}$$

obtained, the column min(`relerr`) *the minimal relative error. In order to get an idea about the average relative accuracy obtained, we computed for each example the arithmetic mean; the range in which these values were found is given in column 'average(`relerr`)'. Finally, in order to compare our results with those given in [13], we computed the average relative accuracy for all examples of each dimension using the arithmetic mean of all examples for each dimension. In [13], these averages are given for dimensions 10, 20, and 40; our results confirm those results.*

$2*n$	$\max(\|M^T JM - J\|_2)$	$\|S^{-1}MS - B\|_2$	condmax	iter
10	$\mathcal{O}(10^{-12})$	$\mathcal{O}(10^{-14}) - \mathcal{O}(10^{-7})$	$\mathcal{O}(10^3)$	0.698
20	$\mathcal{O}(10^{-12})$	$\mathcal{O}(10^{-13}) - \mathcal{O}(10^{-6})$	$\mathcal{O}(10^4)$	0.716
30	$\mathcal{O}(10^{-11})$	$\mathcal{O}(10^{-12}) - \mathcal{O}(10^{-5})$	$\mathcal{O}(10^4)$	0.716
40	$\mathcal{O}(10^{-12})$	$\mathcal{O}(10^{-10}) - \mathcal{O}(10^{-4})$	$\mathcal{O}(10^5)$	0.683
50	$\mathcal{O}(10^{-11})$	$\mathcal{O}(10^{-10}) - \mathcal{O}(10^{-4})$	$\mathcal{O}(10^4)$	0.678
60	$\mathcal{O}(10^{-11})$	$\mathcal{O}(10^{-10}) - \mathcal{O}(10^{-4})$	$\mathcal{O}(10^6)$	0.658
70	$\mathcal{O}(10^{-10})$	$\mathcal{O}(10^{-9}) - \mathcal{O}(10^{-3})$	$\mathcal{O}(10^5)$	0.661
80	$\mathcal{O}(10^{-10})$	$\mathcal{O}(10^{-9}) - \mathcal{O}(10^{-4})$	$\mathcal{O}(10^5)$	0.653
90	$\mathcal{O}(10^{-10})$	$\mathcal{O}(10^{-9}) - \mathcal{O}(10^{-4})$	$\mathcal{O}(10^6)$	0.656
100	$\mathcal{O}(10^{-10})$	$\mathcal{O}(10^{-9}) - \mathcal{O}(10^{-4})$	$\mathcal{O}(10^6)$	0.656

Table 4.4.   Example 4.10 — SR algorithm

$2*n$	$\max(relerr)$	$\min(relerr)$	average($relerr$)	average
10	$\mathcal{O}(10^{-9})$	$\mathcal{O}(10^{-12})$	$\mathcal{O}(10^{-15}) - \mathcal{O}(10^{-9})$	$2.4*10^{-11}$
20	$\mathcal{O}(10^{-6})$	$\mathcal{O}(10^{-10})$	$\mathcal{O}(10^{-14}) - \mathcal{O}(10^{-7})$	$2.8*10^{-9}$
30	$\mathcal{O}(10^{-7})$	$\mathcal{O}(10^{-10})$	$\mathcal{O}(10^{-13}) - \mathcal{O}(10^{-8})$	$2.8*10^{-9}$
40	$\mathcal{O}(10^{-5})$	$\mathcal{O}(10^{-11})$	$\mathcal{O}(10^{-12}) - \mathcal{O}(10^{-6})$	$2.8*10^{-8}$
50	$\mathcal{O}(10^{-5})$	$\mathcal{O}(10^{-11})$	$\mathcal{O}(10^{-12}) - \mathcal{O}(10^{-6})$	$1.6*10^{-8}$
60	$\mathcal{O}(10^{-5})$	$\mathcal{O}(10^{-11})$	$\mathcal{O}(10^{-12}) - \mathcal{O}(10^{-6})$	$4.1*10^{-8}$
70	$\mathcal{O}(10^{-5})$	$\mathcal{O}(10^{-11})$	$\mathcal{O}(10^{-11}) - \mathcal{O}(10^{-6})$	$1.0*10^{-7}$
80	$\mathcal{O}(10^{-5})$	$\mathcal{O}(10^{-11})$	$\mathcal{O}(10^{-11}) - \mathcal{O}(10^{-6})$	$2.5*10^{-8}$
90	$\mathcal{O}(10^{-4})$	$\mathcal{O}(10^{-11})$	$\mathcal{O}(10^{-11}) - \mathcal{O}(10^{-6})$	$5.1*10^{-8}$
100	$\mathcal{O}(10^{-6})$	$\mathcal{O}(10^{-13})$	$\mathcal{O}(10^{-11}) - \mathcal{O}(10^{-7})$	$2.2*10^{-8}$

Table 4.5.   Example 4.10 — SR algorithm

$2*n$	$\|SKQ - \tilde{K}\|_2$	$\|SNQ - \tilde{N}\|_2$	condmax	iter
30	$\mathcal{O}(10^{-11}) - \mathcal{O}(10^{-6})$	$\mathcal{O}(10^{-11}) - \mathcal{O}(10^{-7})$	$\mathcal{O}(10^5)$	0.574
50	$\mathcal{O}(10^{-10}) - \mathcal{O}(10^{-4})$	$\mathcal{O}(10^{-10}) - \mathcal{O}(10^{-5})$	$\mathcal{O}(10^5)$	0.546

Table 4.6.   Example 4.10 — SZ algorithm

$2*n$	$\max(relerr)$	$\min(relerr)$	aver($relerr$)	average
30	$\mathcal{O}(10^{-7})$	$\mathcal{O}(10^{-9})$	$\mathcal{O}(10^{-13}) - \mathcal{O}(10^{-8})$	$9.7*10^{-10}$
50	$\mathcal{O}(10^{-6})$	$\mathcal{O}(10^{-11})$	$\mathcal{O}(10^{-12}) - \mathcal{O}(10^{-7})$	$1.0*10^{-8}$

Table 4.7.   Example 4.10 — SZ algorithm

The same kind of test runs was performed for randomly generated symplectic matrix pencils $K - \lambda N$ where K and N are both symplectic using the SZ algorithm. K and N were generated analogous to M as above. Note that this introduces more difficulties here than above; our SZ algorithm makes use of the fact that $K^T J K = N^T J N = J$; all of these equalities are violated. But despite this, the SZ algorithm performs as well as the SR algorithm. Our implementation of the SZ algorithm first reduces K and N to the butterfly pencil (3.4.19) and than iterates using only quadruple shift steps where the shifts are chosen according to the generalized Rayleigh-quotient strategy.

In the Tables 4.6 and 4.7 we report the same information as in the two tables presented for the SR algorithm. This time we give the data only for dimensions 30 and 50, in order to save some space but to support our claim that the SZ algorithm works as well as the SR algorithm. The SZ algorithm computes symplectic matrices S, Q, $\widetilde{K}$ and $\widetilde{N}$ such that $S(K - \lambda N)Q = \widetilde{K} - \lambda \widetilde{N}$ and the pencil $\widetilde{K} - \lambda \widetilde{N}$ decouples into a number of 2×2 and 4×4 subproblems. The eigenvalues of these small subproblems were computed using the MATLAB *function* eig. *Similar as before, we compare those eigenvalues to the eigenvalues obtained via* eig(K,N).

EXAMPLE 4.11 *A second set of tests was performed to see whether the quadruple shift SR or the quadruple shift SZ algorithm performs better once the symplectic matrix/matrix pencil is reduced to parameterized form (a similar comparison of the double shift SR and the double shift parameterized SR is given in Example 4.15). For this purpose, parameters $a_1, \ldots, a_n, b_1, \ldots, b_n,$ $c_1, \ldots, c_n, d_2, \ldots, d_n \in \mathbf{R}$ were generated, from which a symplectic butterfly pencil $K - \lambda N$ and the corresponding butterfly matrix M were constructed as in (3.4.19), and (3.2.10), respectively.*

The examples generated this way do not suffer from loss of symplecticity, any matrix pencil $K - \lambda N$ of the above form is symplectic. Furthermore no initial reduction to butterfly form is necessary here; K, N, and M are already in parameterized form. For each $n = 5 : 5 : 50$, one hundred sets of parameters were generated, K, N, and M were constructed, and the SR/SZ algorithm was used to compute the eigenvalues. As before, the 2×2 and 4×4 subproblems were solved using eig. *The eigenvalues so obtained were compared to eigenvalues computed via* eig(M). *Table 4.8 reports some of the results so obtained, using the same notation as above.*

As expected, the examples showed the same convergence behavior no matter which algorithm was used. That is, the number of iterations needed for convergence was almost the same, the maximal condition number of the Gauss transformations were the same. The maximal relative error observed for the different examples was bigger for the SR algorithm than for the SZ algo-

$2*n$	SR max($relerr$)	SR average	SZ max($relerr$)	SZ average	SR/SZ iter
10	$\mathcal{O}(10^{-11})$	$1.7*10^{-13}$	$\mathcal{O}(10^{-13})$	$1.6*10^{-15}$	0.60
20	$\mathcal{O}(10^{-9})$	$8.4*10^{-12}$	$\mathcal{O}(10^{-12})$	$5.5*10^{-15}$	0.64
30	$\mathcal{O}(10^{-12})$	$3.5*10^{-14}$	$\mathcal{O}(10^{-13})$	$2.3*10^{-15}$	0.65
40	$\mathcal{O}(10^{-7})$	$6.9*10^{-11}$	$\mathcal{O}(10^{-13})$	$2.7*10^{-15}$	0.65
50	$\mathcal{O}(10^{-11})$	$2.5*10^{-14}$	$\mathcal{O}(10^{-14})$	$2.7*10^{-15}$	0.64
60	$\mathcal{O}(10^{-8})$	$3.6*10^{-12}$	$\mathcal{O}(10^{-11})$	$1.8*10^{-14}$	0.64
70	$\mathcal{O}(10^{-9})$	$2.1*10^{-12}$	$\mathcal{O}(10^{-13})$	$3.4*10^{-15}$	0.63
80	$\mathcal{O}(10^{-9})$	$8.1*10^{-13}$	$\mathcal{O}(10^{-13})$	$3.5*10^{-15}$	0.64
90	$\mathcal{O}(10^{-8})$	$5.4*10^{-12}$	$\mathcal{O}(10^{-13})$	$3.6*10^{-15}$	0.63
100	$\mathcal{O}(10^{-8})$	$2.5*10^{-11}$	$\mathcal{O}(10^{-12})$	$5.3*10^{-15}$	0.63

Table 4.8.   Example 4.11

rithm. *These results indicate that the SZ algorithm computes more accurate eigenvalues than the SR algorithm.*

EXAMPLE 4.12 *Tests with examples from the benchmark collection [22] were performed. None of these examples result in a symplectic pencil $K - \lambda N$ with symplectic K and N matrices. Hence, whenever possible, a symplectic matrix M was formed from the given data. Table 4.9 presents the results obtained applying the SR algorithm to M. Again, the relative error in the eigenvalues was computed by comparing the eigenvalues computed via the SR algorithm with those computed via* eig. *The first column of the table gives the number of the example as given in [22]. The next columns display the dimension of the problem, the maximal and minimal relative errors for the computed eigenvalues, the maximal condition number used, and the total number of iterations needed to achieve convergence.*

For the first two examples, no SR iteration was necessary: after the initial reduction to butterfly form, the problem either decoupled into two 2×2 sub-problems or the eigenvalues could be read off directly. The relative error of the so computed eigenvalues is of order $\mathcal{O}(\mathrm{eps})$. For Example 8 of [22] the eigenvalues computed via the SR algorithm were better than those computed via eig. *This was checked via the smallest singular value $\sigma_{\min}$ of $(M - \lambda I)$ for the eigenvalues λ computed via* eig *as well as via the SR algorithm. It turns out that $\sigma_{\min}(M - \lambda^{SR}I)$ is smaller then $\sigma_{\min}(M - \lambda^{\mathrm{eig}}I)$. For Example 11 from [22] one should note that the matrix M there is only almost symplectic, that is, $\|M^T J M - J\|_2 \approx 1*10^{-10}$ and the condition number of M is given by $\kappa(M) \approx 1.6*10^6$.*

EXAMPLE 4.13 *Flaschka, Mehrmann, and Zywietz report in [53] that the SR algorithm for symplectic J–Hessenberg matrices does not perform satisfactory*

Example Number	$2 * n$	max(*relerr*)	min(*relerr*)	condmax	number of iterations
1	4			2.5	0
2	4			25.3	0
6	8	$1.0 * 10^{-14}$	$4.0 * 10^{-15}$	6.4	4
7	8	$2.2 * 10^{-12}$	$6.8 * 10^{-14}$	$5.4 * 10^2$	3
8	8	$2.4 * 10^{-11}$	$8.5 * 10^{-16}$	8.4	3
9	10	$9.3 * 10^{-13}$	$4.1 * 10^{-16}$	20.4	5
11	18	$1.2 * 10^{-2}$	$7.1 * 10^{-12}$	$7.9 * 10^3$	6

Table 4.9.   Example 4.12

due to roundoff errors. They present two examples to demonstrate the behavior of the SR algorithm for symplectic J–Hessenberg matrices. The first example presented is a symplectic matrix with the eigenvalues $5, 1/5, 3\pm 4i, 0.12\pm 0.16i$; the matrix itself is given in [53]. It is reported in [53] that complete deflation was observed after 19 iteration, but the final iteration matrix was far from being symplectic. The maximal condition number of the symplectic Gauss transformations used during the iterations was $6.4 * 10^3$.

Our algorithm first reduced the symplectic matrix to butterfly form (this is denoted here as iteration step 0), then two iterations were needed for convergence. Moreover, cubic convergence can be observed by monitoring the parameters d_j during the course of the iteration, as they indicate deflation. Table 4.10 reports the values for the d_j's after each iteration.

As can be seen, it takes only two iterations for d_3 to become zero with respect to machine precision. Decoupling is possible and the problem splits into a 2×2 and a 4×4 subproblem. The observed maximal condition number of the symplectic Gauss transformations used during the iterations was 57.39.

The second example discussed in [53] is a 12×12 symplectic matrix with the eigenvalues $1 \pm 1i, 0.5 \pm 0.5i, 2 \pm 2i, 0.25 \pm 0.25i, 3 \pm 4i, 0.12 \pm 0.16i$. Here, a symplectic diagonal matrix with these eigenvalues on the diagonal was constructed and a similarity transformation with a randomly generated orthogonal symplectic matrix was performed to obtain a symplectic matrix M. The implementation presented in [53] first reduces this matrix to J–Hessenberg form, then a double shift SR step with the perfect shift $3 \pm 4i$ is performed. This resulted in deflation and good approximation of these eigenvalues, but symplecticity was lost completely.

Our algorithm again first reduced the symplectic matrix to butterfly form, then six iterations were needed for convergence. As before, cubic convergence can be observed by monitoring the parameters d_j during the course of the iteration. Table 4.11 reports the values for the d_j's after each iteration as well

iteration	d_2	d_3
0	1.8576	$2.389 * 10^{-2}$
1	-0.2783	$-2.117 * 10^{-5}$
2	-4.3422	$2.242 * 10^{-16}$

Table 4.10.   Example 4.13 — first test

iteration	deflation?	size of deflated subproblem	d_3	d_5
0	no		$1.07 * 10^{0}$	$0.91 * 10^{0}$
1	no		$1.29 * 10^{-1}$	$-8.50 * 10^{-2}$
2	no		$-5.30 * 10^{-2}$	$1.37 * 10^{-4}$
3	no		$-1.49 * 10^{-3}$	$-1.26 * 10^{-12}$
4	yes	4×4	$2.18 * 10^{-3}$	$-3.40 * 10^{-24}$
5	no		$-4.36 * 10^{-10}$	
6	yes	4×4	$3.09 * 10^{-25}$	

Table 4.11.   Example 4.13 — second test

as whether deflation occurred and whether a 2×2 or a 4×4 subproblem was deflated.

The observed maximal condition number of the symplectic Gauss transformations used during the iterations was 73.73.

EXAMPLE 4.14 *We also tested an implementation of the SR algorithm using a standard polynomial p instead of a Laurent polynomial q to drive the SR step. In cases where no trouble arose, both algorithms performed similarly. That is, both versions of the algorithm needed the same number of iterations for convergence, and the accuracy of the computed eigenvalues was similar. But, as indicated in our discussion in Section 4.1, using the standard polynomial might sometimes cause some problems. Using the Laurent polynomial to drive the SR step, the algorithm behaved as expected. Convergence of even-dimensional subspaces occurred, which resulted in the convergence of some of the d_k's to zero. But when working with standard polynomials to drive the SR step, one might observe convergence of a_1 to zero and stagnation of the algorithm afterwards. This will be illustrated here by the following example. We generated a 30×30 symplectic matrix using the parameters $a_1, \ldots, a_{15}, b_1, \ldots, b_{15}, c_1, \ldots, c_{15},$ and $d_2, \ldots, d_{15}$ as given in Table 4.12. The resulting symplectic matrix M has only two real eigenvalues:*

$$\mu = 1.97700698420 \quad \text{and} \quad \mu^{-1} = 0.50581510737.$$

The twenty-eight complex eigenvalues occur in pairs $\{\lambda, \overline{\lambda}\}$ where $\lambda \in \mathbb{C}, |\lambda| = 1.

	a	b	c	d
1	0.76880950325	0.82064368228	0.06824661097	
2	0.96970170497	0.97047237460	0.96412426837	0.84800944806
3	0.71479723187	0.48692499554	0.20765658836	0.72860019101
4	0.78196184196	0.81746853554	0.16111822555	0.95509863327
5	0.23756508204	0.64157116784	0.63822138259	0.65635111059
6	0.19573076378	0.30634935951	0.00022817289	0.74230513350
7	0.26321391517	0.66093213223	0.33563294335	0.34496601390
8	0.71378506459	0.35801711338	0.27509982146	0.88402194967
9	0.97759973943	0.93819943010	0.04452752039	0.34724408649
10	0.63712194084	0.48766697476	0.09389649759	0.05947668054
11	0.54592415509	0.09099035774	0.40999739977	0.71841459107
12	0.84805722441	0.67383411686	0.81689231949	0.95821429290
13	0.80209765848	0.51488031898	0.87051707180	0.15683486507
14	0.66830641006	0.22157934638	0.02255512045	0.41635310614
15	0.67098263396	0.72500937095	0.72717698369	0.09403486897

Table 4.12. Example 4.14 — parameters

	a_1	b_1	c_1
1	$1.8 * 10^{-1}$	1.98410591	2.2791
2	$1.6 * 10^{-2}$	1.97948316	30.8977
3	$2.2 * 10^{-3}$	1.97726186	233.4375
4	$6.6 * 10^{-4}$	1.97714161	761.5576
5	$3.6 * 10^{-4}$	1.97718534	1397.3210
6	$4.4 * 10^{-4}$	1.97728897	1149.0454
7	$9.1 * 10^{-5}$	1.97706343	5575.8521
8	$1.0 * 10^{-5}$	1.97701113	49627.9530
9	$7.8 * 10^{-7}$	1.97700726	643154.9327

Table 4.13. Example 4.14 — standard polynomial

	a_1	b_1	c_1
1	0.92647065700	0.85985047660	1.66728006461
2	0.91957849076	0.85402888680	1.76951100126
3	0.91953205232	0.85399738303	1.76962068109
4	0.91952829207	0.85399457780	1.76914546114
5	0.91951299233	0.85398309033	1.76786406820
6	0.91931641917	0.85383550669	1.77172737277
7	0.91929704077	0.85382096040	1.77200623558
8	0.91929694433	0.85382088801	1.77200762831
9	0.91929694382	0.85382088762	1.77200763568

Table 4.14. Example 4.14 — Laurent polynomial

Table 4.13 reports the values of a_1, b_1, and c_1 in the course of the iteration when the SR step is driven by a standard polynomial (the first column indicates the number of iterations). The choice of shifts is as before.

Already after the first iteration the largest eigenvalue μ is emerging as b_1. During the subsequent iterations, b_1 converges towards this eigenvalue while a_1 converges to zero. The growth of c_1 reflects the ill conditioning of the transforming matrices. At the bottom of the matrix, deflations take place: after iteration 5, a 2×2 subproblem is decoupled, after iteration 7 a 4×4, after iteration 11 and 12 a 2×2, and after iteration 16 another 4×4 subproblem is decoupled. At that point a_1 is less than eps so that a 2×2 subproblem can be deflated at the top which corresponds to the pair of real eigenvalues, $b_1 \approx \mu$ and the $(n + 1, n + 1)$ entry of the iteration matrix is approximately equal to μ^{-1}. The resulting 14×14 subproblem has only complex pairs of eigenvalues on the unit circle. Parameterizing this subproblem, one observes that three of the six parameters d_j are of order $\sqrt{eps}$, the other three are of order 1. This does not change during subsequent iterations, no convergence is achieved (the required tolerance for deflation is of order eps).

Using a Laurent polynomial to drive the SR step, the process converges after 22 iterations, a_1 does not converge to zero. All eigenvalues are computed accurately $(\max(relerr) = \mathcal{O}(10^{-15}))$. Table 4.14 gives the same information as Table 4.13 for the case considered here.

EXAMPLE 4.15 *The last set of tests concerned the parameterized SR algorithm. As that algorithm was implemented only for a double shift step, we compared it to a double shift SR algorithm (all the test reported so far used quadruple shift steps). For the tests reported $n \times n$ diagonal matrices D were generated using* MATLAB*'s 'rand' function. Then a symplectic matrix S was constructed such that $S = M^T \mathrm{diag}((\,)D, D^{-1})M$ where $M \in \mathbf{R}^{2n \times 2n}$ are randomly generated symplectic orthogonal matrices. This guarantees that all test matrices have only real-valued pairs of eigenvalues $\{\mu, \mu^{-1}\}, \mu \in \mathbf{R}$. Hence, using only double shift Laurent polynomials to drive the SR step, the corresponding butterfly matrices can be reduced to butterfly matrices such that the $(1, 2)$ and the $(2, 2)$ block is diagonal (that is, all parameters d_j are zero).*

In order to detect deflation in the parameterized SR algorithm, parameters d_j were declared to be zero during the iteration when

$$d_j \leq 10 \cdot n \cdot eps$$

was fulfilled, where the dimension of the problem is $2n \times 2n$ and eps $\approx 2.2204 * 10^{-16}$ is MATLAB*'s floating point relative accuracy. Deflation in the double shift SR algorithm was determined as described in (4.4.39).*

While symplecticity is forced by the parameterized SR algorithm, its has to be enforced after each double shift SR step. Otherwise symplecticity is lost in the double shift SR algorithm.

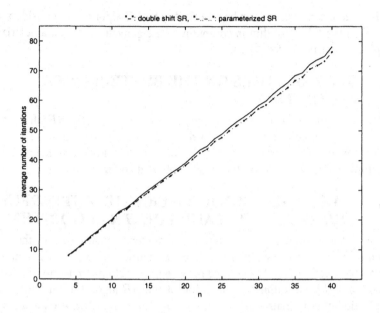

Figure 4.1. average number of iterations, 100 examples for each dimension

All tests showed that the parameterized SR algorithm and the double shift SR algorithm compute the eigenvalues to about the same accuracy. But the parameterized SR algorithm converged slightly faster than the double shift SR algorithm, exhibiting the same cubic convergence behavior. Figure 4.1 shows the average number of iterations needed for convergence using the parameterized SR algorithm and the double shift SR algorithm.

In order to compute the average number of iterations needed for convergence, 100 symplectic matrices S for each of the dimensions $2n \times 2n$ for $n = 4 : 40$ were constructed as described above. It was observed that the parameterized SR algorithm converges typically slightly faster then the double shift SR algorithm. For most of the test examples, the parameterized SR algorithm was as fast or faster than the double shift SR algorithm. Just for very few examples, the parameterized SR algorithm needed more iteration than the double shift SR algorithm; and than only up to 3 iterations more. Mostly this was due to the fact that the deflation criterion for the parameterized SR algorithm is somewhat more strict than the one for the double shift SR algorithm. Similar results were obtained for test matrices $S = M^T \begin{bmatrix} D & F \\ 0 & D^{-1} \end{bmatrix} M$, where D, F are random diagonal $n \times n$ matrices and M is as before.

The examples presented demonstrate that the butterfly SR/SZ algorithm is an efficient structure-preserving algorithm for computing the eigenvalues of symplectic matrices/matrix pencils. Using Laurent polynomials as shift polynomials cubic convergence can be observed. The SZ algorithm is considerably better than the SR algorithm in computing the eigenvalues of a parameterized symplectic matrix/matrix pencil.

4.5   TWO REMARKS ON THE BUTTERFLY SR ALGORITHM

Here we will give two interesting remarks on the butterfly SR algorithm. First we prove a connection between the SR and HR algorithm. Then we discuss how one of the problems that motivated us to study the symplectic SR algorithm can be solved using the results obtained so far.

4.5.1   A CONNECTION BETWEEN THE BUTTERFLY SR ALGORITHM AND THE HR ALGORITHM

The SR algorithm preserves symplectic butterfly matrices, and the HR algorithm preserves D–symmetric tridiagonal matrices where D is a signature matrix (see Section 2.2.3). Here we prove an interesting connection between the SR and HR algorithm: An iteration of the SR algorithm on a $2n \times 2n$ symplectic butterfly matrix using shifts $\mu_i, \mu_i^{-1}, i = 1, \ldots, k$, is equivalent to an iteration of the HR algorithm on a certain $n \times n$ tridiagonal D–symmetric matrix using shifts $\mu_i + \mu_i^{-1}, i = 1, \ldots, k$.

Normally if one speaks of a GR iteration with shifts $\mu_1, \ldots, \mu_k$, one means that the iteration is carried out using the polynomial spectral transformation $p(B) = (B - \mu_1 I) \cdots (B - \mu_k I)$. But as discussed in Section 4.1 in the case of an SR iteration on a symplectic matrix B, it makes sense to do things differently. When we speak of an SR iteration with shifts $\mu_i, \mu_i^{-1}, i = 1, \ldots, k$, applied to a symplectic matrix B, we mean that one forms the Laurent polynomial

$$q(B) = \prod_{i=1}^{k}(B - \mu_i I)(I - \mu_i^{-1}B^{-1}) = \prod_{i=1}^{k}\left([B + B^{-1}] - [\mu_i + \mu_i^{-1}]I\right),$$

next one performs the decomposition $q(B) = SR$, then one performs the similarity transformation $\widehat{B} = S^{-1}BS$. When we speak of an HR iteration with shifts $\mu_i + \mu_i^{-1}$ applied to a D–symmetric tridiagonal matrix T, we mean that one uses the polynomial spectral transformation

$$p(T) = (T - [\mu_1 + \mu_1^{-1}]I) \cdots (T - [\mu_k + \mu_k^{-1}]I).$$

We are restricting ourselves to the nonsingular case, which means that none of the μ_i is allowed to be an eigenvalue of B. This is only for the sake of avoiding complications. As we have seen in Section 4.1, nothing bad happens in the singular case.

We allow complex shifts. However, if μ_i is not real, we insist that $\overline{\mu_i}$ should also appear in the list of shifts, so that $q(B)$ is real. In case that ℓ shifts are used in each SR or HR iteration step we say that the iteration is of *degree* ℓ.

THEOREM 4.16 *Let* $B = \begin{bmatrix} 0 & -D \\ D & T \end{bmatrix}$ *be an unreduced symplectic butterfly matrix in canonical form (3.3.16). Then an SR iteration of degree $2k$ with shifts* $\mu_i,\ \mu_i^{-1},\ i = 1,\ldots,k,$ *on* B *is equivalent to an HR iteration of degree k with shifts* $\mu_i + \mu_i^{-1},\ i = 1,\ldots,k$ *on the D–symmetric matrix T.*

PROOF: The SR iteration has the form

$$q(B) = SR, \quad \widehat{B} = S^{-1}BS,$$

where q is the Laurent polynomial

$$q(\lambda) = \prod_{i=1}^{k} \left([\lambda + \lambda^{-1}] - [\mu_i + \mu_i^{-1}] \right).$$

Notice that $q(B) = p(B + B^{-1})$, where p is the ordinary polynomial

$$p(\nu) = \prod_{i=1}^{k} \left(\nu - [\mu_i + \mu_i^{-1}] \right).$$

Since

$$B = \begin{bmatrix} 0 & -D \\ D & T \end{bmatrix},$$

we have $B + B^{-1} = \mathrm{diag}(T^T, T)$. Therefore

$$q(B) = \begin{bmatrix} p(T^T) & \\ & p(T) \end{bmatrix}. \tag{4.5.40}$$

An HR iteration on T with shifts $\mu_i + \mu_i^{-1},\ i = 1,\ldots,k,$ has the form

$$p(T) = HU, \quad \widehat{T} = H^{-1}TH.$$

U is upper triangular, H satisfies $H^T D H = \widehat{D}$, where $\widehat{D}$ is a signature matrix, and $\widehat{T}$ is $\widehat{D}$–symmetric.

Now let us relate this to the SR iteration on B. Since $Dp(T) = p(T^T)D$, we have

$$p(T^T) = DHUD = H^{-T}\widehat{D}UD.$$

Thus

$$q(B) = \begin{bmatrix} p(T^T) & \\ & p(T) \end{bmatrix} = \begin{bmatrix} H^{-T} & \\ & H \end{bmatrix} \begin{bmatrix} \widehat{D}UD & \\ & U \end{bmatrix}.$$

This is the SR decomposition of $q(B)$, for

$$S = \begin{bmatrix} H^{-T} & \\ & H \end{bmatrix}$$

is symplectic, and

$$R = \begin{bmatrix} \hat{D}UD & \\ & U \end{bmatrix}$$

is J–triangular. Using this SR decomposition to perform the SR iteration, we obtain

$$\hat{B} = S^{-1}BS = \begin{bmatrix} 0 & -\hat{D} \\ \hat{D} & \hat{T} \end{bmatrix}.$$

Thus the HR iteration on T is equivalent to the SR iteration on B. $\checkmark$

In principle we can compute the spectrum of a symplectic butterfly matrix by putting it into canonical form, calculating the eigenvalues of T, then inverting the transformation $\lambda \to \lambda + \lambda^{-1}$. Conversely, we can calculate the eigenvalues of a D–symmetric tridiagonal matrix T by embedding T and D in a symplectic butterfly matrix B, calculating the eigenvalues of B, and applying the transformation $\lambda \to \lambda + \lambda^{-1}$.

These transformations are not necessarily advisable from the standpoint of numerical stability. The first will resolve eigenvalues near ± 1 poorly because, as we already mentioned in Remark 3.17, the inverse transformation is not Lipschitz continuous. The second transformation is perhaps less objectionable. However, any eigenvalues of T that are near zero will have poor relative accuracy, because cancellation will occur in the transformation $\lambda \to \lambda + \lambda^{-1}$.

REMARK 4.17 *a) The decomposition (3.2.7) of the canonical form B is*

$$B = \begin{bmatrix} 0 & -D \\ D & T \end{bmatrix} = \begin{bmatrix} D & 0 \\ 0 & D \end{bmatrix} \begin{bmatrix} 0 & -I \\ I & DT \end{bmatrix}.$$

Thus B is equivalent to the pencil

$$\begin{bmatrix} 0 & -I \\ I & DT \end{bmatrix} - \lambda \begin{bmatrix} D & 0 \\ 0 & D \end{bmatrix}.$$

Now the SZ algorithm on this pencil, driven by

$$q(z) = \prod_{i=1}^{k} (z - \mu_i)(1 - \mu_i^{-1}z^{-1})$$

is the same as the HZ algorithm on the pencil DT − λD driven by

$$p(w) = \prod_{i=1}^{k}(w - [\mu_i + \mu_i^{-1}]).$$

(See [140]).

b) *An iteration of the butterfly SR algorithm is not only equivalent to an iteration of the HR algorithm as proven above, but also to two iterations of the LU algorithm. In [33, Theorem 4.3] it is shown that one step of the HR algorithm applied to a D–symmetric tridiagonal matrix T is essentially equivalent to two steps of the LU algorithm applied to T (if the LU algorithm is constructible and no shifts are used). If T is symmetric (that is, D = I), then two steps of the LU algorithm (if it exists) are equivalent to one QR step (which always exists). For positive definite symmetric T, this was already noted by Wilkinson [144, p. 545], for general symmetric matrices this relation is described in [147].*

4.5.2 DISCRETE TIME ALGEBRAIC RICCATI EQUATIONS AND THE BUTTERFLY SR ALGORITHM

In this section we will briefly discuss how to solve a discrete time algebraic Riccati equation

$$X = F^T(I + XG)^{-1}XF + H \qquad (4.5.41)$$

where $F \in \mathbf{R}^{n \times n}$, G and $H \in \mathbf{R}^{n \times n}$ are symmetric positive semidefinite using the butterfly SR algorithm. Under certain assumptions, the solution of the discrete time algebraic Riccati equation (4.5.41) can be expressed in terms of certain invariant or deflating subspaces of corresponding matrices and matrix pencils (see, e.g., [83, 87, 104], and the references therein). The symmetric solution X of (4.5.41) can be given in terms of the deflating subspace of the symplectic matrix pencil

$$K - \lambda N = \begin{bmatrix} F & 0 \\ H & I \end{bmatrix} - \lambda \begin{bmatrix} I & -G \\ 0 & F^T \end{bmatrix}$$

corresponding to the n eigenvalues $\lambda_1, \ldots, \lambda_n$ inside the unit circle using the relation

$$\begin{bmatrix} F & 0 \\ H & I \end{bmatrix}\begin{bmatrix} I \\ -X \end{bmatrix} = \begin{bmatrix} I & -G \\ 0 & F^T \end{bmatrix}\begin{bmatrix} I \\ -X \end{bmatrix}\Lambda$$

where $\Lambda \in \mathbf{R}^{n \times n}$, $\sigma(\Lambda) = \{\lambda_1, \ldots, \lambda_n\}$. Therefore, if we can compute $Z_1, Z_2 \in \mathbf{R}^{n \times n}$ such that the columns of $\begin{bmatrix} Z_1 \\ Z_2 \end{bmatrix}$ span the desired deflating subspace of $K - \lambda N$, then $X = -Z_2 Z_1^{-1}$ is the desired solution of the Riccati equation (4.5.41).

K and N are not symplectic themselves. Assume that F is regular, then we have two options to use the algorithms presented in order to solve the discrete time algebraic Riccati equation (4.5.41). For the first option, we form the symplectic matrix

$$K^{-1}N = \begin{bmatrix} GF^{-T}H + F & GF^{-T} \\ F^{-T}H & F^{-T} \end{bmatrix} = M.$$

Instead of computing the deflating subspace of $K - \lambda N$ corresponding to the n eigenvalues $\lambda_1, \ldots, \lambda_n$ inside the unit circle, we can compute the invariant subspace of M corresponding to the n eigenvalues inside the unit circle. Using the algorithm given in Table 3.2 M can be reduced to butterfly form

$$B = S_0^{-1} M S_0.$$

Using the SR algorithm for butterfly matrices, a symplectic matrix S_1 is computed such that

$$S_1^{-1} B S_1$$

decouples into 2×2 and 4×4 problems. Solving these subproblems as discussed at the end of Section 4.1, finally a symplectic matrix S_2 is computed such that

$$S_2^{-1} S_1^{-1} B S_1 S_2 = \left[\begin{array}{c|c} \Lambda & \star \\ \hline 0 & \Lambda^{-T} \end{array} \right] \tag{4.5.42}$$

where the eigenvalues of Λ are just the n eigenvalues of M inside the unit circle. Let $S = S_0 S_1 S_2$. Then (4.5.42) is equivalent to

$$KS = NS \left[\begin{array}{c|c} \Lambda & \star \\ \hline 0 & \Lambda^{-T} \end{array} \right]. \tag{4.5.43}$$

Write

$$S = \begin{bmatrix} S_{11} & S_{12} \\ S_{21} & S_{22} \end{bmatrix}. \tag{4.5.44}$$

The first n columns of (4.5.43) are given by

$$K \begin{bmatrix} S_{11} \\ S_{21} \end{bmatrix} = NS \begin{bmatrix} \Lambda \\ 0 \end{bmatrix} = N \begin{bmatrix} S_{11}\Lambda \\ S_{21}\Lambda \end{bmatrix} = N \begin{bmatrix} S_{11} \\ S_{21} \end{bmatrix} \Lambda.$$

Hence, $X = -S_{21} S_{11}^{-1}$ is the desired solution of the Riccati equation (4.5.41) if S_{11} is regular (under usual assumptions, this is always satisfied).

The second option to use the algorithms presented to solve the discrete time algebraic Riccati equation (4.5.41) is to premultiply $K - \lambda N$ by $\begin{bmatrix} I & 0 \\ 0 & F^{-T} \end{bmatrix}$.

This results in a symplectic matrix pencil $K' - \lambda N'$ where K', N' are both symplectic. Hence, the butterfly SZ algorithm can be used to compute the desired solution of the Riccati equation (4.5.41) analogous to the approach just described.

As noted, e.g., in [39, 128], when solving continuous time algebraic Riccati equations via the Hamiltonian SR algorithm it is desirable to combine the SR algorithm with the square reduced algorithm of Van Loan [137] which is used to compute good approximations to the eigenvalues of the Hamiltonian matrix. These eigenvalues are then used as 'ultimate' shifts for an acceleration of the Hamiltonian SR algorithm. Combined with the symmetric updating method of Byers and Mehrmann [40] to compute the solution of the continuous time algebraic Riccati equation and combined with a defect correction method like the one proposed by Mehrmann and Tan [105], the resulting method is a very efficient method and produces highly accurate results.

These ideas can be used here as well. Lin [96] developed an algorithm for computing eigenvalues of symplectic matrices analogous to Van Loan's square reduced method for Hamiltonian matrices [137]. Symmetric updating has been considered by Byers and Mehrmann [40]. Instead of computing the symplectic matrix S as in (4.5.44), one works with $n \times n$ matrices X, Y and Z such that finally $X = -S_{21}S_{11}^{-1}, Y = S_{11}^{-1}S_{12}$ and $Z = S_{11}^{-1}$. Starting from $X = Y = 0, Z = I$, this can be implemented without building the intermediate symplectic transformations used in the butterfly SR algorithm, just using the parameters that determine these transformations. As for every symplectic matrix S written in the form (4.5.44), $S_{21}S_{11}^{-1}$ is symmetric, this approach guarantees that all intermediate (and the final) X are symmetric. The same idea can be used when working with a butterfly pencil.

Using results given by Mehrmann and Tan in [105, Theorem 2.7] (see also [104, Chapter 10]), we have

THEOREM 4.18 *Let X be the symmetric solution of*

$$\mathcal{DR}(X) := -X + F^T(I + XG)^{-1}XF + H = 0$$

where F, G, H as before. Let $\widetilde{X}$ be a symmetric approximation to X. Then the error $V = X - \widetilde{X}$ can be expressed as the solution of

$$-V + \widehat{F}^T + (I + V\widehat{G})^{-1}V\widehat{F} + \widehat{H} = 0$$

where $\widehat{H} = \mathcal{DR}(\widetilde{X}), \widehat{G} = (G^{-1} + \widetilde{X})^{-1} = G(I + \widetilde{X}G)^{-1}$, and $\widehat{F} = (I - \widehat{G}\widetilde{X})F$.

The error V fulfills a discrete time algebraic Riccati equation just like the desired solution X. The defect discrete-time algebraic Riccati equation may be solved by any method for discrete-time algebraic Riccati equations, including the butterfly SR/SZ algorithm. Most frequently, in situations like this, Newton's method is used [71, 104, 16]. Then V can be used to correct $\widetilde{X}$. Iterating this process until $\mathcal{DR}(X)$ is suitably small is called a defect correction method and should always be used if the discrete-time algebraic Riccati equation is solved by the butterfly SR/SZ algorithm.

A flop count shows that the symmetric updating approach is not feasible here. The butterfly SR/SZ algorithm involves updates with symplectic Givens, symplectic Gauss and symplectic Householder transformations. Partition the symplectic transformation matrix S into $n \times n$ blocks

$$S = \begin{bmatrix} S_{11} & S_{12} \\ S_{21} & S_{22} \end{bmatrix},$$

and let $X = -S_{21}S_{11}^{-1}$, $Y = S_{11}^{-1}S_{12}$ and $Z = S_{11}^{-1}$. X and Y are symmetric. Then an update with a symplectic Householder matrix

$$H_k = \begin{bmatrix} \widetilde{P} & 0 \\ 0 & \widetilde{P} \end{bmatrix}$$

where $\widetilde{P} = \text{diag}(I^{k-1,k-1}, I^{n-k+1,n-k+1} - 2\frac{vv^T}{v^Tv})$, $v \in \mathbf{R}^{n-k+1}$ can be expressed as follows:

$$\begin{bmatrix} S_{11} & S_{12} \\ S_{21} & S_{22} \end{bmatrix} \begin{bmatrix} \widetilde{P} & 0 \\ 0 & \widetilde{P} \end{bmatrix} = \begin{bmatrix} S_{11}\widetilde{P} & S_{12}\widetilde{P} \\ S_{21}\widetilde{P} & S_{22}\widetilde{P} \end{bmatrix} = \begin{bmatrix} \widetilde{S}_{11} & \widetilde{S}_{12} \\ \widetilde{S}_{21} & \widetilde{S}_{22} \end{bmatrix},$$

and

$$\begin{aligned} \widetilde{X} &= -\widetilde{S}_{21}\widetilde{S}_{11}^{-1} = -(S_{21}\widetilde{P})(S_{11}\widetilde{P})^{-1} = -S_{21}S_{11}^{-1} = X, \\ \widetilde{Y} &= \widetilde{S}_{11}^{-1}\widetilde{S}_{12} = \widetilde{P}^{-1}S_{11}^{-1}S_{12}\widetilde{P} = \widetilde{P}Y\widetilde{P}, \\ \widetilde{Z} &= \widetilde{S}_{11}^{-1} = \widetilde{P}^{-1}S_{11}^{-1} = \widetilde{P}^{-1}Z. \end{aligned}$$

Hence, a symmetric update with a symplectic Householder matrix H_k requires $2n \cdot (4k - 2) + 2k^3 + k^2 - k$ flops, taking the symmetry of Y into account.

An update with a symplectic Gauss matrix

$$L_k = \begin{bmatrix} D & V \\ 0 & D^{-1} \end{bmatrix},$$

where $D = \text{diag}(I^{k-1,k-1}, c, c, I^{n-k,n-k})$ and $V = d(e_k e_{k+1}^T + e_{k+1}e_k^T)$, can be expressed as follows:

$$\begin{aligned} \begin{bmatrix} S_{11} & S_{12} \\ S_{21} & S_{22} \end{bmatrix} \begin{bmatrix} D & V \\ 0 & D^{-1} \end{bmatrix} &= \begin{bmatrix} S_{11}D & S_{11}V + S_{12}D^{-1} \\ S_{21}D & S_{21}V + S_{22}D^{-1} \end{bmatrix} \\ &= \begin{bmatrix} \widetilde{S}_{11} & \widetilde{S}_{12} \\ \widetilde{S}_{21} & \widetilde{S}_{22} \end{bmatrix}, \end{aligned}$$

and

$$\tilde{X} = -\tilde{S}_{21}\tilde{S}_{11}^{-1} = -(S_{21}D)(S_{11}D)^{-1} = -S_{21}S_{11}^{-1} = X,$$
$$\tilde{Y} = \tilde{S}_{11}^{-1}\tilde{S}_{12} = D^{-1}S_{11}^{-1}(S_{11}V + S_{12}D^{-1}) = D^{-1}V + D^{-1}YD^{-1},$$
$$\tilde{Z} = \tilde{S}_{11}^{-1} = D^{-1}S_{11}^{-1} = D^{-1}Z.$$

Hence, a symmetric update with a symplectic Gauss matrix L_k requires $4n + 8$ flops, taking the symmetry of Y into account.

An update with a symplectic Givens matrix

$$G_k = \begin{bmatrix} C & -S \\ S & C \end{bmatrix},$$

where $C = I + (c-1)e_k e_k^T$, $S = se_k e_k^T$ and $c^2 + s^2 = 1$, can be expressed as follows:

$$\begin{bmatrix} S_{11} & S_{12} \\ S_{21} & S_{22} \end{bmatrix}\begin{bmatrix} C & -S \\ S & C \end{bmatrix} = \begin{bmatrix} S_{11}C + S_{12}S & S_{11}C - S_{11}S \\ S_{21}C + S_{22}S & S_{22}C - S_{21}S \end{bmatrix}$$
$$= \begin{bmatrix} \tilde{S}_{11} & \tilde{S}_{12} \\ \tilde{S}_{21} & \tilde{S}_{22} \end{bmatrix}.$$

A lengthy derivation involving the use of the Sherman-Morrison-Woodbury formula in order to compute the inverse of $S_{11} + ((c-1)S_{11} + sS_{12})e_k e_k^T$ gives

$$\tilde{X} = -\tilde{S}_{21}\tilde{S}_{11}^{-1} = X - \frac{s}{\alpha}S_{11}^{-T}e_k e_k^T S_{11}^{-1},$$
$$\tilde{Y} = \tilde{S}_{11}^{-1}\tilde{S}_{12}$$
$$= Y + \frac{1}{\alpha}\left\{(2y_{kk}(c-1) - s)e_k e_k^T - (c-1)(Ye_k e_k^T + e_k e_k^T Y)\right.$$
$$\left. - sYe_k e_k^T Y\right\},$$
$$\tilde{Z} = \tilde{S}_{11}^{-1} = Z - \frac{1}{\alpha}\left\{(c-1)e_k e_k^T Z - sYe_k e_k^T Z\right\}$$

where $\alpha = c + sy_{kk}$. Hence, a symmetric update with a symplectic Givens matrix G_k requires $4n^2 + 9n + 15$ flops, taking the symmetry of X and Y into account.

For the initial reduction of $B = K^{-1}N$ to butterfly form or of $K - \lambda N$ to butterfly form $n^2 - n$ symplectic Givens transformations, $n - 1$ symplectic Gauss transformations, and 2 symplectic Householder transformations with $v \in \mathbf{R}^j$ for each $j = 2, \ldots, n - 1$ are used (for the reduction of the pencil, we only need to consider the transformations from the right). Hence, $5n^4$ arithmetic operations are needed in order to compute X, Y, and Z. The butterfly SR algorithm itself requires $4n - 4$ symplectic Givens transformations,

$n - 1$ symplectic Gauss transformations, $n - 2$ symplectic Householder transformation with $v \in \mathbf{R}^3$, and $n - 1$ symplectic Householder transformation with $v \in \mathbf{R}^2$; while the butterfly SZ algorithm requires $4n - 6$ symplectic Givens transformations, $n - 1$ symplectic Gauss transformations, $n - 3$ symplectic Householder transformation with $v \in \mathbf{R}^3$, and $n - 1$ symplectic Householder transformation with $v \in \mathbf{R}^2$. Therefore, the butterfly SR/SZ algorithm requires about $11n^4$ arithmetic operations for the symmetric updating. Hence, the symmetric updating requires $\mathcal{O}(n^4)$ arithmetic operations. This increase in computational cost is not rewarded with a significantly better computation of X.

For a discussion of a hybrid method for the numerical solution of discrete-time algebraic Riccati equations which combines the butterfly SZ algorithm and Newton's method see [52].

Chapter 5

THE SYMPLECTIC LANCZOS ALGORITHM

In the previous chapter algorithms for computing the eigenvalues of symplectic matrices have been considered that are based on an elimination process for computing the butterfly form of a symplectic matrix. Unfortunately, this approach is not suitable when dealing with large and sparse symplectic matrices as an elimination process can not make full use of the sparsity. The preparatory step of the SR algorithm involves the initial reduction of the (large and sparse) symplectic matrix to butterfly form. During this reduction process fill-in will occur such that the original sparsity pattern is destroyed. Moreover, in practise, one often does not have direct access to the large and sparse symplectic matrix M itself, one might only have access to the matrix-vector product Mx for any vector x. When considering large and sparse problems, often one is interested in computing only a few of the eigenvalues. The algorithms considered so far compute all $2n$ eigenvalues. Hence, the SR algorithms considered so far are not appropriate for large and sparse symplectic eigenproblems.

In Section 2.2.4 the Lanczos algorithm for general $n \times n$ matrices A was reviewed. It generates a sequence of tridiagonal matrices $T^{j,j} \in \mathbf{R}^{j \times j}$ with the property that the eigenvalues of $T^{j,j}$ are progressively better estimates of A's eigenvalues. Unfortunately, applying that algorithm to a symplectic matrix will ignore the symplectic structure. But, as will be shown in this section, it is fairly easy to derive a structure-preserving Lanczos-like algorithm for the symplectic eigenproblem: Given $s_1 \in \mathbf{R}^{2n}$ and a symplectic matrix $M \in \mathbf{R}^{2n \times 2n}$ the symplectic Lanczos-like algorithm generates a sequence of symplectic butterfly matrices $B^{2k,2k} \in \mathbf{R}^{2k \times 2k}$ such that (if no breakdown occurs)

$$MS^{2n,2k} = S^{2n,2k}B^{2k,2k} + r_{k+1}e_{2k}^T, \qquad (5.0.1)$$

where $S^{2n,2k} \in \mathbf{R}^{2n \times 2k}$, $S^{2n,2k}e_1 = s_1$, and the columns of $S^{2n,2k}$ are orthogonal with respect to the indefinite inner product defined by J as in (2.1.1). That is, the columns of $S^{2n,2k}$ are J–orthogonal. The eigenvalues of the intermediate matrices $B^{2k,2k}$ are progressively better estimates of M's eigenvalues. For $k = n$ the algorithm computes a symplectic matrix S such that S transforms M into butterfly form: $S^{-1}MS = B$.

Such a symplectic Lanczos algorithm was developed by Banse in [13] (using the approach of Freund and Mehrmann [55] for the development of a symplectic Lanczos method for the Hamiltonian eigenproblem). The algorithm is based on the factorization (3.2.15) of a symplectic butterfly matrix. Here we will develop an analogue of that symplectic Lanczos method for the factorization (3.2.7) of a symplectic butterfly matrix. Any such symplectic Lanczos method will suffer from the well-known numerical difficulties inherent to any Lanczos method for unsymmetric matrices. In [13], Banse picks up the look-ahead idea of Parlett, Taylor, and Liu [116]. A look-ahead Lanczos method skips over breakdowns and near-breakdowns. The price paid is that the resulting matrix is no longer of the desired reduced form, but has small bulges in that form to mark each occurrence of a (near) breakdown. Such a symplectic look-ahead Lanczos algorithm is presented in [13] which overcomes breakdown by giving up the strict butterfly form and, to a certain degree, also symplecticity. Unfortunately, so far there do not exist eigenvalue methods that can make use of that special reduced form. Standard eigenvalue methods such as QR or SR have to be employed resulting in a full (symplectic) matrix after only a few iteration steps.

A different approach to deal with the numerical difficulties of the Lanczos process is to modify the starting vectors by an implicitly restarted Lanczos process (see the fundamental work in [41, 132]; for the unsymmetric eigenproblem, the implicitly restarted Arnoldi method has been implemented very successfully, see [94]; for the Hamiltonian eigenproblem the method has been adapted in [18]). Usually only a small subset of the eigenvalues is desired. As the eigenvalues of the symplectic butterfly matrices $B^{2k,2k}$ are estimates for the eigenvalues of M, the length $2k$ symplectic Lanczos factorization (5.0.1) may suffice if the residual vector r_{k+1} is small. The idea of restarted Lanczos algorithms is to fix the number of steps in the Lanczos process at a prescribed value k which is dependent on the required number of approximate eigenvalues. The purpose of the implicit restart is to determine initial vectors such that the associated residual vectors are tiny. Given (5.0.1), an implicit Lanczos restart computes the Lanczos factorization

$$M\breve{S}^{2n,2k} = \breve{S}^{2n,2k}\breve{B}^{2k,2k} + \breve{r}_{k+1}e_{2k}^T$$

which corresponds to the starting vector

$$\breve{s}_1 = q(M)s_1$$

(where $q(M) \in \mathbf{R}^{2n \times 2n}$ is a Laurent polynomial) without explicitly restarting the Lanczos process with the vector $\breve{s}_1$. This process is iterated until the residual vector is tiny. J-orthogonality of the k Lanczos vectors is secured by re-J-orthogonalizing these vectors when necessary. Such an implicit restarting mechanism is derived in Section 5.2 analogous to the technique introduced in [67, 132]. It will be seen that Laurent polynomials should be used to drive the implicit restart.

The symplectic Lanczos method is presented in the following section. Further, that section is concerned with finding conditions for the symplectic Lanczos method terminating prematurely such that an invariant subspace associated with certain desired eigenvalues is obtained. We will also consider the important question of determining stopping criteria. An error analysis of the symplectic Lanczos algorithm in finite-precision arithmetic analogous to the analysis for the unsymmetric Lanczos algorithm presented by Bai [11] will be given. Numerical experiments show that, just like in the conventional Lanczos algorithm, information about the extreme eigenvalues tends to emerge long before the symplectic Lanczos process is completed. The effect of finite-precision arithmetic is discussed. Using Bai's work [11] on the unsymmetric Lanczos algorithm, an analog of Paige's theory [109] on the relationship between the loss of orthogonality among the computed Lanczos vectors and the convergence of a Ritz value is discussed. As to be expected, it follows that (under certain assumptions) the computed symplectic Lanczos vectors loose J-orthogonality when some Ritz values begin to converge. The implicitly restarted symplectic Lanczos method itself is derived in Section 5.2. Numerical properties of the proposed algorithm are discussed. In Section 5.3, we present some numerical examples. As expected, they demonstrate that re-J-orthogonalizing is necessary as the computed symplectic Lanczos vectors loose J-orthogonality when some Ritz values begin to converge. Moreover, the observed behavior of the implicitly restarted symplectic Lanczos algorithm corresponds to the reported behavior of the implicitly restarted Arnoldi method of Sorensen [132].

Some of the results discussed in this chapter appeared in [19, 17, 51].

For most of the discussion in this chapter, in order to simplify the notation, we use permuted versions of M, B, S, and J as in the previous chapter:

$$M_P = PMP^T, \quad B_P = PBP^T, \quad S_P = PSP^T, \quad J_P = PJP^T$$

with the permutation matrix P (2.1.2). For ease of reference, let us recall the definitions of B_P, $(K_u)_P^{-1}$, and $(N_u)_P^{-1}$. $(K_u)_P^{-1}$ and $(N_u)_P^{-1}$ are given by

$$(K_u)_P^{-1} = \begin{bmatrix} a_1^{-1} & b_1 & & & \\ 0 & a_1 & & & \\ & & \ddots & & \\ & & & a_n^{-1} & b_n \\ & & & 0 & a_n \end{bmatrix}, \tag{5.0.2}$$

and

$$(N_u)_P^{-1} = \begin{bmatrix} c_1 & 1 & d_2 & 0 & & & & \\ -1 & 0 & 0 & 0 & & & & \\ d_2 & 0 & c_2 & 1 & \ddots & & & \\ 0 & 0 & -1 & 0 & & \ddots & & \\ & & \ddots & & \ddots & & d_n & 0 \\ & & & \ddots & & \ddots & 0 & 0 \\ & & & & d_n & 0 & c_n & 1 \\ & & & & 0 & 0 & -1 & 0 \end{bmatrix}, \qquad (5.0.3)$$

while B_P is given by

$$\begin{bmatrix} b_1 & b_1c_1 - a_1^{-1} & 0 & b_1d_2 & & & \\ a_1 & a_1c_1 & 0 & a_1d_2 & & & \\ 0 & b_2d_2 & b_2 & b_2c_2 - a_2^{-1} & \ddots & & \\ 0 & a_2d_2 & a_2 & a_2c_2 & & \ddots & \\ & \ddots & & \ddots & & 0 & b_{n-1}d_n \\ & & & \ddots & & 0 & a_{n-1}d_n \\ & & 0 & b_nd_n & b_n & b_nc_n - a_n^{-1} \\ & & 0 & a_nd_n & a_n & a_nc_n \end{bmatrix}. \qquad (5.0.4)$$

5.1   THE SYMPLECTIC LANCZOS FACTORIZATION

We want to compute a symplectic matrix S in a Lanczos-like style such that S transforms the symplectic matrix M to a symplectic butterfly matrix B. In the permuted version, $MS = SB$ yields

$$M_P S_P = S_P B_P.$$

Equivalently, as $B = K_u^{-1}N_u$ (3.2.7), we can consider

$$M_P S_P (N_u)_P^{-1} = S_P (K_u)_P^{-1}.$$

The structure-preserving Lanczos method generates a sequence of permuted symplectic matrices $S_P^{2n,2k}$ (that is, the columns of $S_P^{2n,2k}$ are J_P–orthogonal) which we partition columnwise as

$$S_P^{2n,2k} = [v_1, w_1, v_2, w_2, \dots, v_k, w_k] \in \mathbf{R}^{2n \times 2k}.$$

These symplectic matrices satisfy

$$M_P S_P^{2n,2k} = S_P^{2n,2k} B_P^{2k,2k} + d_{k+1}(b_{k+1}v_{k+1} + a_{k+1}w_{k+1})e_{2k}^T. \qquad (5.1.5)$$

Equivalently, as

$$B_P^{2k,2k} = (K_u^{2k,2k})_P^{-1}(N_u^{2k,2k})_P$$

and

$$e_{2k}^T (N_u^{2k,2k})_P^{-1} = -e_{2k-1}^T,$$

we have

$$M_P S_P^{2n,2k} (N_u^{2k,2k})_P^{-1} \qquad\qquad (5.1.6)$$
$$= S_P^{2n,2k} (K_u^{2k,2k})_P^{-1} - d_{k+1}(b_{k+1}v_{k+1} + a_{k+1}w_{k+1})e_{2k-1}^T.$$

Here, $B_P^{2k,2k} = P^{2k,2k} B^{2k,2k} (P^{2k,2k})^T$ is a permuted $2k \times 2k$ symplectic butterfly matrix as in (5.0.4) and

$$(X_u^{2k,2k})_P = P^{2k,2k}(X_u^{2k,2k})(P^{2k,2k})^T, \qquad X = K, \text{ or } X = N,$$

is a permuted $2k \times 2k$ symplectic matrix of the form (5.0.2), resp. (5.0.3). The space spanned by the columns of

$$S^{2n,2k} = (P^{2n,2n})^T S_P^{2n,2k} P^{2k,2k}$$

is J–orthogonal, since

$$S_P^{2n,2k^T} J_P^{2n,2n} S_P^{2n,2k} = J_P^{2k,2k},$$

where $P^{2j,2j} J^{2j,2j} (P^{2j,2j})^T = J_P^{2j,2j}$ and $J^{2j,2j}$ is a $2j \times 2j$ matrix of the form (2.1.1).

The *residual vector*

$$r_{k+1} := b_{k+1}v_{k+1} + a_{k+1}w_{k+1}$$

is J_P–orthogonal to the columns of $S_P^{2n,2k}$ which are called *Lanczos vectors*. The matrix

$$B_P^{2k,2k} = J_P^{2k,2k}(S_P^{2n,2k})^T J_P^{2n,2n} M_P S_P^{2n,2k}$$

is the J_P–orthogonal projection of M_P onto the range of $S_P^{2n,2k}$. Equation (5.1.5) (resp. (5.1.6)) defines a *length $2k$ Lanczos factorization* of M_P. If the residual vector r_{k+1} is the zero vector, then equation (5.1.5) (resp. (5.1.6)) is called a *truncated Lanczos factorization* if $k < n$. Note that theoretically, r_{n+1} must vanish since

$$(S_P^{2n,2n})^T J_P^{2n,2n} r_{n+1} = 0,$$

and the columns of $S_P^{2n,2n}$ form a J_P–orthogonal basis for $\mathbf{R}^{2n}$. In this case the symplectic Lanczos method computes a reduction to butterfly form.

Before developing the symplectic Lanczos method itself, we state the following theorem which explains that the symplectic Lanczos factorization is completely specified by the starting vector v_1.

THEOREM 5.1 *Let two length 2k Lanczos factorizations be given by*

$$M_P S_P^{2n,2k} = S_P^{2n,2k} B_P^{2k,2k} + d_{k+1}(b_{k+1}v_{k+1} + a_{k+1}w_{k+1})e_{2k}^T,$$
$$M_P \widehat{S}_P^{2n,2k} = \widehat{S}_P^{2n,2k} \widehat{B}_P^{2k,2k} + \widehat{d}_{k+1}(\widehat{b}_{k+1}\widehat{v}_{k+1} + \widehat{a}_{k+1}\widehat{w}_{k+1})e_{2k}^T,$$

where $S_P^{2n,2k}$, $\widehat{S}_P^{2n,2k}$ *have* J_P*-orthogonal columns,* $B_P^{2k,2k}$, $\widehat{B}_P^{2k,2k}$ *are permuted unreduced symplectic butterfly matrices with*

$$(B_P^{2k,2k})_{jj} = (\widehat{B}_P^{2k,2k})_{jj} = 1,$$
$$|(B_P^{2k,2k})_{j+1,j}| = |(\widehat{B}_P^{2k,2k})_{j+1,j}| = 1,$$

for $j = 1, 3, 5, \ldots, 2k - 1$,

$$\mathrm{sign}((B_P^{2k,2k})_{j+1,j-1}) = \mathrm{sign}((\widehat{B}_P^{2k,2k})_{j+1,j-1}) = 1,$$

for $j = 3, 5, \ldots, 2k - 1$, *and*

$$0 = J_P^{2k,2k}(S_P^{2n,2k})^T J_P(b_{k+1}v_{k+1} + a_{k+1}w_{k+1}),$$
$$0 = J_P^{2k,2k}(\widehat{S}_P^{2n,2k})^T J_P(\widehat{b}_{k+1}\widehat{v}_{k+1} + \widehat{a}_{k+1}\widehat{w}_{k+1}).$$

If the first column of $S_P^{2n,2k}$ *and* $\widehat{S}_P^{2n,2k}$ *are equal, then*

$$B_P^{2k,2k} = \widehat{B}_P^{2k,2k}, \qquad S_P^{2n,2k} = \widehat{S}_P^{2n,2k},$$

and

$$d_{k+1}(b_{k+1}v_{k+1} + a_{k+1}w_{k+1}) = \widehat{d}_{k+1}(\widehat{b}_{k+1}\widehat{v}_{k+1} + \widehat{a}_{k+1}\widehat{w}_{k+1}).$$

PROOF: This is a direct consequence of Theorem 3.7 c) and Remark 3.9. $\checkmark$

Next we will see how the factorization (5.1.5) (resp. (5.1.6)) may be computed. As this reduction is strongly dependent on the first column of the transformation matrix that carries out the reduction and might not exist, we must expect breakdown or near-breakdown in the Lanczos process. Assuming for the moment that no such breakdowns occur, a symplectic Lanczos method can be derived as follows.

Let $S_P = [v_1, w_1, v_2, w_2, \ldots, v_n, w_n]$. For a given vector v_1, our symplectic Lanczos method constructs the matrix S_P columnwise from the equations

$$M_P S_P (N_u)_P^{-1} e_j = S_P (K_u)_P^{-1} e_j, \qquad j = 1, 2, \ldots .$$

That is, for even numbered columns we have

$$M_P v_m = b_m v_m + a_m w_m.$$

This implies

$$
\begin{aligned}
a_m w_m &= M_P v_m - b_m v_m \\
&=: \tilde{w}_m.
\end{aligned}
\tag{5.1.7}
$$

Similar, for odd numbered columns we have

$$a_m^{-1} v_m = M_P(d_m v_{m-1} + c_m v_m - w_m + d_{m+1} v_{m+1})$$

which implies

$$
\begin{aligned}
d_{m+1} v_{m+1} &= -d_m v_{m-1} - c_m v_m + w_m + a_m^{-1} M_P^{-1} v_m \\
&=: \tilde{v}_{m+1}.
\end{aligned}
\tag{5.1.8}
$$

Note that $M_P^{-1} = -J_P M_P^T J_P$, since M is symplectic. Thus $M_P^{-1} v_m$ is just a matrix-vector-product with the transpose of M_P.

Now we have to choose the parameters a_m, b_m, c_m, d_{m+1} such that

$$S_P^T J_P S_P = J_P$$

is satisfied. That is, we have to choose the parameters such that

$$v_{m+1}^T J_P w_{m+1} = 1.$$

One possibility is to choose

$$
\begin{aligned}
d_{m+1} &= \|\tilde{v}_{m+1}\|_2, \\
a_{m+1} &= v_{m+1}^T J_P M_P v_{m+1}.
\end{aligned}
$$

Premultiplying $\tilde{v}_{m+1}$ by $w_m^T J_P$ and using $S_P^T J_P S_P = J_P$ yields

$$c_m = -a_m^{-1} w_m^T J_P M_P^{-1} v_m = a_m^{-1} v_m^T J_P M_P w_m.$$

Thus we obtain the algorithm given in Table 5.1.

REMARK 5.2 *Using the derived formulae (5.1.7) for w_m, the residual term r_{k+1} can be expressed as*

$$r_{k+1} = M_P v_{k+1}.$$

There is still some freedom in the choice of the parameters that occur in this algorithm. Essentially, the parameters b_m can be chosen freely. Here we set $b_m = 1$. Likewise a different choice of the parameters a_m, d_m is possible.

REMARK 5.3 *Choosing $b_m = 0$, a different interpretation of the algorithm in Table 5.1 can be given. The resulting butterfly matrix $B = S^{-1}MS$ is of the form*

$$
\begin{bmatrix} 0 & -A \\ A & T \end{bmatrix} = \begin{bmatrix} 0 & \diagdown \\ \diagdown & \diagdown\diagdown \end{bmatrix},
$$

Algorithm: Symplectic Lanczos method

Given an initial vector $\tilde{v}_1 \in \mathbf{R}^{2n}, \tilde{v}_1 \neq 0$ and a symplectic matrix $M \in \mathbf{R}^{2n \times 2n}$, this algorithm computes the parameters $a_1, \ldots, a_m, b_1, \ldots, b_m,$ $c_1, \ldots, c_m, d_2, \ldots, d_m$ that determine a $2m \times 2m$ symplectic butterfly matrix $B^{2m,2m}$ with the property $\sigma(B^{2m,2m}) \subset \sigma(M)$ if breakdown is caused by $d_{m+1} = 0$. How to interpret the results obtained, if the breakdown is caused by $a_m = 0$, is discussed in the text.

> $v_0 = 0 \in \mathbf{R}^{2n}$
> $d_1 = \|\tilde{v}_1\|_2$
> $v_1 = \frac{1}{d_1}\tilde{v}_1$
> **for m = 1, 2, ... until breakdown do**
>   (update of w_m)
> $\tilde{w}_m = M_P v_m - b_m v_m$
> $a_m = v_m^T J_P M_P v_m$
> $w_m = \frac{1}{a_m}\tilde{w}_m$
>   (computation of c_m)
> $c_m = a_m^{-1} v_m^T J_P M_P w_m$
>   (update of v_{m+1})
> $\tilde{v}_{m+1} = -d_m v_{m-1} - c_m v_m + w_m + a_m^{-1} M_P^{-1} v_m$
> $d_{m+1} = \|\tilde{v}_{m+1}\|_2$
> $v_{m+1} = \frac{1}{d_{m+1}}\tilde{v}_{m+1}$
> **end**

Table 5.1.   Symplectic Lanczos Method

---

where A is a diagonal matrix and T is an unsymmetric tridiagonal matrix. As $S^{-1}MS = B$, *we have* $S^{-1}M^{-1}S = B^{-1}$ *and*

$$S^{-1}(M + M^{-1})S = B + B^{-1} = \begin{bmatrix} T^T & 0 \\ 0 & T \end{bmatrix}.$$

Obviously there is no need to compute both T and T^T. It is sufficient to compute the first n columns of S. This corresponds to computing the v_m in our algorithm. This case is not considered here any further. See also the discussion in Section 4.5.1.

Note that M_P is not altered during the entire Lanczos process. As $M_P^{-1} = -J_P M_P^T J_P$, the algorithm can be rewritten such that only one matrix-vector product is required for each computed Lanczos vector w_m or v_m. Thus an efficient implementation of this algorithm requires $6n + (4nz + 32n)k$ flops, where nz is the number of nonzero elements in M_P and $2k$ is the number of

Lanczos vectors computed (that is, the loop is executed k times). The algorithm as given in Table 5.1 computes an odd number of Lanczos vectors, for a practical implementation one has to omit the computation of the last vector v_{k+1} (or one has to compute an additional vector w_{k+1}).

In the symplectic Lanczos method as given above we have to divide by parameters that may be zero or close to zero. If the normalization parameter d_{m+1} is zero, the corresponding vector $\tilde{v}_{m+1}$ is the zero vector. In this case, a J_P–orthogonal invariant subspace of M_P or equivalently, a symplectic invariant subspace of M is detected. By redefining $\tilde{v}_{m+1}$ to be any vector satisfying

$$
\begin{aligned}
v_j^T J_P \tilde{v}_{m+1} &= 0, \\
w_j^T J_P \tilde{v}_{m+1} &= 0,
\end{aligned}
$$

for $j = 1, \ldots, m$, the algorithm can be continued. The resulting butterfly matrix is no longer unreduced; the eigenproblem decouples into two smaller subproblems. In case $d_{m+1} \approx 0$, a good approximation to a symplectic invariant subspace of M may have been found (if $\|S_P^{2n,2m}\|$ is large, then d_{m+1} can not be trusted, see Section 5.1.2 for a discussion); then one can proceed as described above. In case $\tilde{w}_m$ is zero (close to zero), an invariant subspace of M_P with dimension $2m - 1$ is found (may be found). From (5.1.7) it is easy to see that the parameter a_m will be (close to) zero if $\tilde{w}_m = 0$ ($\tilde{w}_m \approx 0$). We further obtain from (5.1.7) that in this case $M_P v_m = b_m v_m$, i.e., b_m is an eigenvalue of M_P with corresponding eigenvector v_m. (In case $\tilde{w}_m \approx 0$, we have $M_P v_m \approx b_m v_m$). Due to the symmetry of the spectrum of M, we also have that $1/b_m$ is an eigenvalue of M. Computing an eigenvector y of M_P corresponding to $1/b_m$, we can try to augment the $(2m - 1)$–dimensional invariant subspace to an M_P–invariant subspace of even dimension. If this is possible, the space can be made J_P–orthogonal by J_P–orthogonalizing y against $\{ v_1, w_1, \ldots, v_{m-1}, w_{m-1} \}$ and normalizing such that $y^T J_P v_m = 1$.

Thus, if either v_{m+1} or w_{m+1} vanishes, the breakdown is benign. If $v_{m+1} \neq 0$ and $w_{m+1} \neq 0$ but $a_{m+1} = 0$, then the breakdown is *serious*. No reduction of the symplectic matrix to a symplectic butterfly matrix with v_1 as first column of the transformation matrix exists. On the other hand, an initial vector v_1 exists so that the symplectic Lanczos process does not encounter serious breakdown. However, determining this vector requires knowledge of the minimal polynomial of M. Thus, no algorithm for successfully choosing v_1 at the start of the computation yet exists.

Furthermore, in theory, the above recurrences for v_m and w_m are sufficient to guarantee the J–orthogonality of theses vectors. Yet, in practice, the J–orthogonality will be lost, re–J–orthogonalization is necessary, increasing the computational cost significantly. We re–J_P–orthogonalize each symplectic

Lanczos vector as soon as it is computed against the previous ones via

$$w_m = w_m + S_P^{2n,2m-2} J_P^{2m-2,2m-2} S_P^{2n,2m-2^T} J_P^{2n,2n} w_m,$$

$$v_{m+1} = v_{m+1} + S_P^{2n,2m} J_P^{2m,2m} S_P^{2n,2m^T} J_P^{2n,2n} v_{m+1}.$$

A different way to write this re–J_P–orthogonalization is

$$w_m \leftarrow w_m + \sum_{j=1}^{m-1} (<v_j, w_m>_{J_P} w_j - <w_j, w_m>_{J_P} v_j),$$

$$v_{m+1} \leftarrow v_{m+1} + \sum_{j=1}^{m} (<v_j, v_{m+1}>_{J_P} w_j - <w_j, v_{m+1}>_{J_P} v_j),$$

where for $x, y \in \mathbf{R}^{2n}$, $<x, y>_{J_P} := x^T J_P^{2n,2n} y$ defines the indefinite inner product implied by $J_P^{2n,2n}$.

This re–J_P–orthogonalization is costly, it requires $16n(m-1)$ flops for the vector w_m and $16nm$ flops for v_{m+1}. Thus, if $2k$ Lanczos vectors $v_1, w_1, \ldots, v_k, w_k$ are computed, the re–J_P–orthogonalization adds a computational cost of the order of $16nk^2$ flops to the overall cost of the symplectic Lanczos method. For standard Lanczos algorithms, different reorthogonalization techniques have been studied (for references see, e.g., [58]). Those ideas can be used to design analogous re–J_P–orthogonalizations for the symplectic Lanczos method. It should be noted that if k is small, the cost for re–J_P–orthogonalization is not too expensive.

REMARK 5.4 *The discussion given so far (and in the following) assumes that we are interested in solving a standard symplectic eigenproblem $Mx = \lambda x$. But, as for any Lanczos-like algorithm, it is easy to modify the algorithm such that a generalized symplectic eigenvalue problem $Kx = \lambda Nx$ can be tackled. This implies that for each symplectic Lanczos vector a linear system of equations has to be solved. See Example 5.22 for a discussion.*

REMARK 5.5 *The usual unsymmetric Lanczos algorithm generates two sequences of vectors. Recall the unsymmetric Lanczos algorithm as discussed in Section 2.2.4: Given $p_1, q_1 \in \mathbf{R}^n$ and an unsymmetric matrix $A \in \mathbf{R}^{n \times n}$, the standard unsymmetric Lanczos algorithm produces matrices $P^{n,k} = [p_1, \ldots, p_k] \in \mathbf{R}^{n \times k}$ and $Q^{n,k} = [q_1, \ldots, q_k] \in \mathbf{R}^{n \times k}$ which satisfy*

$$(P^{n,k})^T Q^{n,k} = I^{k,k},$$

and

$$AQ^{n,k} = Q^{n,k} T^{k,k} + \beta_{k+1} q_{k+1} e_k^T, \qquad (5.1.9)$$

$$A^T P^{n,k} = P^{n,k} (T^{k,k})^T + \gamma_{k+1} p_{k+1} e_k^T, \qquad (5.1.10)$$

where $T^{k,k}$ is an unsymmetric tridiagonal matrix.

Adapted to the situation considered here, this implies that the symplectic Lanczos process should have been stated as follows: Given $v_1, t_1 \in \mathbf{R}^{2n}$ and a symplectic matrix $M \in \mathbf{R}^{2n \times 2n}$, the symplectic Lanczos algorithm produces matrices $S_P^{2n,2k} = [v_1, \ w_1, \ \ldots, \ v_k, \ w_k] \in \mathbf{R}^{2n \times 2k}$ and $W_P^{2n,2k} = [t_1, \ldots, t_{2k}] \in \mathbf{R}^{2n \times 2k}$ with J_P–orthogonal columns which satisfy

$$(W_P^{2n,2k})^T S_P^{2n,2k} = I^{2k,2k},$$

and

$$M_P S_P^{2n,2k} = S_P^{2n,2k} B_P^{2k,2k} + d_{k+1} r_{k+1} e_{2k}^T,$$
$$M_P^T W_P^{2n,2k} = W_P^{2n,2k} (B_P^{2k,2k})^T + d_{k+1} \check{r}_{k+1} e_{2k}^T.$$

As S_P is symplectic, we obtain from $(W_P^{2n,2k})^T S_P^{2n,2k} = I^{2k,2k}$ that

$$W_P^{2n,2k} = J_P^{2n,2n} S_P^{2n,2k} J_P^{2k,2k} = [-J_P w_1, \ J_P v_1, \ \ldots, \ -J_P w_k, \ J_P v_k].$$

Moreover,

$$r_{k+1} = M_P v_{k+1}, \quad \text{and} \quad \check{r}_{k+1} = J_P v_{k+1}.$$

Substituting the expressions for $W_P^{2n,2k}$ and $\check{r}_{k+1}$ into the second recursion equation and pre- and postmultiplying with J_P yields that the two recursions are equivalent. Hence one of the two sequences can be eliminated here and thus work and storage can essentially be halved. (This property is valid for a broader class of matrices, see [56].)

The numerical difficulties of the symplectic Lanczos method described above are inherent to all Lanczos-like methods for unsymmetric matrices. Different approaches to overcome these difficulties have been proposed. Taylor [134] and Parlett, Taylor, and Liu [116] were the first to propose a look-ahead Lanczos algorithm that skips over breakdowns and near-breakdowns. Freund, Gutknecht, and Nachtigal present in [57] a look-ahead Lanczos code that can handle look-ahead steps of any length. Banse adapted this method to the symplectic Lanczos method given in [13]. The price paid is that the resulting matrix is no longer of butterfly form, but has a small bulge in the butterfly form to mark each occurrence of a (near) breakdown. Unfortunately, so far there exists no eigenvalue method that can make use of that special reduced form.

A different approach to deal with the numerical difficulties of Lanczos-like algorithms is to implicitly restart the symplectic Lanczos factorization. This was first introduced by Sorensen [132] in the context of unsymmetric matrices and the Arnoldi process. Usually only a small subset of the eigenvalues is desired. As the eigenvalues of the symplectic butterfly matrices $B^{2k,2k}$ are estimates for the eigenvalues of M, the length $2k$ symplectic Lanczos factorization (5.1.5) may suffice if the residual vector r_{k+1} is small. The idea of restarted Lanczos algorithms is to fix the number of steps in the Lanczos process at a prescribed value k which is dependent on the required number of approximate eigenvalues. The purpose of the implicit restart is to determine initial vectors such that the associated residual vectors are tiny. Given (5.1.5), an implicit Lanczos restart computes the Lanczos factorization

$$M \breve{S}^{2n,2k} = \breve{S}^{2n,2k} \breve{B}^{2k,2k} + \breve{r}_{k+1}e_{2k}^T$$

which corresponds to the starting vector

$$\breve{s}_1 = q(M)s_1$$

(where $q(M) \in \mathbf{R}^{2n\times 2n}$ is a Laurent polynomial) without having to explicitly restart the Lanczos process with the vector $\breve{s}_1$. This process is iterated until the residual vector r_{k+1} is tiny. J–orthogonality of the k Lanczos vectors is secured by re–J–orthogonalizing these vectors when necessary. This idea will be investigated in Section 5.2.

5.1.1   TRUNCATED SYMPLECTIC LANCZOS FACTORIZATIONS

This section is concerned with finding conditions for the symplectic Lanczos method terminating prematurely. This is a welcome event since in this case we have found an invariant symplectic subspace $S^{2n,2k}$ and the eigenvalues of $B^{2k,2k}$ are a subset of those of M. We will first discuss the conditions under which the residual vector of the symplectic Lanczos factorization will vanish at some step k. Then we will show how the residual vector and the starting vector are related. Finally a result indicating when a particular starting vector generates an exact truncated factorization is given.

First the conditions under which the residual vector of the symplectic Lanczos factorization will vanish at some step k will be discussed. From the derivation of the algorithm it is immediately clear that if no breakdown occurs, then

$$\text{span}\{v_1,\ldots,v_{k+1},w_1,\ldots,w_k\}$$
$$= \text{span}\{v_1,M_P^{-1}v_1,\ldots,M_P^{-k}v_1,M_Pv_1,\ldots,M_P^kv_1\}$$
$$= \text{span}\{\mathcal{L}(M_P,v_1,k)\cup\{M_P^{-k}v_1\}\},$$
$$\text{span}\{v_1,\ldots,v_{k+1},w_1,\ldots,w_{k+1}\}$$
$$= \text{span}\{v_1,M_P^{-1}v_1,\ldots,M_P^{-k}v_1,M_Pv_1,\ldots,M_P^{k+1}v_1\}$$
$$= \mathcal{L}(M_P,v_1,k+1),$$

where

$$\mathcal{L}(X, v, k) = \text{span}\{v, X^{-1}v, X^{-2}v \ldots, X^{-(k-1)}v, Xv, X^2v, \ldots, X^kv\}$$

defines the linear space spanned by the columns of the generalized Krylov matrix $L(X, v, k)$ (see Definition 2.3). If $\dim(\mathcal{L}(M_P, v_1, n)) = 2n$, then there will be no breakdown in the symplectic Lanczos process. The first zero residuum will be $r_{n+1} = 0$. What happens if $\dim(\mathcal{L}(M_P, v_1, n)) < 2n$? First let us note that in that case, extending the generalized Krylov sequence will not increase the dimension of the generalized Krylov space.

LEMMA 5.6 *If*

$$\dim(\mathcal{L}(M_P, v_1, k)) = d < 2k,$$

then

$$\dim(\mathcal{L}(M_P, v_1, j)) = d$$

for all $j > k$.

Let us assume that

$$\dim(\mathcal{L}(M_P, v_1, k)) = 2k,$$

and

$$\dim(\mathcal{L}(M_P, v_1, k + 1)) < 2k + 2.$$

If $\dim(\mathcal{L}(M_P, v_1, k + 1)) = 2k + 1$, then

$$M_P v_{k+1} \in \text{span}\{v_1, \ldots, v_{k+1}, w_1, \ldots, w_k\}.$$

Hence, there exist real scalars $\alpha_1, \ldots, \alpha_{k+1}$ and $\beta_1, \ldots, \beta_k$ such that

$$M_P v_{k+1} = \alpha_1 v_1 + \ldots + \alpha_{k+1} v_{k+1} + \beta_1 w_1 + \ldots + \beta_k w_k.$$

Using the definition of a_{k+1} as given in Table 5.1 and the above expression we obtain due to J–orthogonality,

$$\begin{aligned} a_{k+1} &= v_{k+1}^T J_P M_P v_{k+1} \\ &= \alpha_1 v_{k+1}^T J_P v_1 + \ldots + \alpha_{k+1} v_{k+1}^T J_P v_{k+1} \\ &\quad + \beta_1 v_{k+1}^T J_P w_1 + \ldots + \beta_k v_{k+1}^T J_P w_k \\ &= 0. \end{aligned}$$

As $\widetilde{w}_{k+1} = a_{k+1} w_{k+1} = M_P v_{k+1} - b_{k+1} v_{k+1}$ (see (5.1.7)) it follows that

$$\widetilde{w}_{k+1} = 0.$$

This implies that an invariant subspace of M_P of dimension $2k + 1$ is found.

If $\dim(\mathcal{L}(M_P, v_1, k+1)) = 2k$, then

$$M_P^{-1} v_k \in \text{span}\{v_1, \ldots, v_k, w_1, \ldots, w_k\}.$$

Hence

$$a_k^{-1} M_P^{-1} v_k = \alpha_1 v_1 + \ldots + \alpha_k v_k + \beta_1 w_1 + \ldots + \beta_k w_k,$$

for properly chosen α_j, β_j and from the algorithm in Table 5.1

$$
\begin{aligned}
\tilde{v}_{k+1} &= \alpha_1 v_1 + \ldots + \alpha_{k-2} v_{k-2} + (\alpha_{k-1} - d_k) v_{k-1} + (\alpha_k - c_k) v_k \\
&\quad + \beta_1 w_1 + \ldots + \beta_{k-1} w_{k-1} + (\beta_k + 1) w_k.
\end{aligned}
$$

Since $[v_1, w_1, \ldots, v_k, w_k]^T J_P \tilde{v}_{k+1} = [0, \ldots, 0]$ we obtain for $j < k$ and $\ell < k - 2$

$$
\begin{aligned}
v_j^T J_P \tilde{v}_{k+1} &= \beta_j v_j^T J_P w_j = \beta_j = 0, \\
v_k^T J_P \tilde{v}_{k+1} &= (\beta_k + 1) v_k^T J_P w_k = \beta_k + 1 = 0, \\
w_\ell^T J_P \tilde{v}_{k+1} &= \alpha_\ell w_\ell^T J_P v_\ell = -\alpha_\ell = 0, \\
w_{k-1}^T J_P \tilde{v}_{k+1} &= (\alpha_{k-1} - d_k) w_{k-1}^T J_P v_{k-1} = d_k - \alpha_{k-1} = 0, \\
w_k^T J_P \tilde{v}_{k+1} &= (\alpha_k - c_k) w_k^T J_P v_k = c_k - \alpha_k = 0.
\end{aligned}
$$

Therefore

$$\tilde{v}_{k+1} = 0,$$

and further

$$d_{k+1} = 0.$$

This implies that the residual vector of the symplectic Lanczos factorization will vanish at the first step k such that the dimension of $\mathcal{L}(M, v_1, k+1)$ is equal to $2k$ and hence is guaranteed to vanish for some $k \leq n$.

The above discussion is summarized in the following proposition.

PROPOSITION 5.7 *Assume that* $\dim(\mathcal{L}(M_P, v_1, k)) = 2k.$

a) *If* $\dim(\mathcal{L}(M_P, v_1, k+1)) = 2k + 1$, *then* $\tilde{w}_{k+1} = 0$.

b) *If* $\dim(\mathcal{L}(M_P, v_1, k+1)) = 2k$, *then* $d_{k+1} = 0$.

Hence, in any case an invariant subspace of M_P is found. For

$$\dim(\mathcal{L}(M_P, v_1, k+1)) = 2k$$

the residual vector r_{k+1} will vanish.

Now we will discuss the relation between the residual term and the starting vector. As the discussion states the result for the not permuted symplectic

Lanczos recursion, we use $\widehat{v}_1$ denote the back permuted first column v_1 from $S_P^{2n,2k}$. If $\dim(\mathcal{L}(M, \widehat{v}_1, n)) = 2n$ then

$$ML(M, \widehat{v}_1, n) = L(M, \widehat{v}_1, n)C^{2n,2n},$$

where $C^{2n,2n}$ is a generalized companion matrix of the form

$$
C^{2n,2n} =
\begin{bmatrix}
0 & 1 & & & & c_1 \\
& \ddots & \ddots & & & \vdots \\
& & \ddots & 1 & & \vdots \\
& & & 0 & & c_n \\
\hline
1 & & & 0 & & c_{n+1} \\
& & & 1 & \ddots & \vdots \\
& & & & \ddots & 0 & c_{2n-1} \\
& & & & & 1 & c_{2n}
\end{bmatrix}
$$

(see proof of Theorem 3.7). Thus

$$
\begin{aligned}
ML(M, \widehat{v}_1, k) &= L(M, \widehat{v}_1, k)C^{2k,2k} \\
&\quad + (M^{k+1}\widehat{v}_1 - L(M, \widehat{v}_1, k)C^{2k,2k}e_{2k})e_{2k}^T.
\end{aligned}
\tag{5.1.11}
$$

Define the residual in (5.1.11) by

$$f_{k+1} := M^{k+1}\widehat{v}_1 - L(M, \widehat{v}_1, k)C^{2k,2k}e_{2k}.$$

Note that

$$f_{k+1} = p_k(M)\widehat{v}_1, \tag{5.1.12}$$

where

$$p_k(\lambda) := \lambda^{k+1} - \sum_{j=0}^{k-1}(c_{k+j+1}\lambda^{j+1} + c_{j+1}\lambda^{-j}).$$

We will now show that f_{k+1} is up to scaling the residual of the length $2k$ symplectic Lanczos iteration with starting vector $\widehat{v}_1$. Together with (5.1.12) this reveals the relation between residual and starting vectors.

Since

$$\det(C^{2k,2k} - \lambda I) = \lambda^{2k} - \sum_{j=0}^{k-1}(c_{k-j}\lambda^j + c_{k+j+1}\lambda^{k+j}),$$

we obtain

$$p_k(\lambda) = \lambda^{-(k-1)}\det(C^{2k,2k} - \lambda I).$$

Let $L(M, \widehat{v}_1, k) = S^{2n,2k} R$ where $S^{2n,2k} \in \mathbf{R}^{2n \times 2k}$ with J–orthogonal columns (that is $(S^{2n,2k})^T J^{2n,2n} S^{2n,2k} = J^{2k,2k}$) and $R \in \mathbf{R}^{2k \times 2k}$ is a J–triangular matrix. Then $S^{2n,2k} e_1 = \widehat{v}_1$. The diagonal elements of R are nonzero if and only if the columns of $L(M, \widehat{v}_1, k)$ are linear independent. Choosing

$$c = \begin{bmatrix} c_1 \\ \vdots \\ c_{2k} \end{bmatrix} = R^{-1}(-J^{2k,2k}(S^{2n,2k})^T J^{2n,2n}) M^{k+1} \widehat{v}_1$$

assures that $(-J^{2k,2k}(S^{2n,2k})^T J^{2n,2n}) f_{k+1} = 0$. Now multiplying (5.1.11) from the right by R^{-1} yields

$$M L(M, \widehat{v}_1, k) R^{-1} - L(M, \widehat{v}_1, k) C^{2k,2k} R^{-1} = f_{k+1} e_{2k}^T R^{-1}.$$

This implies

$$M S^{2n,2k} - S^{2n,2k} B = f_{k+1} e_{2k}^T / r_{2k,2k}, \qquad (5.1.13)$$

where $B = R C^{2k,2k} R^{-1}$ is an unreduced butterfly matrix (see proof of Theorem 3.7) with the same characteristic polynomial as $C^{2k,2k}$. Equation (5.1.13) is a valid symplectic Lanczos recursion with starting vector $\widehat{v}_1 = S^{2n,2k} e_1$ and residual vector $f_{k+1}/r_{2k,2k}$. By (5.1.12) and due to the essential uniqueness of the symplectic Lanczos recursion any symplectic Lanczos recursion with starting vector $\widehat{v}_1$ yields a residual vector that can be expressed as a polynomial in M times the starting vector $\widehat{v}_1$.

REMARK 5.8 *From (5.1.12) it follows that if* $\dim(\mathcal{L}(M, \widehat{v}_1, k+1)) \leq 2k$, *then we can choose* $c_1, \ldots, c_{2k}$ *such that* $f_{k+1} = 0$. *Hence, if the generalized Krylov subspace* $\mathcal{L}(M, \widehat{v}_1, k + 1)$ *forms a 2k–dimensional M–invariant subspace, the residual of the symplectic Lanczos recursion will be zero after k Lanczos steps such that the columns of* $S^{2n,2k}$ *span a symplectic basis for the subspace* $\mathcal{L}(M, \widehat{v}_1, k + 1)$.

The final result of this section will give necessary and sufficient conditions for a particular starting vector to generate an exact truncated factorization. This is desirable since then the columns of $S^{2n,2k}$ form a basis for an invariant symplectic subspace of M and the eigenvalues of $B^{2k,2k}$ are a subset of those of M. As the theorem states the result for the not permuted symplectic Lanczos recursion, we use $\widehat{v}_{k+1}$ and $\widehat{w}_{k+1}$ to denote the back permuted columns v_{k+1} and w_{k+1} from $S_P^{2n,2k}$.

THEOREM 5.9 *Let*

$$M S^{2n,2k} - S^{2n,2k} B^{2k,2k} = d_{k+1}(b_{k+1} \widehat{v}_{k+1} + a_{k+1} \widehat{w}_{k+1}) e_{2k}^T$$

be the symplectic Lanczos factorization after k steps, with $B^{2k,2k}$ unreduced. Then $d_{k+1} = 0$ if and only if $\widehat{v}_1 = Xy$ where $MX = XY$ with rank $(X) = 2k$ and Y is a Jordan matrix of order $2k$.

PROOF: If $d_{k+1} = 0$, let $B^{2k,2k}\widetilde{X} = \widetilde{X}Y$ be the Jordan canonical form of $B^{2k,2k}$ and put $X = S^{2n,2k}\widetilde{X}$. Then

$$MX = S^{2n,2k}B^{2k,2k}\widetilde{X} = S^{2n,2k}\widetilde{X}Y = XY$$

and

$$\widehat{v}_1 = S^{2n,2k}e_1 = S^{2n,2k}\widetilde{X}\widetilde{X}^{-1}e_1 = Xy$$

with $y = \widetilde{X}^{-1}e_1$.

Suppose now that $MX = XY$, rank$(X) = 2k$, and $\widehat{v}_1 = Xy$. Then $M^m X = XY^{2m,2m}$ for $m \in \mathbf{N}$ and it follows that

$$M^m\widehat{v}_1 = M^m Xy = XY^{2m,2m}y \in \text{ran}(X)$$

for $m \in \mathbf{N}$. Hence by Lemma 5.6, $\dim(\mathcal{L}(M,\widehat{v}_1, k+1)) \le \text{rank}(X) = 2k$. Since $B^{2k,2k}$ is unreduced, $\dim(\mathcal{L}(M,\widehat{v}_1, j)) = 2j$ for $j = 1, ..., k$. This implies $\dim(\mathcal{L}(M,\widehat{v}_1, k+1)) = 2k$ and therefore, $d_{k+1} = 0$. $\checkmark$

A similar result may be formulated in terms of symplectic Schur vectors instead of generalized eigenvectors. It is known (see, e.g., Theorem 2.12) that for any symplectic matrix $M \in \mathbf{R}^{2n \times 2n}$ which has no eigenvalues of modulus 1, there exists an orthogonal and symplectic matrix Q such that

$$Q^T M Q = \begin{bmatrix} T & N \\ 0 & T^{-T} \end{bmatrix}, \qquad T, N \in \mathbf{R}^{n \times n}, \tag{5.1.14}$$

where T is quasi upper triangular. Q can be chosen such that T has only eigenvalues inside the (closed) unit circle. Such a symplectic Schur decomposition exists for a broader class of symplectic matrices, see Theorem 2.12.

THEOREM 5.10 *Let M be a symplectic matrix having a symplectic Schur decomposition as in Theorem 2.12 a). Let*

$$MS^{2n,2k} - S^{2n,2k}B^{2k,2k} = d_{k+1}(b_{k+1}\widehat{v}_{k+1} + a_{k+1}\widehat{w}_{k+1})e_{2k}^T$$

be the symplectic Lanczos factorization after k steps, with $B^{2k,2k}$ unreduced. Then $d_{k+1} = 0$ if and only if $\widehat{v}_1 = Q^{2k}y$ where

$$MQ^{2n,2k} = Q^{2n,2k}\begin{bmatrix} T & N \\ 0 & T^{-T} \end{bmatrix} = Q^{2n,2k}R$$

with $(Q^{2n,2k})^T Q^{2n,2k} = I^{2k,2k}$, *the columns of* $Q^{2n,2k}$ *are* J–*orthogonal, and* T *quasi upper triangular of order* k.

PROOF: If $d_{k+1} = 0$, then $MS^{2n,2k} = S^{2n,2k} B^{2k,2k}$. Let $B^{2k,2k} Z = ZR$ be a real symplectic Schur decomposition where $Z \in \mathbf{R}^{2k,2k}$ is orthogonal and symplectic and R is of the form (5.1.14). Then

$$\widehat{v}_1 = S_P^{2n,2k} e_1 = S_P^{2n,2k} ZZ^T e_1 =: Q^{2n,2k} y$$

where $y = Z^T e_1$ and $Q^{2n,2k} = S_P^{2n,2k} Z \in \mathbf{R}^{2n \times 2k}$. Note that $MQ^{2n,2k} = Q^{2n,2k} R$.

Suppose now that $MQ^{2n,2k} = Q^{2n,2k} R$ with $(Q^{2n,2k})^T Q^{2n,2k} = I^{2k,2k}$, the columns of $Q^{2n,2k}$ are J–orthogonal and R is of the form (5.1.14). Let $\widehat{v}_1 = Q^{2n,2k} y$ with $y \in \mathbf{R}^{2n,2k}$ arbitrary. Now, for any $m \in \mathbf{N}$, $M^m Q^{2n,2k} = Q^{2n,2k} R^m$ and thus

$$M^m \widehat{v}_1 = M^m Q^{2n,2k} y = Q^{2n,2k} R^m y \in \operatorname{ran}(Q^{2n,2k}).$$

Hence $\dim(\mathcal{L}(M, \widehat{v}_1, k+1)) \leq \operatorname{rank}(X) = 2k$. Since $B^{2k,2k}$ is unreduced, $\dim(\mathcal{L}(M, \widehat{v}_1, j)) = 2j$ for $j = 1, ..., k$. This implies $\dim(\mathcal{L}(M, \widehat{v}_1, k+1)) = 2k$ and therefore, $d_{k+1} = 0$. $\sqrt{}$

These theorems provide the motivation for the implicit restart developed in the next section. Theorem 5.9 suggests that one might find an invariant subspace by iteratively replacing the starting vector with a linear combination of approximate eigenvectors corresponding to eigenvalues of interest. Such approximations are readily available through the Lanczos factorization.

5.1.2   STOPPING CRITERIA

Now assume that we have performed k steps of the symplectic Lanczos method and thus obtained the identity (after permuting back)

$$MS^{2n,2k} = S^{2n,2k} B^{2k,2k} + d_{k+1}(b_{k+1} \widehat{v}_{k+1} + a_{k+1} \widehat{w}_{k+1}) e_{2k}^T.$$

If the norm of the residual vector is small, the $2k$ eigenvalues of $B^{2k,2k}$ are approximations to the eigenvalues of M. Numerical experiments indicate that the norm of the residual rarely becomes small by itself. Nevertheless, some eigenvalues of $B^{2k,2k}$ may be good approximations to eigenvalues of M. Let λ be an eigenvalue of $B^{2k,2k}$ with the corresponding eigenvector y. Then the vector $x = S^{2n,2k} y$ satisfies

$$
\begin{aligned}
\|Mx - \lambda x\|_2 &= \|(MS^{2n,2k} - S^{2n,2k} B^{2k,2k})y\|_2 \\
&= |d_{k+1}| \, |e_{2k}^T y| \, \|b_{k+1} \widehat{v}_{k+1} + a_{k+1} \widehat{w}_{k+1}\|_2. \quad (5.1.15)
\end{aligned}
$$

The vector x is referred to as *Ritz vector* and λ as *Ritz value* of M. If the last component of the eigenvector y is sufficiently small, the right-hand side of (5.1.15) is small and the pair $\{\lambda, x\}$ is a good approximation to an eigenvalue-eigenvector pair of M. Note that by Lemma 3.11 $|e_{2k}^T y| > 0$ if $B^{2k,2k}$ is unreduced. The pair $\{\lambda, x\}$ is exact for the nearby problem

$$(M + E)x = \lambda x$$

where

$$E = -d_{k+1}(b_{k+1}\widehat{v}_{k+1} + a_{k+1}\widehat{w}_{k+1})e_k^T (S^{2n,2k})^T J^{2n,2n}.$$

In an actual implementation, typically the *Ritz estimate*

$$|d_{k+1}|\, |e_{2k}^T y|\, \|b_{k+1}\widehat{v}_{k+1} + a_{k+1}\widehat{w}_{k+1}\|_2$$

is used in order to decide about the numerical accuracy of an approximate eigenpair. This avoids the explicit formation of the residual $(M S^{2n,2k} - S^{2n,2k}B^{2k,2k})y$ when deciding about the numerical accuracy of an approximate eigenpair.

A small Ritz estimate is not sufficient for the Ritz pair $\{\lambda, x\}$ to be a good approximation to an eigenvalue-eigenvector pair of M. It does not guarantee that λ is a good approximation to an eigenvalue of M. That is

$$\min_j |\lambda - \mu_j|, \quad \text{where } \mu_j \in \sigma(M)$$

is not necessarily small when the Ritz estimate is small (see, e.g., [76, Section 3]). For nonnormal matrices the norm of the residual of an approximate eigenvector is not by itself sufficient information to bound the error in the approximate eigenvalue. It is sufficient however to give a bound on the distance to the nearest matrix to which the Ritz triplet $\{\lambda, x, y\}$ is exact [76] (here y denotes the left Ritz vector of M corresponding to the Ritz value λ). In the following, we will give a computable expression for the error using the results of Kahan, Parlett, and Jiang [76]. Assume that $B^{2k,2k}$ is diagonalizable, i.e., there exists Y such that

$$Y^{-1}B^{2k,2k}Y = \begin{bmatrix} \lambda_1 & & & & & \\ & \ddots & & & & \\ & & \lambda_k & & & \\ \hline & & & \lambda_1^{-1} & & \\ & & & & \ddots & \\ & & & & & \lambda_k^{-1} \end{bmatrix} = \Lambda;$$

Y can be chosen symplectic. Let $X = S^{2n,2k}Y = [x_1 \ldots, x_{2k}]$ and denote $b_{k+1}\widehat{v}_{k+1} + a_{k+1}\widehat{w}_{k+1}$ by $\widehat{r}_{k+1}$. Since

$$M S^{2n,2k} = S^{2n,2k} B^{2k,2k} + d_{k+1}\widehat{r}_{k+1}e_{2k}^T,$$

it follows that

$$MS^{2n,2k}Y = S^{2n,2k}YY^{-1}B^{2k,2k}Y + d_{k+1}\widehat{r}_{k+1}e_{2k}^T Y$$

or

$$MX = X\Lambda + d_{k+1}\widehat{r}_{k+1}e_{2k}^T Y.$$

Thus

$$Mx_i = \lambda_i x_i + y_{2k,i}d_{k+1}\widehat{r}_{k+1},$$

and

$$Mx_{k+i} = \lambda_i^{-1}x_{k+i} + y_{2k,k+i}d_{k+1}\widehat{r}_{k+1}$$

for $i = 1, \ldots, k$. The last equation can be rewritten as

$$(Jx_{k+i})^T M = \lambda_i(Jx_{k+i})^T + y_{2k,k+i}\lambda_i d_{k+1}\widehat{r}_{k+1}^T JM.$$

Using Theorem 2' of [76] we obtain that $\{\lambda_i, x_i, (Jx_{k+i})^T\}$ is an eigen-triplet of $M - F_{\lambda_i}$ where

$$\|F_{\lambda_i}\|_2 = |d_{k+1}| \max_i \left\{ \frac{\|\widehat{r}_{k+1}\|_2 |y_{2k,i}|}{\|x_i\|_2}, \frac{\|\widehat{r}_{k+1}^T JM\|_2 |y_{2k,k+i}\lambda_i|}{\|Jx_{k+i}\|_2} \right\}.$$

Furthermore, if $\|F_{\lambda_i}\|_2$ is small enough, then

$$|\theta_i - \lambda_j| \leq \text{cond}(\lambda_j)\|F_{\lambda_i}\|_2 + \mathcal{O}(\|F_{\lambda_i}\|_2^2),$$

where θ_i is an eigenvalue of M and $\text{cond}(\lambda_j)$ is the condition number of the Ritz value λ_j

$$\text{cond}(\lambda_j) = \frac{\|x_i\|_2 \|Jx_{k+i}\|_2}{|x_{k+i}^T Jx_i|} = \|x_i\|_2 \|x_{k+i}\|_2.$$

Similarly, we obtain that $\{\lambda_i^{-1}, x_{k+i}, (Jx_i)^T\}$ is an eigen-triplet of $M - F_{\lambda_i^{-1}}$ where

$$\|F_{\lambda_i^{-1}}\|_2 = |d_{k+1}| \max_i \left\{ \frac{\|\widehat{r}_{k+1}\|_2 |y_{2k,k+i}|}{\|x_{k+i}\|_2}, \frac{\|\widehat{r}_{k+1}^T JM\|_2 |y_{2k,i}\lambda_i^{-1}|}{\|Jx_i\|_2} \right\}.$$

Consequently, as λ_i and λ_i^{-1} should be treated alike, the symplectic Lanczos algorithm should be continued until $\|F_{\lambda_i}\|_2$ and $\|F_{\lambda_i^{-1}}\|_2$ are small, and until $\text{cond}(\lambda_j)\|F_{\lambda_i}\|_2$ and $\text{cond}(\lambda_j)\|F_{\lambda_i^{-1}}\|_2$ are below a given threshold for accuracy. Note that as in the Ritz estimate, in the criteria derived here the essential quantities are $|d_{k+1}|$ and the last component of the desired eigenvectors $|y_{2k,i}|$ and $|y_{2k,k+i}|$.

5.1.3 THE SYMPLECTIC LANCZOS ALGORITHM IN FINITE-PRECISION ARITHMETIC

In this section, we present a rounding error analysis of the symplectic Lanczos algorithm in finite-precision arithmetic. Our analysis will follow the lines of Bai's analysis of the unsymmetric Lanczos algorithm [11]. It is in the spirit of Paige's analysis for the symmetric Lanczos algorithm [109], except that we (as Bai) carry out the analysis componentwise rather than normwise. The componentwise analysis allows to measure each element of a perturbation relative to its individual tolerance, so that, unlike in the normwise analysis, the sparsity pattern of the problem under consideration can be exploited.

We use the usual model of floating-point arithmetic, as, e.g., in [58, 72]:

$$fl(x \circ y) = (x \circ y)(1 + \varepsilon)$$

where $\circ$ denotes any of the four basic arithmetic operations $+, -, *, /$ and $|\varepsilon| \leq$ u with u denoting the *unit roundoff*.

We summarize (as in [11]) all the results for basic linear algebra operations of sparse vectors and/or matrices that we need for our analysis:

Saxpy operation:
$$fl(\alpha x + y) = \alpha x + y + e, \qquad |e| \leq \mathbf{u}\, (2|\alpha x| + |y|) + \mathcal{O}(\mathbf{u}^2),$$

Inner product:
$$fl(x^T y) = x^T y + e, \qquad |e| \leq k\,\mathbf{u}\,|x|^T|y| + \mathcal{O}(\mathbf{u}^2),$$

Matrix-vector multiplication:
$$fl(Ax) = Ax + e, \qquad |e| \leq m\,\mathbf{u}\,|A|\,|x| + \mathcal{O}(\mathbf{u}^2),$$

where k is the number of overlapping nonzero components in the vectors x and y, and m is the maximal number of nonzero elements of the matrix A in any row. For a vector $x = [x_1, \ldots, x_n]^T$, $|x|$ denotes the vector $[|x_1|, \ldots, |x_n|]^T$. Similar, for a matrix $A = [a_{ij}]_{i,j=1}^n$, $|A|$ denotes the $n \times n$ matrix $[|a_{ij}|]_{i,j=1}^n$.

We will now analyze one step of the symplectic Lanczos algorithm to see the effects of the finite-precision arithmetic. Any computed quantity will be denoted by a hat, e.g., $\widehat{\alpha}$ will denote a computed quantity that is affected by rounding errors. (Please note, that in the previous section, we used hatted quantities to denote the not permuted symplectic Lanczos vectors.) After $j - 1$ steps of the symplectic Lanczos algorithm, we have computed $\widehat{a}_{j-1}, \widehat{w}_{j-1}, \widehat{b}_{j-1}, \widehat{c}_{j-1}, \widehat{d}_j, \widehat{v}_j$. During the jth step we will compute $\widehat{a}_j, \widehat{w}_j, \widehat{b}_j, \widehat{c}_j, \widehat{d}_{j+1}$ and $\widehat{v}_{j+1}$. Recall that we set $b_j = 1$ in the symplectic Lanczos algorithm. As a different choice is possible, we will treat b_j as a computed quantity without considering its actual computation in the following analysis.

At first we have to compute $a_j = v_j^T J_P M_P v_j$. Due to its special structure, multiplication by J_P does not cause any roundoff-error; hence it will not influence our analysis. Let M_P have at most m nonzero entries in any row or column. Then for the matrix-vector multiplication $J_P M_P v_j$ we have

$$\widehat{s}_1 = fl(J_P M_P \widehat{v}_j) = J_P M_P \widehat{v}_j + \widehat{e}_1,$$

where

$$|\widehat{e}_1| \leq m \, \mathbf{u} \, |J_P M_P| \, |\widehat{v}_j| + \mathcal{O}(\mathbf{u}^2).$$

Then a_j is computed by an inner product

$$\widehat{s}_2 = fl(\widehat{v}_j^T \widehat{s}_1) = \widehat{v}_j^T \widehat{s}_1 + \widehat{e}_2,$$

where

$$|\widehat{e}_2| \leq 2n \, \mathbf{u} \, |\widehat{v}_j|^T |\widehat{s}_1| + \mathcal{O}(\mathbf{u}^2),$$

assuming that $\widehat{v}_j$ and $\widehat{s}_1$ are full vectors. Overall, we have

$$\widehat{a}_j = \widehat{v}_j^T J_P M_P \widehat{v}_j + \widehat{f}_j^{[1]}, \tag{5.1.16}$$

where the roundoff error $\widehat{f}_j^{[1]} = \widehat{v}_j^T \widehat{e}_1 + \widehat{e}_2$ is bounded by

$$
\begin{aligned}
|\widehat{f}_j^{[1]}| &\leq m \, \mathbf{u} \, |\widehat{v}_j|^T |J_P M_P| \, |\widehat{v}_j| + 2n \, \mathbf{u} \, |\widehat{v}_j|^T |\widehat{s}_1| + \mathcal{O}(\mathbf{u}^2) \\
&\leq (m + 2n) \, \mathbf{u} \, |\widehat{v}_j|^T |J_P M_P| \, |\widehat{v}_j| + \mathcal{O}(\mathbf{u}^2).
\end{aligned}
$$

Next we have to compute $w_j = (M_P v_j - b_j v_j)/a_j$. For the matrix-vector multiplication $M_P v_j$ we obtain

$$\widehat{s}_3 = fl(M_P \widehat{v}_j) = M_P \widehat{v}_j + \widehat{e}_3,$$

where

$$|\widehat{e}_3| \leq m \, \mathbf{u} \, |M_P| \, |\widehat{v}_j| + \mathcal{O}(\mathbf{u}^2).$$

The saxpy operation $\widetilde{w}_j = M_P v_j - b_j v_j$ yields

$$\widehat{s}_4 = fl(\widehat{s}_3 - \widehat{b}_j \widehat{v}_j) = \widehat{s}_3 - \widehat{b}_j \widehat{v}_j + \widehat{e}_4,$$

with

$$|\widehat{e}_4| \leq \mathbf{u} \, (2|\widehat{b}_j \widehat{v}_j| + |\widehat{s}_3|) + \mathcal{O}(\mathbf{u}^2).$$

Thus overall we have

$$\widehat{\widetilde{w}}_j = M_P \widehat{v}_j - \widehat{b}_j \widehat{v}_j + \widehat{f}_j^{[2]}, \tag{5.1.17}$$

where the rounding error vector $\widehat{f}_j^{[2]} = \widehat{e}_3 + \widehat{e}_4$ is bounded by

$$
\begin{aligned}
|\widehat{f}_j^{[2]}| &\leq m \, \mathbf{u} \, |M_P| \, |\widehat{v}_j| + \mathbf{u} \, (2|\widehat{b}_j \widehat{v}_j| + |\widehat{s}_3|) + \mathcal{O}(\mathbf{u}^2) \\
&\leq m \, \mathbf{u} \, |M_P| \, |\widehat{v}_j| + \mathbf{u} \, (2|\widehat{b}_j \widehat{v}_j| + |M_P| \, |\widehat{v}_j|) + \mathcal{O}(\mathbf{u}^2) \\
&\leq (m + 1) \, \mathbf{u} \, |M_P| \, |\widehat{v}_j| + 2 \, \mathbf{u} \, |\widehat{b}_j \widehat{v}_j| + \mathcal{O}(\mathbf{u}^2).
\end{aligned}
$$

The computation of w_j is completed by

$$\widehat{w}_j = fl(\widehat{\widetilde{w}}_j/\widehat{a}_j) = \widehat{\widetilde{w}}_j/\widehat{a}_j + \widehat{f}_j^{[3]} \qquad (5.1.18)$$

where the rounding error vector $\widehat{f}_j^{[3]}$ is bounded by

$$|\widehat{f}_j^{[3]}| \leq \mathbf{u}\, |\widehat{\widetilde{w}}_j \widehat{a}_j^{-1}| + \mathcal{O}(\mathbf{u}^2).$$

The analysis of the computation of $c_j = v_j^T J_P M_P w_j / a_j$ is entirely analogous to the analysis of the computation of a_j. We start with the matrix-vector multiplication $J_P M_P w_j$

$$\widehat{s}_5 = fl(J_P M_P \widehat{w}_j) = J_P M_P \widehat{w}_j + \widehat{e}_5,$$

where

$$|\widehat{e}_5| \leq m\, \mathbf{u}\, |J_P M_P|\, |\widehat{w}_j| + \mathcal{O}(\mathbf{u}^2).$$

This is followed by an inner product $v_j^T J_P M_P w_j$

$$\widehat{s}_6 = fl(\widehat{v}_j^T \widehat{s}_5) = \widehat{v}_j^T \widehat{s}_5 + \widehat{e}_6,$$

with

$$|\widehat{e}_6| \leq 2n\, \mathbf{u}\, |\widehat{v}_j|^T |\widehat{s}_5| + \mathcal{O}(\mathbf{u}^2).$$

Finally, the computation is completed by

$$\widehat{s}_7 = fl(\widehat{s}_6/\widehat{a}_j) = \widehat{s}_6/\widehat{a}_j + \widehat{e}_7,$$

where

$$|\widehat{e}_7| \leq \mathbf{u}\, |\widehat{s}_6 \widehat{a}_j^{-1}| + \mathcal{O}(\mathbf{u}^2).$$

Overall, we have

$$\widehat{c}_j = \widehat{v}_j^T J_P M_P \widehat{w}_j / \widehat{a}_j + \widehat{f}_j^{[4]},$$

where the roundoff error $\widehat{f}_j^{[4]} = \widehat{a}_j^{-1} \widehat{v}_j^T \widehat{e}_5 + \widehat{a}_j^{-1} \widehat{e}_6 + \widehat{e}_7$ is bounded by

$$\begin{aligned}
|\widehat{f}_j^{[4]}| &\leq m\, \mathbf{u}\, |\widehat{v}_j|^T |J_P M_P|\, |\widehat{w}_j|\, |\widehat{a}_j^{-1}| + 2n\, \mathbf{u}\, |\widehat{v}_j|^T |J_P M_P|\, |\widehat{w}_j|\, |\widehat{a}_j^{-1}| \\
&\quad + \mathbf{u}\, |\widehat{v}_j|^T |J_P M_P|\, |\widehat{w}_j|\, |\widehat{a}_j^{-1}| + \mathcal{O}(\mathbf{u}^2) \\
&\leq (m + 2n + 1)\, \mathbf{u}\, |\widehat{a}_j^{-1}|\, |\widehat{v}_j|^T |J_P M_P|\, |\widehat{w}_j| + \mathcal{O}(\mathbf{u}^2).
\end{aligned}$$

Finally, we have to compute $\widetilde{v}_{j+1} = -d_j v_{j-1} - c_j v_j + w_j + a_j^{-1} M_P^{-1} v_j$, $d_{j+1} = \sqrt{\widetilde{v}_{j+1}^T \widetilde{v}_{j+1}}$ and $v_{j+1} = \widetilde{v}_{j+1}/d_{j+1}$. Recall that, as M is symplectic, the inverse of M_P is given by $M_P^{-1} = -J_P M_P^T J_P$. Let us start us with the matrix-vector multiplication $M_P^{-1} v_j$

$$\hat{s}_8 = fl(M_P^{-1}\hat{v}_j) = M_P^{-1}\hat{v}_j + \hat{e}_8$$

where

$$|\hat{e}_8| \le m\,\mathbf{u}\,|M_P^{-1}|\,|\hat{v}_j| + \mathcal{O}(\mathbf{u}^2).$$

Next three saxpy operations are used to finish the computation of $\tilde{v}_{j+1}$:

$$\hat{s}_9 = fl(\hat{s}_8\hat{a}_j^{-1} + \hat{w}_j) = \hat{s}_8\hat{a}_j^{-1} + \hat{w}_j + \hat{e}_9,$$

where

$$|\hat{e}_9| \le \mathbf{u}\,(2|\hat{s}_8\hat{a}_j^{-1}| + |\hat{w}_j|) + \mathcal{O}(\mathbf{u}^2),$$

and

$$\hat{s}_{10} = fl(\hat{s}_9 - \hat{c}_j\hat{v}_j) = \hat{s}_9 - \hat{c}_j\hat{v}_j + \hat{e}_{10},$$

where

$$|\hat{e}_{10}| \le \mathbf{u}\,(2|\hat{c}_j\hat{v}_j| + |\hat{s}_9|) + \mathcal{O}(\mathbf{u}^2),$$

and

$$\hat{s}_{11} = fl(\hat{s}_{10} - \hat{d}_j\hat{v}_{j-1}) = \hat{s}_{10} - \hat{d}_j\hat{v}_{j-1} + \hat{e}_{11},$$

where

$$|\hat{e}_{11}| \le \mathbf{u}\,(2|\hat{d}_j\hat{v}_{j-1}| + |\hat{s}_{10}|) + \mathcal{O}(\mathbf{u}^2).$$

Overall, we have for $\tilde{v}_{j+1}$

$$\hat{\tilde{v}}_{j+1} = -\hat{d}_j\hat{v}_{j-1} - \hat{c}_j\hat{v}_j + \hat{w}_j + \hat{a}_j^{-1}M_P^{-1}\hat{v}_j + \hat{f}_{j+1}^{[5]}, \qquad (5.1.19)$$

where the roundoff error vector $\hat{f}_{j+1}^{[5]} = \hat{a}_j^{-1}\hat{e}_8 + \hat{e}_9 + \hat{e}_{10} + \hat{e}_{11}$ is bounded by

$$
\begin{aligned}
|\hat{f}_{j+1}^{[5]}| &\le m\,\mathbf{u}\,|\hat{a}_j^{-1}|\,|M_P^{-1}|\,|\hat{v}_j| + \mathbf{u}\,(2|\hat{s}_8\hat{a}_j^{-1}| + |\hat{w}_j|) \\
&\quad + \mathbf{u}\,(2|\hat{c}_j\hat{v}_j| + |\hat{s}_9|) + \mathbf{u}\,(2|\hat{d}_j\hat{v}_{j-1}| + |\hat{s}_{10}|) + \mathcal{O}(\mathbf{u}^2) \\
&\le (m+4)\,\mathbf{u}\,|\hat{a}_j^{-1}|\,|M_P^{-1}|\,|\hat{v}_j| + 3\,\mathbf{u}\,|\hat{w}_j| + 3\,\mathbf{u}\,|\hat{c}_j\hat{v}_j| \\
&\quad + 2\,\mathbf{u}\,|\hat{d}_j\hat{v}_{j-1}| + \mathcal{O}(\mathbf{u}^2).
\end{aligned}
$$

Next we compute $d_{j+1} = \sqrt{\tilde{v}_{j+1}^T \tilde{v}_{j+1}}$.

$$\hat{s}_{12} = fl(\hat{\tilde{v}}_{j+1}^T \hat{\tilde{v}}_{j+1}) = \hat{\tilde{v}}_{j+1}^T \hat{\tilde{v}}_{j+1} + \hat{e}_{12},$$

with

$$|\hat{e}_{12}| \le 2n\,\mathbf{u}\,|\hat{\tilde{v}}_{j+1}|^T|\hat{\tilde{v}}_{j+1}| + \mathcal{O}(\mathbf{u}^2).$$

Hence,

$$\widehat{d}_{j+1} = fl(\sqrt{\widehat{s}_{12}}) = \sqrt{\widetilde{\widehat{v}}_{j+1}^T \widehat{\widetilde{v}}_{j+1}} + \widehat{f}_{j+1}^{[6]},$$

where the roundoff error $\widehat{f}_{j+1}^{[6]}$ is bounded by

$$|\widehat{f}_{j+1}^{[6]}| \leq \mathbf{u}\sqrt{\widehat{s}_{12}} \leq \mathbf{u}\sqrt{\widetilde{\widehat{v}}_{j+1}^T \widehat{\widetilde{v}}_{j+1}} + \mathcal{O}(\mathbf{u}^2).$$

The symplectic Lanczos step is completed by computing $v_{j+1} = \widetilde{v}_{j+1}/d_{j+1}$:

$$\widehat{v}_{j+1} = fl(\widehat{\widetilde{v}}_{j+1}\widehat{d}_{j+1}^{-1}) = \widehat{\widetilde{v}}_{j+1}\widehat{d}_{j+1}^{-1} + \widehat{f}_{j+1}^{[7]}, \tag{5.1.20}$$

with

$$|\widehat{f}_{j+1}^{[7]}| \leq \mathbf{u}\,|\widehat{\widetilde{v}}_{j+1}|\,|\widehat{d}_{j+1}^{-1}| + \mathcal{O}(\mathbf{u}^2).$$

From (5.1.20) and (5.1.19) we know that

$$\widehat{d}_{j+1}\widehat{v}_{j+1} = -\widehat{d}_j\widehat{v}_{j-1} - \widehat{c}_j\widehat{v}_j + \widehat{w}_j + \widehat{a}_j^{-1}M_P^{-1}\widehat{v}_j + g_{j+1}, \tag{5.1.21}$$

where g_{j+1} is the sum of roundoff errors in computing the intermediate vector $\widetilde{v}_{j+1}$ and the symplectic Lanczos vector v_{j+1}

$$g_{j+1} = \widehat{f}_{j+1}^{[5]} + \widehat{d}_{j+1}\widehat{f}_{j+1}^{[7]}.$$

Using the bounds for the rounding errors $\widehat{f}_{j+1}^{[5]}$ and $\widehat{f}_{j+1}^{[7]}$ we have

$$\begin{aligned}
|g_{j+1}| &\leq (m+4)\,\mathbf{u}\,|\widehat{a}_j^{-1}|\,|M_P^{-1}|\,|\widehat{v}_j| + 3\,\mathbf{u}\,|\widehat{w}_j| \tag{5.1.22}\\
&\quad + 3\,\mathbf{u}\,|\widehat{c}_j\widehat{v}_j| + 2\,\mathbf{u}\,|\widehat{d}_j\widehat{v}_{j-1}| + \mathbf{u}\,|\widehat{\widetilde{v}}_{j+1}| + \mathcal{O}(\mathbf{u}^2)\\
&\leq (m+5)\,\mathbf{u}\,|\widehat{a}_j^{-1}|\,|M_P^{-1}|\,|\widehat{v}_j| + 4\,\mathbf{u}\,|\widehat{w}_j|\\
&\quad + 4\,\mathbf{u}\,|\widehat{c}_j\widehat{v}_j| + 3\,\mathbf{u}\,|\widehat{d}_j\widehat{v}_{j-1}| + \mathcal{O}(\mathbf{u}^2). \tag{5.1.23}
\end{aligned}$$

Similar, from (5.1.18) and (5.1.17) we know that

$$\widehat{a}_j\widehat{w}_j = M_P\widehat{v}_j - \widehat{b}_j\widehat{v}_j + h_j, \tag{5.1.24}$$

where

$$h_j = \widehat{f}_j^{[2]} + \widehat{a}_j\widehat{f}_j^{[3]},$$

and

$$\begin{aligned}
|h_j| &\leq (m+1)\,\mathbf{u}\,|M_P|\,|\widehat{v}_j| + 2\,\mathbf{u}\,|\widehat{b}_j\widehat{v}_j| + \mathbf{u}\,|\widehat{\widetilde{w}}_j| + \mathcal{O}(\mathbf{u}^2)\\
&\leq (m+2)\,\mathbf{u}\,|M_P|\,|\widehat{v}_j| + 3\,\mathbf{u}\,|\widehat{b}_j\widehat{v}_j| + \mathcal{O}(\mathbf{u}^2). \tag{5.1.25}
\end{aligned}$$

While the equation $a_j w_j = M_P v_j - b_j v_j$ is given by the $(2j)$th column of $M_P S_P (N_u)_P^{-1} = S_P (K_u)_P^{-1}$, the equation $d_{j+1} v_{j+1} = -d_j v_{j_1} - c_j v_j + w_j + a_j^{-1} M_P^{-1} v_j$ corresponds to the $(2j - 1)$th column of $S_P (N_u)_P^{-1} = M_P^{-1} S_P (K_u)_P^{-1}$. Hence, in order to summarize the results obtained so far into one single equation, let

$$E_k = [M_P g_2, \ -h_1, \ M_P g_3, \ -h_2, \ \ldots, \ M_P g_{k+1}, \ -h_k]. \qquad (5.1.26)$$

Then we have from (5.1.21) and (5.1.24)

$$M_P [\widehat{v}_1, \widehat{w}_1, \ldots, \widehat{v}_k, \widehat{w}_k]
\begin{bmatrix}
\widehat{c}_1 & 1 & \widehat{d}_2 & 0 & & & & \\
-1 & 0 & 0 & 0 & & & & \\
\widehat{d}_2 & 0 & \widehat{c}_2 & 1 & \ddots & & & \\
0 & 0 & -1 & 0 & & \ddots & & \\
& & \ddots & & \ddots & & \widehat{d}_k & 0 \\
& & & \ddots & & \ddots & 0 & 0 \\
& & & & \widehat{d}_k & 0 & \widehat{c}_k & 1 \\
& & & & 0 & 0 & -1 & 0
\end{bmatrix}$$

$$= [\widehat{v}_1, \widehat{w}_1, \ldots, \widehat{v}_k, \widehat{w}_k]
\begin{bmatrix}
\widehat{a}_1^{-1} & \widehat{b}_1 & & \\
0 & \widehat{a}_1 & & \\
& & \ddots & \\
& & & \widehat{a}_k^{-1} & \widehat{b}_k \\
& & & 0 & \widehat{a}_k
\end{bmatrix}$$

$$- \widehat{d}_{k+1} M_P \widehat{v}_{k+1} e_{2k-1}^T + E_k,$$

or, even shorter,

$$M_P \widehat{S}_P^{2n,2k} (\widehat{N_u}^{2k,2k})_P^{-1} = \widehat{S}_P^{2n,2k} (\widehat{K_u}^{2k,2k})_P^{-1} \\
- \widehat{d}_{k+1} M_P \widehat{v}_{k+1} e_{2k-1}^T + E_k. \qquad (5.1.27)$$

Using the componentwise upper bounds for $|g_{j+1}|$ and $|h_j|$, let us derive an upper bound for $\|E_k\|_F$. Clearly,

$$\|E_k\|_F \le \| [h_1, h_2, \ldots, h_k] \|_F + \|M_P\|_F \| [g_2, g_3, \ldots, g_{k+1}] \|_F.$$

From (5.1.25) we have

$$|[h_1, h_2, \ldots, h_k]| \le (m + 2) \, \mathbf{u} \, |M_P| \, |\widehat{S}_P^{2n,2k}| \\
+ 3 \, \mathbf{u} \, |\widehat{S}_P^{2n,2k}| \, |(\widehat{K_u}^{2k,2k})_P| + \mathcal{O}(\mathbf{u}^2),$$

and

$$\| [h_1, h_2, \ldots, h_k] \|_F \leq (m+2) \mathbf{u} \| \widehat{S}^{2n,2k} \|_F \| M \|_F$$
$$+ 3 \mathbf{u} \| \widehat{S}^{2n,2k} \|_F \| \widehat{K_u}^{2k,2k} \|_F \quad (5.1.28)$$
$$+ \mathcal{O}(\mathbf{u}^2).$$

Using (5.1.23) we obtain

$$| [g_2, g_3, \ldots, g_{k+1}] | \leq \mathbf{u} \left[(m+5) |(\widehat{K_u}^{2k,2k})_P| |M_P^{-1}| |\widehat{S}_P^{2n,2k}| \right.$$
$$+ 4 |\widehat{S}_P^{2n,2k}| + 4 |\widehat{S}_P^{2n,2k}| |(\widehat{N_u}^{2k,2k})_P| \right]$$
$$+ \mathcal{O}(\mathbf{u}^2),$$

and

$$\| [g_2, g_3, \ldots, g_{k+1}] \|_F \leq \mathbf{u} \| \widehat{S}^{2n,2k} \|_F \left[(m+5) \| \widehat{K_u}^{2k,2k} \|_F \| M \|_F \right.$$
$$+ 4 + 4 \| \widehat{N_u}^{2k,2k} \|_F \right] + \mathcal{O}(\mathbf{u}^2). \quad (5.1.29)$$

Hence, summarizing we obtain as an upper bound for the error matrix E_k of (5.1.27)

$$\| E_k \|_F \leq \mathbf{u} \| \widehat{S}^{2n,2k} \|_F \left[(m+5) \| \widehat{K_u}^{2k,2k} \|_F \| M \|_F^2 \right.$$
$$+ 4 \| \widehat{N_u}^{2k,2k} \|_F \| M \|_F + (m+6) \| M \|_F$$
$$+ 3 \| \widehat{K_u}^{2k,2k} \|_F \right] + \mathcal{O}(\mathbf{u}^2).$$

Summarizing the analysis of one step of the symplectic Lanczos algorithm to see the effects of the finite-precision arithmetic we obtain the following theorem.

THEOREM 5.11 *Let $M \in \mathbf{R}^{2n \times 2n}$ be a symplectic matrix with at most m nonzero entries in any row or column. If no breakdown occurs during the execution of k steps of the symplectic Lanczos algorithm as given in Table 5.7, the computed Lanczos vectors satisfy*

$$\widehat{a}_j \widehat{w}_j = M_P \widehat{v}_j - \widehat{v}_j + h_j, \quad (5.1.30)$$
$$\widehat{d}_{j+1} \widehat{v}_{j+1} = -\widehat{d}_j \widehat{v}_{j-1} - \widehat{c}_j \widehat{v}_j + \widehat{w}_j + \widehat{a}_j^{-1} M_P^{-1} \widehat{v}_j + g_{j+1}, \quad (5.1.31)$$

where

$$|h_j| \leq (m+2) \mathbf{u} |M_P| |\widehat{v}_j| + 2 \mathbf{u} |\widehat{v}_j| + \mathcal{O}(\mathbf{u}^2), \quad (5.1.32)$$
$$|g_{j+1}| \leq (m+5) \mathbf{u} |\widehat{a}_j^{-1}| |M_P^{-1}| |\widehat{v}_j| + 4 \mathbf{u} |\widehat{w}_j|$$
$$+ 4 \mathbf{u} |\widehat{c}_j \widehat{v}_j| + 3 \mathbf{u} |\widehat{d}_j \widehat{v}_{j-1}| + \mathcal{O}(\mathbf{u}^2). \quad (5.1.33)$$

The computed matrices $\widehat{S}_P^{2k}$, $\widehat{N}_P^{2k,2k}$, and $\widehat{K}_P^{2k,2k}$ satisfy

$$M_P \widehat{S}_P^{2k} (\widehat{N}_P^{2k,2k})^{-1} = \widehat{S}_P^{2k} (\widehat{K}_P^{2k,2k})^{-1} - \widehat{d}_{k+1} M_P \widehat{v}_{k+1} e_{2k-1}^T + E_k, \quad (5.1.34)$$

where

$$\begin{aligned}
\|E_k\|_F \ \leq \ & \mathbf{u} \, \|\widehat{S}^{2k}\|_F \left[(m+5) \, \|\widehat{K}^{2k,2k}\|_F \|M\|_F^2 \right. \\
& + 4 \, \|\widehat{N}^{2k,2k}\|_F \|M\|_F + (m+6) \, \|M\|_F \qquad (5.1.35) \\
& \left. + 3 \, \|\widehat{K}^{2k,2k}\|_F \right] + \mathcal{O}(\mathbf{u}^2).
\end{aligned}$$

This indicates that the recursion equation (5.1.34)

$$M_P S_P^{2k} (N_P^{2k,2k})^{-1} = S_P^{2k} (K_P^{2k,2k})^{-1} - d_{k+1} M_P v_{k+1} e_{2k-1}^T$$

is satisfied to working precision, if

$$\|M\|_F^2, \|\widehat{S}^{2k}\|_F, \|\widehat{K}^{2k,2k}\|_F, \ \text{and} \ \|\widehat{N}^{2k,2k}\|_F \|M\|_F$$

are of moderate size. But, unfortunately, $\|\widehat{S}^{2k}\|_F$ may grow unboundedly in the case of near-breakdown.

While the equation (5.1.30) is given by the $(2j)$th column of

$$M_P S_P N_P^{-1} = S_P K_P^{-1},$$

the equation (5.1.31) corresponds to the $(2j - 1)$th column of

$$S_P N_P^{-1} = M_P^{-1} S_P K_P^{-1}.$$

The upper bounds associated with (5.1.30) and (5.1.31) involve only $\|M\|_F$ as to be expected, see (5.1.32) and (5.1.33). Recall that $M_P^{-1} = -J_P M_P^T J_P$, since M is symplectic. Thus $|M_P^{-1}|$ does not introduce any problems usually involved by forming the inverse of a matrix. In order to summarize these results into one single equation, we define E_k in (5.1.26), then (5.1.34) holds. Using the component-wise upper bounds for $|h_j|$ and $|g_{j+1}|$, we obtain the upper bound for E_k as given in (5.1.35). As we summarize our results in terms of the equation $M_P S_P N_P^{-1} = S_P K_P^{-1}$, we have to premultiply the error bound associated with (5.1.31) by M_P, resulting in an artificial $\|M\|_F^2$ term here. Hence combining all our findings into one single equation forces the $\|M\|_F^2$ term.

For the unsymmetric Lanczos algorithm, Bai obtains a similar result in [11]. The equations corresponding to our equations (5.1.30) and (5.1.31) are (see (5.1.9) and (5.1.10))

$$\begin{aligned}
\widehat{\beta}_{j+1} \widehat{q}_{j+1} &= A \widehat{q}_j - \widehat{\alpha}_j \widehat{q}_j - \widehat{\gamma}_j \widehat{q}_j + h_j^{nonsymLan}, \\
\widehat{\gamma}_{j+1} \widehat{p}_{j+1} &= A^T \widehat{p}_j - \widehat{\alpha}_j \widehat{p}_j - \widehat{\beta}_j \widehat{p}_{j-1} + g_{j+1}^{nonsymLan}.
\end{aligned}$$

The errors associated are given by

$$
\begin{aligned}
|h_j^{nonsymLan}| &\leq (3+m)\,\mathbf{u}\,|A|\,|\widehat{q}_j| + 4\,\mathbf{u}\,|\widehat{\alpha}_j|\,|\widehat{q}_j| \\
&\quad + 3\,\mathbf{u}\,|\widehat{\gamma}_j|\,|\widehat{q}_{j-1}| + \mathcal{O}(\mathbf{u}^2), \\
|g_{j+1}^{nonsymLan}| &\leq (3+m)\,\mathbf{u}\,|A|\,|\widehat{p}_j| + 4\,\mathbf{u}\,|\widehat{\alpha}_j|\,|\widehat{p}_j| \\
&\quad + 3\,\mathbf{u}\,|\widehat{\gamma}_j|\,|\widehat{p}_{j-1}| + \mathcal{O}(\mathbf{u}^2).
\end{aligned}
$$

Hence, the symplectic Lanczos algorithms behaves essentially like the unsymmetric Lanczos algorithm. The additional restriction of preserving the symplectic structure does not pose any additional problems concerning the rounding error analysis, the results of the analysis are essentially the same.

REMARK 5.12 *In Remark 5.5 we have noted that the usual unsymmetric Lanczos algorithm generates two sequences of vectors, but that due to the symplectic structure, the two recurrence relations of the standard unsymmetric Lanczos algorithm are equivalent for the situation discussed here. It was noted that the equation which is not used is given by*

$$
M_P^T W_P^{2n,2k}(K_u^{2k,2k})_P^T = W_P^{2n,2k}(N_u^{2k,2k})_P^T + d_{k+1}J_P v_{k+1}e_{2k}^T,
$$

where

$$
W_P^{2n,2k} = J_P^{2n,2n} S_P^{2n,2k} J_P^{2k,2k} = [-J_P w_1,\ J_P v_1,\ \ldots,\ -J_P w_k,\ J_P v_k].
$$

Instead of summarizing our findings into equation (5.1.34), we could have summarized

$$
\begin{aligned}
M_P^T \widehat{W}_P^{2n,2k}(\widehat{K}_u^{2k,2k})_P^T &= \widehat{W}_P^{2n,2k}(\widehat{N}_u^{2k,2k})_P^T \\
&\quad + \widehat{d}_{k+1}J_P \widehat{v}_{k+1}e_{2k}^T + F_k \quad (5.1.36)
\end{aligned}
$$

where

$$
\begin{aligned}
\widehat{W}_P^{2n,2k} &= J_P^{2n,2n}\widehat{S}_P^{2n,2k}J_P^{2k,2k}, \\
F_k &= \left[M_P^T J_P h_1,\ J_P g_2,\ \ldots,\ M_P^T J_P h_k,\ J_P g_{k+1}\right]. \quad (5.1.37)
\end{aligned}
$$

Using (5.1.28) and (5.1.29) we obtain as an upper bound for $\|F_k\|_F$

$$
\|F_k\|_F \leq \mathbf{u}\,\|\widehat{S}^{2n,2k}\|_F \left[(m+2)\,\|M\|_F^2 + (m+8)\,\|\widehat{K}_u^{2k,2k}\|_F\|M\|_F + 4\,\|\widehat{N}_u^{2k,2k}\|_F + 4\right] + \mathcal{O}(\mathbf{u}^2)
$$

As before, the term $\|M\|_F^2$ is introduced because we summarize all our findings into one single equation.

It is well-known, that in finite-precision arithmetic, orthogonality between the computed Lanczos vectors in the symmetric Lanczos process is lost. This loss of orthogonality is due to cancellation and is not the result of the gradual accumulation of roundoff error (see, e.g., [114, 131]). What can we say about the J–orthogonality of the computed symplectic Lanczos vectors? Obviously, rounding errors, once introduced into some computed Lanczos vectors, are propagated to future steps. Such error propagation for the unsymmetric Lanczos process is analyzed by Bai [11]. Here we will show that J–orthogonality between the computed symplectic Lanczos vectors is lost, following the lines of the proof of Corollary 3.1 in [11].

J–orthogonality between the symplectic Lanczos vectors implies that we should have

$$
\begin{aligned}
w_j^T J_P v_j &= -1, & \text{for all } j, \\
v_j^T J_P v_m &= 0, \\
w_j^T J_P w_m &= 0, \\
v_j^T J_P w_m &= 0, & \text{for all } j \neq m.
\end{aligned}
$$

Let us take a closer look at these relations for the computed symplectic Lanczos vectors. Define

$$
H = [\widehat{v}_1, \widehat{w}_1, \ldots, \widehat{v}_k, \widehat{w}_k]^T J_P [\widehat{v}_1, \widehat{w}_1, \ldots, \widehat{v}_k, \widehat{w}_k].
$$

That is,

$$
\begin{aligned}
h_{2j-1,2m-1} &= \widehat{v}_j^T J_P \widehat{v}_m, \\
h_{2j-1,2m} &= \widehat{v}_j^T J_P \widehat{w}_m, \\
h_{2j,2m-1} &= \widehat{w}_j^T J_P \widehat{v}_m, \\
h_{2j,2m} &= \widehat{w}_j^T J_P \widehat{w}_m.
\end{aligned}
$$

In exact arithmetic we would have $H = J_P$. As $x^T J_P x = 0$ for any vector x, we have

$$
h_{2j,2j} = h_{2j-1,2j-1} = 0,
$$

not depending on the loss of J_P-orthogonality between the computed symplectic Lanczos vectors. Moreover, as $h_{2m,2j-1} = -h_{2j-1,2m}$, we only need to examine $h_{2j,2j-1}$ for $j = 1, \ldots, k$, and $h_{2j,2m-1}, h_{2j-1,2m-1}$ and $h_{2j,2m}$ for $j, m = 1, \ldots, k, j < m$. Examining these elements of H we obtain the following lemma.

LEMMA 5.13 *The elements h_{jm} of H satisfy the following equations*

$$
\begin{aligned}
h_{jj} &= 0 & j = 1, \ldots, 2k \\
-h_{j,j+1} = h_{j+1,j} &= -1 + \kappa_j & j = 1, \ldots, 2k-1
\end{aligned}
\tag{5.1.38}
$$

where

$$|\kappa_j| \;\le\; \mathbf{u}\, \frac{|\widehat{\widehat{v}}_j|^T |J_P| \left\{ 2(m+n+2)\,|M_P| + 5 \right\} |\widehat{v}_j|}{|\widehat{w}_j^T J_P \widehat{\widehat{v}}_j|} + \mathcal{O}(\mathbf{u}^2), \quad (5.1.39)$$

and

$$
\begin{aligned}
\widehat{d}_m \widehat{a}_m h_{2j,2m} \;=\;& \widehat{a}_j^{-1} h_{2j-1,2m-2} - \widehat{d}_m \widehat{b}_m h_{2j,2m-1} \\
& - \widehat{c}_{m-1} \widehat{a}_{m-1} h_{2j,2m-2} \\
& - \widehat{c}_{m-1} \widehat{b}_{m-1} h_{2j,2m-3} + \widehat{a}_{m-1}^{-1} h_{2j,2m-3} \\
& - \widehat{d}_{m-1} \widehat{a}_{m-2} h_{2j,2m-4} - \widehat{d}_{m-1} \widehat{b}_{m-2} h_{2j,2m-5} \\
& + \widehat{b}_j \widehat{d}_j h_{2m-2,2j-3} + \widehat{b}_j \widehat{c}_j h_{2m-2,2j-1} \\
& + \widehat{b}_j \widehat{d}_{j+1} h_{2m-2,2j+1} + \widehat{b}_j h_{2m-2,2j} \\
& + \widehat{d}_{m-1} \widehat{w}_j^T J_P h_{m-2} + \widehat{c}_{m-1} \widehat{w}_j^T J_P h_{m-1} \\
& + \widehat{d}_m \widehat{w}_j^T J_P h_m + \widehat{a}_j^{-1} h_j^T J_P M_P \widehat{w}_{m-1} \\
& - \widehat{b}_j \widehat{w}_{m-1}^T J_P g_{j+1} + \widehat{w}_j^T J_P M_P g_m.
\end{aligned}
\quad (5.1.40)
$$

Similar expressions can be derived for $h_{2j,2m-1}$, and $h_{2j-1,2m-1}$.

PROOF: Let us start our analysis with $h_{2j,2j-1}$. Using (5.1.18) and (5.1.20) we have

$$
\begin{aligned}
h_{2j,2j-1} \;=\;& \widehat{w}_j^T J_P \widehat{v}_j \quad\quad\quad\quad\quad\quad\quad\quad\quad\quad (5.1.41)\\
\;=\;& \left(\frac{\widehat{\widehat{w}}_j}{\widehat{a}_j} + \widehat{f}_j^{[3]} \right)^T J_P \left(\frac{\widehat{\widehat{v}}_j}{\widehat{d}_j} + \widehat{f}_j^{[7]} \right) \\
\;=\;& \frac{\widehat{\widehat{w}}_j^T J_P \widehat{\widehat{v}}_j + \widehat{a}_j (\widehat{f}_j^{[3]})^T J_P \widehat{\widehat{v}}_j + \widehat{d}_j \widehat{\widehat{w}}_j^T J_P \widehat{f}_j^{[7]}}{\widehat{a}_j \widehat{d}_j} + \mathcal{O}(\mathbf{u}^2) \\
\;=:\;& \frac{\widehat{\widehat{w}}_j^T J_P \widehat{\widehat{v}}_j + \zeta_1}{\widehat{a}_j \widehat{d}_j} + \mathcal{O}(\mathbf{u}^2), \quad\quad\quad (5.1.42)
\end{aligned}
$$

where

$$
\begin{aligned}
|\zeta_1| \;\le\;& |\widehat{a}_j (\widehat{f}_j^{[3]})^T J_P \widehat{\widehat{v}}_j| + |\widehat{d}_j \widehat{\widehat{w}}_j^T J_P \widehat{f}_j^{[7]}| \\
\;\le\;& 2\,\mathbf{u}\, |\widehat{\widehat{w}}_j|^T |J_P|\, |\widehat{\widehat{v}}_j| \\
\;\le\;& 2\,\mathbf{u}\, |\widehat{v}_j|^T |J_P|\, (|M_P|\, |\widehat{v}_j| - |\widehat{b}_j|\, |\widehat{v}_j|).
\end{aligned}
$$

We would like to be able to rewrite $h_{2j,2j-1} = \widehat{w}_j^T J_P \widehat{v}_j$ as $-1+$ some small error. In order to do so, we rewrite $\widehat{a}_j \widehat{d}_j$ suitably. From (5.1.16) and (5.1.20) we have

$$
\begin{aligned}
\hat{a}_j \hat{d}_j &= (\widetilde{v}_j^T J_P M_P \widehat{v}_j + \widehat{f}_j^{[1]}) \hat{d}_j \\
&= \left[(\frac{\widehat{\widetilde{v}}_j}{\hat{d}_j} + \widehat{f}_j^{[7]})^T J_P M_P \widehat{v}_j + \widehat{f}_j^{[1]} \right] \hat{d}_j \\
&= \widehat{\widetilde{v}}_j^T J_P M_P \widehat{v}_j + \hat{d}_j ((\widehat{f}_j^{[7]})^T J_P M_P \widehat{v}_j + \widehat{f}_j^{[1]}).
\end{aligned}
$$

Using (5.1.17) we obtain

$$
\begin{aligned}
\hat{a}_j \hat{d}_j &= \widehat{\widetilde{v}}_j^T J_P (\widehat{\widetilde{w}}_j + \widehat{b}_j \widehat{v}_j - \widehat{f}_j^{[2]}) + \hat{d}_j ((\widehat{f}_j^{[7]})^T J_P M_P \widehat{v}_j + \widehat{f}_j^{[1]}) \\
&= \widehat{\widetilde{v}}_j^T J_P \widehat{\widetilde{w}}_j + \widehat{b}_j \widehat{\widetilde{v}}_j^T J_P \widehat{v}_j - \widehat{\widetilde{v}}_j^T J_P \widehat{f}_j^{[2]} + \hat{d}_j ((\widehat{f}_j^{[7]})^T J_P M_P \widehat{v}_j + \widehat{f}_j^{[1]}) \\
&= \widehat{\widetilde{v}}_j^T J_P \widehat{\widetilde{w}}_j + \widehat{b}_j \hat{d}_j (\widehat{v}_j - \widehat{f}_j^{[7]})^T J_P \widehat{v}_j - \widehat{\widetilde{v}}_j^T J_P \widehat{f}_j^{[2]} \\
&\quad + \hat{d}_j ((\widehat{f}_j^{[7]})^T J_P M_P \widehat{v}_j + \widehat{f}_j^{[1]}).
\end{aligned}
$$

For the last equation we used again (5.1.20). This rewriting allows us to make us of the fact that $x^T J x = 0$ for any vector x. Thus

$$
\begin{aligned}
\hat{a}_j \hat{d}_j &= \widehat{\widetilde{v}}_j^T J_P \widehat{\widetilde{w}}_j - \widehat{\widetilde{v}}_j^T J_P \widehat{f}_j^{[2]} + \hat{d}_j (\widehat{f}_j^{[7]})^T J_P (M_P \widehat{v}_j - \widehat{b}_j \widehat{v}_j) + \hat{d}_j \widehat{f}_j^{[1]} \\
&= \widehat{\widetilde{v}}_j^T J_P \widehat{\widetilde{w}}_j - \widehat{\widetilde{v}}_j^T J_P \widehat{f}_j^{[2]} + \hat{d}_j (\widehat{f}_j^{[7]})^T J_P (\widehat{\widetilde{w}}_j - \widehat{f}_j^{[2]}) + \hat{d}_j \widehat{f}_j^{[1]} \\
&=: \widehat{\widetilde{v}}_j^T J_P \widehat{\widetilde{w}}_j + \zeta_2 \\
&= -\widehat{\widetilde{w}}_j^T J_P \widehat{\widetilde{v}}_j + \zeta_2, \tag{5.1.43}
\end{aligned}
$$

where we used (5.1.17). The roundoff error is bounded by

$$
\begin{aligned}
|\zeta_2| &\leq |\widehat{\widetilde{v}}_j^T J_P \widehat{f}_j^{[2]}| + |\hat{d}_j| \, |\widehat{f}_j^{[7]}|^T |J_P| \, (|\widehat{\widetilde{w}}_j| + |\widehat{f}_j^{[2]}|) + |\hat{d}_j| \, |\widehat{f}_j^{[1]}| \\
&\leq (m+1) \, \mathbf{u} \, |\widehat{\widetilde{v}}_j|^T |J_P| \, |M_P| \, |\widehat{v}_j| + 2 \, \mathbf{u} \, |\widehat{b}_j| \, |\widehat{\widetilde{v}}_j|^T |J_P| \, |\widehat{v}_j| \\
&\quad + (m+2n) \, \mathbf{u} \, |\hat{d}_j| \, |\widehat{v}_j|^T |J_P M_P| \, |\widehat{v}_j| \\
&\quad + \mathbf{u} \, |\widehat{\widetilde{v}}_j|^T |J_P| \, |\widehat{\widetilde{w}}_j| + \mathcal{O}(\mathbf{u}^2) \\
&\leq (m+1) \, \mathbf{u} \, |\widehat{\widetilde{v}}_j|^T |J_P| \, |M_P| \, |\widehat{v}_j| + 2 \, \mathbf{u} \, |\widehat{b}_j| \, |\widehat{\widetilde{v}}_j|^T |J_P| \, |\widehat{v}_j| \\
&\quad + \mathbf{u} \, |\widehat{\widetilde{v}}_j|^T |J_P| \, (|M_P| \, |\widehat{v}_j| + |\widehat{b}_j| \, |\widehat{v}_j|) \\
&\quad + (m+2n) \, \mathbf{u} \, |\widehat{\widetilde{v}}_j|^T |J_P| \, |M_P| \, |\widehat{v}_j| + \mathcal{O}(\mathbf{u}^2) \\
&\leq (2m+2n+2) \, \mathbf{u} \, |\widehat{\widetilde{v}}_j|^T |J_P| \, |M_P| \, |\widehat{v}_j| \\
&\quad + 3 \, \mathbf{u} \, |\widehat{b}_j| \, |\widehat{\widetilde{v}}_j|^T |J_P| \, |\widehat{v}_j| + \mathcal{O}(\mathbf{u}^2).
\end{aligned}
$$

Combining (5.1.42) and (5.1.43) we have

$$h_{2j,2j-1} = \frac{\widehat{w}_j^T J_P \widehat{v}_j + \zeta_1}{-\widehat{w}_j^T J_P \widehat{v}_j + \zeta_2} + \mathcal{O}(\mathbf{u}^2)$$

$$= -1 + \frac{\zeta_1 + \zeta_2}{-\widehat{w}_j^T J_P \widehat{v}_j + \zeta_2} + \mathcal{O}(\mathbf{u}^2)$$

$$=: -1 + \kappa_j + \mathcal{O}(\mathbf{u}^2).$$

Using the Taylor expansion of $f(x) = \frac{\zeta_1 + \zeta_2}{x + \zeta_2}$ at $t = x - \zeta_2$,

$$f(x) = f(t) + f'(t)(x - t) + \frac{f''(t)}{2}(x - t)^2 + higher\ order\ terms$$

$$= \frac{\zeta_1 + \zeta_2}{x} - \frac{\zeta_1 + \zeta_2}{x^2}\zeta_2 + \frac{\zeta_1 + \zeta_2}{x^3}\zeta_2^2 + higher\ order\ terms,$$

we obtain

$$|\kappa_j| \leq \frac{|\zeta_1| + |\zeta_2|}{|\widehat{w}_j^T J_P \widehat{v}_j|} + \mathcal{O}(\mathbf{u}^2)$$

$$\leq \frac{2(m + n + 2)\,\mathbf{u}\,|\widehat{v}_j|^T |J_P|\,|M_P|\,|\widehat{v}_j|}{|\widehat{w}_j^T J_P \widehat{v}_j|} \qquad (5.1.44)$$

$$+ \frac{5\,\mathbf{u}\,|\widehat{b}_j|\,|\widehat{v}_j|^T |J_P|\,|\widehat{v}_j|}{|\widehat{w}_j^T J_P \widehat{v}_j|} + \mathcal{O}(\mathbf{u}^2).$$

Next we turn our attention to the terms $h_{2j,2m-1}, h_{2j-1,2m-1}$, and $h_{2j,2m}$. The analysis of these three terms will be demonstrated by considering $h_{2j,2m} = \widehat{w}_j^T J_P \widehat{w}_m$. Let us assume that we have already analyzed all previous terms, that is, all the terms in the $2m \times 2m$ leading principal submatrix of H, printed in bold face,

$$\begin{bmatrix}
\mathbf{h_{11}} & \cdots & \mathbf{h_{1,2j-1}} & \mathbf{h_{1,2j}} & \cdots & \mathbf{h_{1,2m-1}} & \mathbf{h_{1,2m}} \\
\vdots & & \vdots & \vdots & & \vdots & \vdots \\
\mathbf{h_{2j-1,1}} & \cdots & \mathbf{h_{2j-1,2j-1}} & \mathbf{h_{2j-1,2j}} & \cdots & \mathbf{h_{2j-1,2m-1}} & \mathbf{h_{2j-1,2m}} \\
\mathbf{h_{2j,1}} & \cdots & \mathbf{h_{2j,2j-1}} & \mathbf{h_{2j,2j}} & \cdots & \mathbf{h_{2j,2m-1}} & h_{2j,2m} \\
\mathbf{h_{2j+1,1}} & \cdots & \mathbf{h_{2j+1,2j-1}} & \mathbf{h_{2j+1,2j}} & \cdots & \mathbf{h_{2j+1,2m-1}} & \mathbf{h_{2j+1,2m}} \\
\vdots & & \vdots & \vdots & & \vdots & \vdots \\
\mathbf{h_{2m-1,1}} & \cdots & \mathbf{h_{2m-1,2j-1}} & \mathbf{h_{2m-1,2j}} & \cdots & \mathbf{h_{2m-1,2m-1}} & \mathbf{h_{2m-1,2m}} \\
\mathbf{h_{2m,1}} & \cdots & \mathbf{h_{2m,2j-1}} & h_{2m,2j} & \cdots & h_{2m,2m-1} & h_{2m,2m}
\end{bmatrix}$$

Our goal is to rewrite $h_{2j,2m}$ in terms of any of these already analyzed terms. First of all, note that for $j = m$ we have $h_{2m,2m} = 0$. Hence for the following discussion we assume $j < m$. From (5.1.24) we have

$$
\begin{aligned}
\widehat{a}_m h_{2j,2m} &= \widehat{w}_j^T J_P(M_P \widehat{v}_m - \widehat{b}_m \widehat{v}_m + h_m) \\
&= \widehat{w}_j^T J_P M_P \widehat{v}_m - \widehat{b}_m h_{2j,2m-1} + \widehat{w}_j^T J_P h_m.
\end{aligned}
$$

Using (5.1.21) we obtain for $\widehat{w}_j^T J_P M_P \widehat{v}_m$

$$
\begin{aligned}
\widehat{d}_m \widehat{w}_j^T J_P M_P \widehat{v}_m &= -\widehat{d}_{m-1} \widehat{w}_j^T J_P M_P \widehat{v}_{m-2} - \widehat{c}_{m-1} \widehat{w}_j^T J_P M_P \widehat{v}_{m-1} \\
&\quad + \widehat{w}_j^T J_P M_P \widehat{w}_{m-1} + \widehat{a}_{m-1}^{-1} \widehat{w}_j^T J_P \widehat{v}_{m-1} \\
&\quad + \widehat{w}_j^T J_P M_P g_m.
\end{aligned}
$$

Using (5.1.24) twice yields

$$
\begin{aligned}
\widehat{d}_m \widehat{w}_j^T J_P M_P \widehat{v}_m &= -\widehat{d}_{m-1} \widehat{w}_j^T J_P(\widehat{a}_{m-2} \widehat{w}_{m-2} + \widehat{b}_{m-2} \widehat{v}_{m-2} - h_{m-2}) \\
&\quad - \widehat{c}_{m-1} \widehat{w}_j^T J_P(\widehat{a}_{m-1} \widehat{w}_{m-1} + \widehat{b}_{m-1} \widehat{v}_{m-1} - h_{m-1}) \\
&\quad + \widehat{w}_j^T J_P M_P \widehat{w}_{m-1} + \widehat{a}_{m-1}^{-1} h_{2j,2m-3} \\
&\quad + \widehat{w}_j^T J_P M_P g_m \\
&= \widehat{w}_j^T J_P M_P \widehat{w}_{m-1} - \widehat{d}_{m-1} \widehat{a}_{m-2} h_{2j,2m-4} \\
&\quad - \widehat{d}_{m-1} \widehat{b}_{m-2} h_{2j,2m-5} - \widehat{c}_{m-1} \widehat{a}_{m-1} h_{2j,2m-2} \\
&\quad - \widehat{c}_{m-1} \widehat{b}_{m-1} h_{2j,2m-3} + \widehat{a}_{m-1}^{-1} h_{2j,2m-3} \\
&\quad + \widehat{d}_{m-1} \widehat{w}_j^T J_P h_{m-2} + \widehat{c}_{m-1} \widehat{w}_j^T J_P h_{m-1} \\
&\quad + \widehat{w}_j^T J_P M_P g_m.
\end{aligned}
$$

The last term that needs our attention here is $\widehat{w}_j^T J_P M_P \widehat{w}_{m-1}$. From (5.1.24) we have

$$
\begin{aligned}
\widehat{a}_j \widehat{w}_j^T J_P M_P \widehat{w}_{m-1} &= (M_P \widehat{v}_j - \widehat{b}_j \widehat{v}_j + h_j)^T J_P M_P \widehat{w}_{m-1} \\
&= \widehat{v}_j^T J_P \widehat{w}_{m-1} + \widehat{b}_j \widehat{w}_{m-1}^T J_P M_P^{-1} \widehat{v}_j \\
&\quad + h_j^T J_P M_P \widehat{w}_{m-1}.
\end{aligned}
$$

as M is symplectic. Using (5.1.21) yields

$$
\begin{aligned}
\widehat{a}_j \widehat{w}_j^T J_P M_P \widehat{w}_{m-1} &= h_{2j-1,2m-2} + h_j^T J_P M_P \widehat{w}_{m-1} \\
&\quad + \widehat{a}_j \widehat{b}_j \widehat{w}_{m-1}^T J_P(\widehat{d}_{j+1} \widehat{v}_{j+1} + \widehat{d}_j \widehat{v}_{j-1} \\
&\quad + \widehat{c}_j \widehat{v}_j - \widehat{w}_j - g_{j+1}) \\
&= h_{2j-1,2m-2} + h_j^T J_P M_P \widehat{w}_{m-1} \\
&\quad - \widehat{a}_j \widehat{b}_j \widehat{w}_{m-1}^T J_P g_{j+1} \\
&\quad + \widehat{a}_j \widehat{b}_j (\widehat{d}_{j+1} h_{2m-2,2j+1} + \widehat{d}_j h_{2m-2,2j-3} \\
&\quad + \widehat{c}_j h_{2m-2,2j-1} + h_{2m-2,2j}).
\end{aligned}
$$

Therefore we obtain the expression (5.1.40) for $\widehat{d}_m\widehat{a}_m h_{2j,2m}$. A similar analysis can be done for $h_{2j,2m-1}$ and $h_{2j-1,2m-1}$.   √

EXAMPLE 5.14 *In order to illustrate the findings of Lemma 5.13 some numerical experiments were done using a* 100×100 *symplectic block-diagonal matrix*

$$M = \text{diag}(D, D^{-1}), \qquad (5.1.45)$$

with

$$D = \text{diag}(200, 100, 50, 47, \ldots, 4, 3, \left[\begin{smallmatrix} 2 & 1 \\ -1 & 2 \end{smallmatrix}\right]). \qquad (5.1.46)$$

All computations were done using MATLAB *Version 5.3 on a Sun Ultra 1 with IEEE double-precision arithmetic and machine precision* $\epsilon = 2.2204 \times 10^{-16}$. *Our code implements exactly the algorithm as given in Table 5.7.*

The symplectic Lanczos process generates a sequence of symplectic butterfly matrices $B^{2k,2k}$ *whose eigenvalues are increasingly better approximates to eigenvalues of* M. *The largest Ritz value approximates the largest eigenvalue* $\lambda_1 = 200$ *of* M.

For the first set of tests a random starting vector v_1 *was used. Table 5.2 lists the upper bound* κ_m *for the deviation of* $h_{m+1,m}$ *from 1 and of* $h_{m,m+1}$ *from* -1 *for* $m = 1, \ldots, 13$. *Due to roundoff errors in the computation of* H *these deviations are not the same, as they should be theoretically. The bound for* $|\kappa_m|$ *is typically one order of magnitude larger than the computed values of* κ_m.

m	bound for $|\kappa_m|$	$|h_{m+1,m} + 1|$	$|h_{m,m+1} - 1|$
1	$2.8599 * 10^{-14}$	$2.2204 * 10^{-16}$	$4.4409 * 10^{-16}$
2	$8.3207 * 10^{-14}$	$2.2204 * 10^{-16}$	$1.1102 * 10^{-16}$
3	$7.0045 * 10^{-14}$	$2.2204 * 10^{-16}$	$1.1102 * 10^{-16}$
4	$5.5521 * 10^{-14}$	$2.2204 * 10^{-16}$	$2.2204 * 10^{-16}$
5	$6.6807 * 10^{-14}$	$2.2204 * 10^{-16}$	$2.2204 * 10^{-16}$
6	$6.4046 * 10^{-14}$	$2.2204 * 10^{-16}$	$2.2204 * 10^{-16}$
7	$6.8944 * 10^{-14}$	$2.2204 * 10^{-16}$	$2.2204 * 10^{-16}$
8	$5.2126 * 10^{-14}$	$2.2204 * 10^{-16}$	$2.2204 * 10^{-16}$
9	$1.0334 * 10^{-13}$	$2.2204 * 10^{-16}$	$3.3307 * 10^{-16}$
10	$2.0037 * 10^{-13}$	$3.3307 * 10^{-16}$	$3.3307 * 10^{-16}$
11	$2.7005 * 10^{-11}$	$3.3307 * 10^{-16}$	$3.5527 * 10^{-15}$
12	$3.6116 * 10^{-14}$	$3.5527 * 10^{-15}$	$5.3291 * 10^{-15}$
13	$4.3575 * 10^{-14}$	$3.5527 * 10^{-15}$	$5.3291 * 10^{-15}$

Table 5.2.   upper bound for $|\kappa_m|$ from Lemma 5.13, random starting vector

The propagation of the roundoff error in H, *described by (5.1.40), can nicely be seen in Figure 5.1. In order to follow the error propagation we have computed* $Z = H - J_P$. *In each step of the symplectic Lanczos method*

two symplectic Lanczos vectors are computed. Hence looking at the principal submatrices of Z of dimension $2m, m = 2, 3, 4, \ldots$ we can follow the error propagation as these submatrices grow by two additional rows and columns representing the error associated with the two new Lanczos vectors. In Figure 5.1 the absolute values of the entries of the principal submatrices of Z of dimension $2m, m = 3, 4, 5, 6$ are shown. For $m = 3$, the entries of the 6×6 principal submatrix of Z are of the order of 10^{-16}. The same is true for the entries of the 8×8 principal submatrix, but it can be seen that the error in the newly computed entries is slightly larger than before. The next two figures for $m = 5$ and $m = 6$ show that the error associated with the new computed entries is increasing slowly.

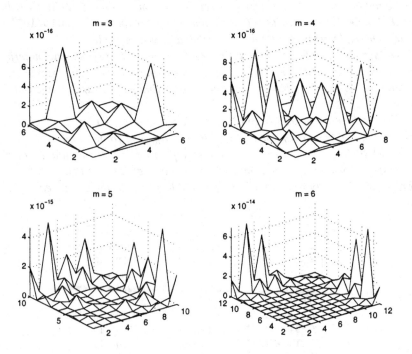

Figure 5.1.   error propagation, random starting vector

The next test reported here was done using the starting vector

$$v_1 = [1, 1, 10^{-11}, \ldots, 10^{-11}]^T \in \mathbf{R}^{100}.$$

This starting vector is close to the sum of the eigenvectors corresponding to the largest and the smallest eigenvalue of M_P. Hence, it can be expected that an invariant subspace corresponding to these eigenvalues is detected soon. Table 5.3 and Figure 5.2 give the same information as the Table 5.2 and Figure 5.1.

j	bound for $|\kappa_j|$	$|h_{j+1,j} + 1|$	$|h_{j,j+1} - 1|$
1	$4.4787 * 10^{-14}$	$1.1102 * 10^{-16}$	$1.1102 * 10^{-16}$
2	$7.1240 * 10^{-04}$	$2.2204 * 10^{-16}$	$1.1102 * 10^{-16}$
3	$1.2347 * 10^{-13}$	$3.3307 * 10^{-16}$	$1.1102 * 10^{-16}$
4	$5.4242 * 10^{-14}$	$3.3307 * 10^{-16}$	$4.4409 * 10^{-16}$
5	$6.6991 * 10^{-14}$	$3.3307 * 10^{-16}$	$4.4409 * 10^{-16}$
6	$6.6742 * 10^{-14}$	$3.3307 * 10^{-16}$	$4.4409 * 10^{-16}$
7	$1.1444 * 10^{-13}$	$3.3307 * 10^{-16}$	$4.4409 * 10^{-16}$
8	$4.9490 * 10^{-14}$	$3.3307 * 10^{-16}$	$4.4409 * 10^{-16}$
9	$6.0268 * 10^{-14}$	$3.3307 * 10^{-16}$	$4.4409 * 10^{-16}$
10	$8.5459 * 10^{-14}$	$3.3307 * 10^{-16}$	$4.4409 * 10^{-16}$
11	$1.4171 * 10^{-13}$	$3.3307 * 10^{-16}$	$4.4409 * 10^{-16}$
12	$2.0743 * 10^{-10}$	$2.1316 * 10^{-14}$	$4.4409 * 10^{-16}$
13	$3.5934 * 10^{-14}$	$4.9738 * 10^{-14}$	$5.6843 * 10^{-14}$

Table 5.3.   upper bound for $|\kappa_m|$ from Lemma 5.13, special starting vector

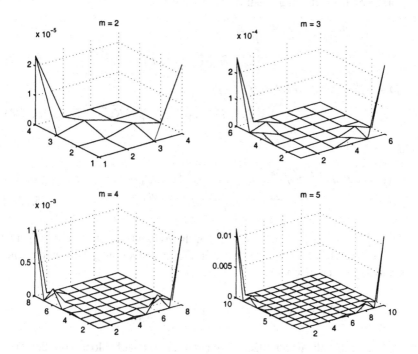

Figure 5.2.   error propagation, special starting vector

Obviously, the J-orthogonality is lost already after two steps of the symplectic Lanczos method. This has almost no effect on the computed values for κ_m.

The next section will discuss loss of J-orthogonality versus convergence. It will be seen that under certain conditions, loss of J-orthogonality is accompanied by the convergence of Ritz values, just as in the last example.

5.1.4   CONVERGENCE VERSUS LOSS OF J–ORTHOGONALITY

It is well-known that in the symmetric Lanczos procedure, loss of orthogonality between the computed Lanczos vectors implies convergence of a Ritz pair to an eigenpair, see, e.g., [113]. Here we will discuss the situation for the symplectic Lanczos algorithm, following the lines of Section 4 of Bai's analysis of the unsymmetric Lanczos algorithm in [11]. We will see that a conclusion similar to the one for the symmetric Lanczos process holds here, subject to a certain condition.

From the previous section, we know that the computed symplectic Lanczos vectors obey the following equalities:

$$
\begin{aligned}
M_P \widehat{S}_P^{2n,2k} &= \widehat{S}_P^{2n,2k} \widehat{B}_P^{2k,2k} \\
&\quad - \left[\widehat{d}_{k+1}\widehat{r}_{k+1}e_{2k-1}^T - E_k \right](\widehat{N_u}^{2k,2k})_P, \quad (5.1.47)
\end{aligned}
$$

$$
\begin{aligned}
M_P^T \widehat{W}_P^{2n,2k} &= \widehat{W}_P^{2n,2k}(\widehat{B}_P^{2k,2k})_P^T \\
&\quad + \left[\widehat{d}_{k+1}J_P\widehat{v}_{k+1}e_{2k}^T + F_k \right](\widehat{K_u}^{2k,2k})_P^{-T}, \quad (5.1.48)
\end{aligned}
$$

with

$$
(\widehat{S}_P^{2n,2k})^T J_P^{2n,2n} \widehat{S}_P^{2n,2k} = K = J_P^{2k,2k} + C_k + \Delta_k - C_k^T, \quad (5.1.49)
$$

$$
\widehat{W}_P^{2n,2k} = J_P^{2n,2n} \widehat{S}_P^{2n,2k} J_P^{2k,2k}, \quad (5.1.50)
$$

where $\widehat{B}_P^{2k,2k} = (\widehat{K_u}^{2k,2k})_P^{-1}(\widehat{N_u}^{2k,2k})_P$, the rounding error matrices E_k and F_k are as in (5.1.26) and, resp., (5.1.37), Δ_k is a block diagonal matrix with 2×2 block on the diagonal,

$$
\Delta_k = \text{diag}\left(\begin{bmatrix} 0 & \kappa_1 \\ -\kappa_1 & 0 \end{bmatrix}, \dots, \begin{bmatrix} 0 & \kappa_k \\ -\kappa_k & 0 \end{bmatrix} \right),
$$

and C_k is a strictly lower block triangular matrix with block size 2. That is $(C_k)_{\ell,j} = 0$ for $\ell = 1, \dots, 2k, j = \ell, \dots, 2k$, and $(C_k)_{2\ell,2\ell-1} = 0$ for $\ell = 1, \dots, k$.

To simplify our discussion, we make two assumptions, which are also used in the analysis of the symmetric Lanczos process [114, p. 265] and in the analysis of the unsymmetric Lanczos process [11]. The first assumption is *local J–orthogonality*, that is, the computed symplectic Lanczos vectors are

J–orthogonal to their neighboring Lanczos vectors:

$$\left[\begin{array}{c} \widehat{v}_{j-1}^T \\ \widehat{w}_{j-1}^T \end{array}\right] \left[\begin{array}{cc} 0 & 1 \\ -1 & 0 \end{array}\right] [\widehat{v}_j \; \widehat{w}_j] = \left[\begin{array}{cc} 0 & 0 \\ 0 & 0 \end{array}\right]. \tag{5.1.51}$$

This implies that the 2×2 block on the subdiagonal of C_k are zero

$$C_k = \left[\begin{array}{ccccccccc} 0 & 0 & 0 & \cdots & \cdots & 0 & 0 & 0 \\ 0 & 0 & 0 & \cdots & \cdots & 0 & 0 & 0 \\ X & 0 & 0 & \cdots & \cdots & 0 & 0 & 0 \\ X & X & 0 & \cdots & \cdots & 0 & 0 & 0 \\ X & X & X & \ddots & & 0 & 0 & 0 \\ & \vdots & & \ddots & \ddots & & \vdots & \\ X & X & X & \cdots & X & 0 & 0 & 0 \\ X & X & X & \cdots & X & X & 0 & 0 \end{array}\right],$$

where the X denote 2×2 blocks.

The second assumption is that the eigenvalue problem for the $2k \times 2k$ butterfly matrix $\widehat{B}_P^{2k,2k} = (\widehat{K}_u^{2k,2k})_P^{-1}(\widehat{N}_u^{2k,2k})_P$ is solved exactly, that is,

$$Y_k^{-1} \widehat{B}_P^{2k,2k} Y_k = \text{diag}(\lambda_1, \lambda_1^{-1}, \ldots, \lambda_k, \lambda_k^{-1}). \tag{5.1.52}$$

This implies that the computed Ritz vector for λ_j is given by

$$z_j = \widehat{S}_P^{2n,2k} y_{2j-1},$$

while the computed Ritz vector for λ_j^{-1} is given by

$$x_j = \widehat{S}_P^{2n,2k} y_{2j}.$$

Our goal is to derive expressions for $z_j^T J_P \widehat{r}_{k+1}$ and $x_j^T J_P \widehat{r}_{k+1}$ that describe the way in which J–orthogonality is lost. In exact arithmetic, these expressions are zero. Our approach follows Bai's derivations in [11, Proof of Theorem 4.1]. Premultiplying (5.1.48) by $(\widehat{S}_P^{2n,2k})^T$ and taking the transpose yields

$$(\widehat{W}_P^{2n,2k})^T M_P \widehat{S}_P^{2n,2k}$$
$$= \widehat{B}_P^{2k,2k} (\widehat{W}_P^{2n,2k})^T \widehat{S}_P^{2n,2k}$$
$$+ (\widehat{K}_u^{2k,2k})_P^{-1} \left[\widehat{d}_{k+1} J_P \widehat{v}_{k+1} e_{2k}^T + F_k\right]^T \widehat{S}_P^{2n,2k}.$$

Premultiplying (5.1.47) by $(\widehat{W}_P^{2n,2k})^T$ we obtain

$$(\widehat{W}_P^{2n,2k})^T M_P \widehat{S}_P^{2n,2k}$$
$$= (\widehat{W}_P^{2n,2k})^T \widehat{S}_P^{2n,2k} \widehat{B}_P^{2k,2k}$$
$$- (\widehat{W}_P^{2n,2k})^T \left[\widehat{d}_{k+1} \widehat{r}_{k+1} e_{2k-1}^T - E_k\right] (\widehat{N}_u^{2k,2k})_P.$$

Subtracting these two equations, we obtain

$$
\begin{aligned}
(\widehat{W}_P^{2n,2k})^T \widehat{S}_P^{2n,2k} &\widehat{B}_P^{2k,2k} - \widehat{B}_P^{2k,2k} (\widehat{W}_P^{2n,2k})^T \widehat{S}_P^{2n,2k} \\
&= \widehat{d}_{k+1} (\widehat{K}_u^{2k,2k})_P^{-1} e_{2k} \widehat{v}_{k+1}^T J_P^T \widehat{S}_P^{2n,2k} \\
&\quad - \widehat{d}_{k+1} (\widehat{W}_P^{2n,2k})^T \widehat{r}_{k+1} e_{2k}^T \\
&\quad + (\widehat{K}_u^{2k,2k})_P^{-1} F_k^T \widehat{S}_P^{2n,2k} \\
&\quad - (\widehat{W}_P^{2n,2k})^T E_k (\widehat{N}_u^{2k,2k})_P .
\end{aligned} \tag{5.1.53}
$$

We are most interested in deriving an expression for

$$
(\widehat{S}_P^{2n,2k})^T J_P \widehat{r}_{k+1} e_{2k-1}^T \quad (\text{or } (\widehat{W}_P^{2n,2k})^T J_P \widehat{r}_{k+1} e_{2k-1}^T)
$$

from the above equation. From this we can easily obtain expressions for $z_j^T J_P r_{k+1}$ or $x_j^T J_P r_{k+1}$ as desired. In order to do so, we note that most of the matrices in (5.1.53) have a very special form. Let us start with the left-hand side. From (5.1.49) we have

$$
(\widehat{S}_P^{2n,2k})^T J_P \widehat{S}_P^{2n,2k} = K = J_P^{2k,2k} + C_k + \Delta_k - C_k^T .
$$

This implies

$$
\begin{aligned}
(\widehat{W}_P^{2n,2k})^T \widehat{S}_P^{2n,2k} &= J_P^{2k,2k} (\widehat{S}_P^{2n,2k})^T J_P^{2n,2n} \widehat{S}_P^{2n,2k} \\
&= J_P^{2k,2k} K \\
&= -I^{2k,2k} + J_P^{2k,2k} C_k + J_P^{2k,2k} \Delta_k - J_P^{2k,2k} C_k^T ,
\end{aligned}
$$

where $J_P^{2k,2k} C_k$ and $(J_P^{2k,2k} C_k^T)^T$ have the same form as C_k, and $J_P^{2k,2k} \Delta_k$ is a diagonal matrix,

$$
J_P^{2k,2k} \Delta_k = \mathrm{diag}\left(\begin{bmatrix} -\kappa_1 & 0 \\ 0 & -\kappa_1 \end{bmatrix}, \dots, \begin{bmatrix} -\kappa_k & 0 \\ 0 & -\kappa_k \end{bmatrix} \right).
$$

Therefore, we can rewrite the left-hand side of (5.1.53) as

$$
\begin{aligned}
(\widehat{W}_P^{2n,2k})^T &\widehat{S}_P^{2n,2k} \widehat{B}_P^{2k,2k} - \widehat{B}_P^{2k,2k} (\widehat{W}_P^{2n,2k})^T \widehat{S}_P^{2n,2k} \\
&= \left[-I^{2k,2k} + J_P^{2k,2k} C_k + J_P^{2k,2k} \Delta_k + J_P^{2k,2k} C_k^T \right] \widehat{B}_P^{2k,2k} \\
&\quad - \widehat{B}_P^{2k,2k} \left[-I^{2k,2k} + J_P^{2k,2k} C_k + J_P^{2k,2k} \Delta_k + J_P^{2k,2k} C_k^T \right] \\
&= \left[J_P^{2k,2k} C_k \widehat{B}_P^{2k,2k} - \widehat{B}_P^{2k,2k} J_P^{2k,2k} C_k \right] \\
&\quad + \left[J_P^{2k,2k} \Delta_k \widehat{B}_P^{2k,2k} - \widehat{B}_P^{2k,2k} J_P^{2k,2k} \Delta_k \right] \\
&\quad + \left[J_P^{2k,2k} C_k^T \widehat{B}_P^{2k,2k} - \widehat{B}_P^{2k,2k} J_P^{2k,2k} C_k^T \right].
\end{aligned}
$$

By the local J–orthogonality assumption (and, therefore, by the special form of $J_P^{2k,2k} C_k$), it follows that

$$L_k^{(1)} := J_P^{2k,2k} C_k \widehat{B}_P^{2k,2k} - \widehat{B}_P^{2k,2k} J_P^{2k,2k} C_k$$

is a strictly lower block triangular matrix with block size 2. With the same argument we have that

$$U_k^{(1)} := J_P^{2k,2k} C_k^T \widehat{B}_P^{2k,2k} - \widehat{B}_P^{2k,2k} J_P^{2k,2k} C_k^T$$

is a strictly upper block triangular matrix with block size 2. Since the 2×2 diagonal blocks of $J_P^{2k,2k} \Delta_k \widehat{B}_P^{2k,2k} - \widehat{B}_P^{2k,2k} J_P^{2k,2k} \Delta_k$ are zero, we can write

$$J_P^{2k,2k} \Delta_k \widehat{B}_P^{2k,2k} - \widehat{B}_P^{2k,2k} J_P^{2k,2k} \Delta_k = L_k^{(2)} + U_k^{(2)},$$

where $L_k^{(2)}$ is strictly lower block triangular and $U_k^{(2)}$ strictly upper block triangular. Hence,

$$(\widehat{W}_P^{2n,2k})^T \widehat{S}_P^{2n,2k} \widehat{B}_P^{2k,2k} - \widehat{B}_P^{2k,2k} (\widehat{W}_P^{2n,2k})^T \widehat{S}_P^{2n,2k}$$
$$= L_k^{(1)} + L_k^{(2)} + U_k^{(1)} + U_k^{(2)}.$$

Now let us turn our attention to the right-hand side of (5.1.53). The row vector

$$\widehat{v}_{k+1}^T J_P^T \widehat{S}_P^{2n,2k} = [* \ldots * \; 0 \; 0]$$

has nonzero elements in its first $(2n - 2)$ positions. As $(\widehat{K_u}^{2k,2k})_P^{-1} e_{2k} = b_k e_{2k-1} + a_k e_{2k}$ we have that

$$L_k^{(3)} := \widehat{d}_{k+1} (\widehat{K_u}^{2k,2k})_P^{-1} e_{2k} \widehat{v}_{k+1}^T J_P^T \widehat{S}_P^{2n,2k}$$

is a strictly lower block triangular matrix with block size 2. Similarly we have that

$$U_k^{(3)} := \widehat{d}_{k+1} (\widehat{W}_P^{2n,2k})^T \widehat{r}_{k+1} e_{2k}^T$$

is a strictly upper block triangular matrix with block size 2. Hence, we can rewrite (5.1.53) as

$$L_k^{(1)} + L_k^{(2)} - L_k^{(3)} + U_k^{(1)} + U_k^{(2)} - U_k^{(3)}$$
$$= (\widehat{K_u}^{2k,2k})_P^{-1} F_k^T \widehat{S}_P^{2n,2k} - (\widehat{W}_P^{2n,2k})^T E_k (\widehat{N_u}^{2k,2k})_P. \quad (5.1.54)$$

This implies that the diagonal blocks of

$$(\widehat{K_u}^{2k,2k})_P^{-1} F_k^T \widehat{S}_P^{2n,2k} - (\widehat{W}_P^{2n,2k})^T E_k (\widehat{N_u}^{2k,2k})_P$$

must be zero. Therefore, we can write

$$(\widehat{K_u}^{2k,2k})_P^{-1} F_k^T \widehat{S}_P^{2n,2k} - (\widehat{W}_P^{2n,2k})^T E_k (\widehat{N_u}^{2k,2k})_P = L_k^{(4)} + U_k^{(4)}$$

where $L_k^{(4)}$ is strictly lower block triangular and $U_k^{(4)}$ is strictly upper block triangular. By writing down only the strictly upper block triangular part of (5.1.54) we have

$$U_k^{(3)} = U_k^{(1)} + U_k^{(2)} - U_k^{(4)}$$

or

$$\widehat{d}_{k+1}(\widehat{W}_P^{2n,2k})^T \widehat{r}_{k+1} e_{2k}^T$$
$$= J_P^{2k,2k} C_k^T \widehat{B}_P^{2k,2k} - \widehat{B}_P^{2k,2k} J_P^{2k,2k} C_k^T + U_k^{(2)} - U_k^{(4)}.$$

This is equivalent to

$$\widehat{d}_{k+1}(\widehat{S}_P^{2n,2k})^T J_P^{2n,2n} \widehat{r}_{k+1} e_{2k}^T$$
$$= - J_P^{2k,2k} \left[J_P^{2k,2k} C_k^T \widehat{B}_P^{2k,2k} - \widehat{B}_P^{2k,2k} J_P^{2k,2k} C_k^T \right.$$
$$\left. + U_k^{(2)} - U_k^{(4)} \right]$$
$$= C_k^T \widehat{B}_P^{2k,2k} - (\widehat{B}_P^{2k,2k})^{-T} C_k^T - J_P^{2k,2k} \left[U_k^{(2)} - U_k^{(4)} \right], \quad (5.1.55)$$

where we have used the fact that $\widehat{B}_P^{2k,2k}$ is symplectic.

From (5.1.52) we get

$$\widehat{B}_P^{2k,2k} y_{2j-1} = \lambda_j y_{2j-1} \quad \text{and} \quad \widehat{B}_P^{2k,2k} y_{2j} = \lambda_j^{-1} y_{2j}.$$

This implies

$$y_{2j-1}^T (\widehat{B}_P^{2k,2k})^{-T} = \lambda_j^{-1} y_{2j-1}^T \quad \text{and} \quad y_{2j}^T (\widehat{B}_P^{2k,2k})^{-T} = \lambda_j y_{2j}^T.$$

Hence, premultiplying (5.1.55) by y_{2j}^T and postmultiplying by y_{2j-1} yields

$$\widehat{d}_{k+1} y_{2j}^T (\widehat{S}_P^{2n,2k})^T J_P^{2n,2n} \widehat{r}_{k+1} (e_{2k}^T y_{2j-1})$$
$$= y_{2j}^T C_k^T \widehat{B}_P^{2k,2k} y_{2j-1} - y_{2j}^T (\widehat{B}_P^{2k,2k})^{-T} C_k^T y_{2j-1}$$
$$\quad - y_{2j}^T J_P^{2k,2k} \left[U_k^{(2)} - U_k^{(4)} \right] y_{2j-1}$$
$$= \lambda_j y_{2j}^T C_k^T y_{2j-1} - \lambda_j y_{2j}^T C_k^T y_{2j-1}$$
$$\quad - y_{2j}^T J_P^{2k,2k} \left[U_k^{(2)} - U_k^{(4)} \right] y_{2j-1}$$
$$= y_{2j}^T J_P^{2k,2k} \left[U_k^{(4)} - U_k^{(2)} \right] y_{2j-1}.$$

Similarly, premultiplying (5.1.55) by y_{2j-1}^T and postmultiplying by y_{2j} yields

$$\widehat{d}_{k+1}y_{2j-1}^T(\widehat{S}_P^{2n,2k})^T J_P^{2n,2n}\widehat{r}_{k+1}e_{2k}^T y_{2j} = y_{2j-1}^T J_P^{2k,2k}\left[U_k^{(4)} - U_k^{(2)}\right]y_{2j}.$$

Therefore, with the assumptions (5.1.51) and (5.1.52) we have

THEOREM 5.15 *Assume that the symplectic Lanczos algorithm in finite-precision arithmetic satisfies (5.1.47) – (5.1.50). Let*

$$L_k^{(2)} + U_k^{(2)} = J_P^{2k,2k}\Delta_k\widehat{B}_P^{2k,2k} - \widehat{B}_P^{2k,2k}J_P^{2k,2k}\Delta_k$$

$$L_k^{(4)} + U_k^{(4)} = (\widehat{K}_u^{2k,2k})_P^{-1}F_k^T\widehat{S}_P^{2n,2k} - (\widehat{W}_P^{2n,2k})^T E_k(\widehat{N}_u^{2k,2k})_P$$

where $L_k^{(2)}$ and $L_k^{(4)}$ are strictly lower block triangular matrices, and $U_k^{(2)}$ and $U_k^{(4)}$ are strictly upper block triangular matrices with block size 2. Then the computed symplectic Lanczos vectors $x_j = \widehat{S}_P^{2n,2k}y_{2j}$ and $z_j = \widehat{S}_P^{2n,2k}y_{2j-1}$ satisfy

$$x_j^T J_P^{2n,2n}\widehat{r}_{k+1} = \frac{y_{2j}^T J_P^{2k,2k}\left[U_k^{(4)} - U_k^{(2)}\right]y_{2j-1}}{\widehat{d}_{k+1}(e_{2k}^T y_{2j-1})}$$

$$=: \frac{\psi_1}{\widehat{d}_{k+1}(e_{2k}^T y_{2j-1})}, \qquad (5.1.56)$$

$$z_j^T J_P^{2n,2n}\widehat{r}_{k+1} = \frac{y_{2j-1}^T J_P^{2k,2k}\left[U_k^{(4)} - U_k^{(2)}\right]y_{2j}}{\widehat{d}_{k+1}(e_{2k}^T y_{2j})}$$

$$=: \frac{\psi_2}{\widehat{d}_{k+1}(e_{2k}^T y_{2j})}. \qquad (5.1.57)$$

The derived equations are similar to those obtained by Bai for the unsymmetric Lanczos process. Hence we can interpret our findings analogously: Equations (5.1.56) and (5.1.57) describe the way in which the J–orthogonality is lost. Recall that the scalar d_{k+1} and the last eigenvector components $(e_{2k}^T y_{2j-1})$ and $(e_{2k}^T y_{2j})$ are also essential quantities used as the backward error criteria for the computed Ritz triplets $\{\lambda_i, z_i, (Jx_i)^T\}$ and $\{\lambda_i^{-1}, x_i, (Jz_i)^T\}$ discussed in Section 5.1.2. (Also recall that by Lemma 3.11 $|e_{2k}^T y_\ell| > 0$ if $B^{2k,2k}$ is unreduced.) Hence, if the quantities $|\psi_1|$ and $|\psi_2|$ are bounded and bounded away from zero, then (5.1.56) and (5.1.57) reflect the reciprocal relation between the convergence of the symplectic Lanczos process (i.e., tiny $\widehat{d}_{k+1}(e_{2k}^T y_{2j-1})$ and $\widehat{d}_{k+1}(e_{2k}^T y_{2j})$) and the loss of J–orthogonality (i.e., large $\widehat{r}_{k+1}^T J_P x_j$ and $\widehat{r}_{k+1}^T J_P z_j$).

EXAMPLE 5.16 *Here we continue the numerical tests with the test matrix (5.1.45). The first test reported was done using a random starting vector v_1. Table 5.4 illustrates the loss of J–orthogonality among the symplectic Lanczos vectors in terms of $z_1^T J_P \widehat{r}_{k+1}$ and the convergence of a Ritz value in terms of the residual $\widehat{d}_{k+1}(e_{2k}^T y_2)$. As predicted by Theorem 5.15, the loss of J–orthogonality accompanies the convergence of a Ritz value to the largest eigenvalue λ_1 (and the convergence of a Ritz value to the smallest eigenvalue λ_1^{-1}) in terms of small residuals. When the symplectic Lanczos process is*

Lanczos step	$z_1^T J_P \widehat{r}_{k+1}$	$\widehat{d}_{k+1}(e_{2k}^T y_2)$
1	$-9.1290 * 10^{-17}$	$-6.4278 * 10^{-01}$
2	$1.6751 * 10^{-17}$	$3.5949 * 10^{-01}$
3	$-8.4297 * 10^{-18}$	$-8.1016 * 10^{-02}$
4	$2.6983 * 10^{-17}$	$-1.7984 * 10^{-02}$
5	$2.8513 * 10^{-16}$	$-1.3822 * 10^{-03}$
6	$4.7089 * 10^{-15}$	$-8.3119 * 10^{-05}$
7	$6.8569 * 10^{-14}$	$-5.7074 * 10^{-06}$
8	$-8.3995 * 10^{-13}$	$-4.6590 * 10^{-07}$
9	$-9.3850 * 10^{-12}$	$-4.1698 * 10^{-08}$
10	$9.0525 * 10^{-11}$	$-4.3229 * 10^{-09}$
11	$-4.1822 * 10^{-10}$	$9.3571 * 10^{-10}$
12	$6.8361 * 10^{-09}$	$-5.7230 * 10^{-11}$
13	$-2.9881 * 10^{-07}$	$1.3010 * 10^{-12}$
14	$5.7946 * 10^{-06}$	$6.5210 * 10^{-14}$
15	$-1.0299 * 10^{-04}$	$2.6478 * 10^{-15}$
16	$1.5128 * 10^{-03}$	$-1.0915 * 10^{-15}$

Table 5.4.   loss of J–orthogonality versus convergence of Ritz value, random starting vector

stopped at $k = 16$, the computed largest Ritz value λ_1 has the relative accuracy

$$\frac{|200 - \lambda_1|}{200} \approx 1.5632 * 10^{-15}.$$

We note that in this example, the Ritz value corresponding to the largest eigenvalue of M is well conditioned, while the condition number for all eigenvalues of M is one, the condition number of the largest Ritz value is ≈ 1.08. The results for $w_1^T J_P \widehat{r}_{k+1}$ and $\widehat{d}_{k+1}(e_{2k}^T y_1)$ are almost the same.

Using the special starting vector $v_1 = [1, 1, 10^{-11}, \ldots, 10^{-11}]^T$, the results presented in Table 5.5 are obtained. As already seen in Example 5.14, J–orthogonality is lost fast. This is accompanied by the convergence of a Ritz value to the largest eigenvalue λ_1 (and the convergence of a Ritz value to the smallest eigenvalue λ_1^{-1}) in terms of small residuals. When the symplectic Lanczos process is stopped at $k = 10$, the computed butterfly matrix has two eigenvalues close to 200 and one eigenvalue close to 100. Hence the loss of J-orthogonality results, as in the standard unsymmetric Lanczos algorithm,

Lanczos step	$z_1^T J_P \widehat{r}_{k+1}$	$\widehat{d}_{k+1}(e_{2k}^T y_2)$
1	$1.2839 * 10^{-08}$	$-5.4906 * 10^{-11}$
2	$1.1915 * 10^{-07}$	$-5.9165 * 10^{-12}$
3	$4.0409 * 10^{-07}$	$1.7423 * 10^{-12}$
4	$-2.3952 * 10^{-06}$	$-2.9338 * 10^{-13}$
5	$-3.0405 * 10^{-05}$	$2.3785 * 10^{-14}$
6	$-4.0010 * 10^{-04}$	$3.2861 * 10^{-15}$
7	$1.5373 * 10^{-03}$	$-1.1173 * 10^{-14}$

Table 5.5. loss of *J*–orthogonality versus convergence of Ritz value, special starting vector

in ghost eigenvalues. That is, multiple eigenvalues of $B^{2k,2k}$ correspond to simple eigenvalues of M. The eigenvalues close to 200 have relative accuracy $1.4211 * 10^{-16}$, resp. $4.0767 * 10^{-07}$ for its ghost, the one close to 100 has the relative accuracy $9.4163 * 10^{-10}$.

Let us conclude our analysis by estimating $|\psi_1|$ and $|\psi_2|$. Let us assume (again analogous to Bai's analysis) that $\Delta_k = 0$, i.e., $\widehat{w}_j^T J_P \widehat{v}_j = -1$, which simplifies the technical details of the analysis and appears to be the case in practice, up to the order of machine precision. Under this assumption, we have $U_k^{(2)} = 0$. Moreover, we have

$$|\psi_\ell| \leq ||U_k^{(4)}||_F ||y_{2j}||_2 ||y_{2j-1}||_2,$$

for $\ell = 1, 2$. Let us derive an estimate for $||U_k^{(4)}||_F$. $U_k^{(4)}$ is the strictly upper block triangular part of

$$(\widehat{K_u}^{2k,2k})_P^{-1} F_k^T \widehat{S}_P^{2n,2k} - (\widehat{W}_P^{2n,2k})^T E_k (\widehat{N_u}^{2k,2k})_P. \qquad (5.1.58)$$

A generous upper bound is therefore given by

$$
\begin{aligned}
||U_k^{(4)}||_F \ \leq \ & ||\widehat{K_u}^{2k,2k}||_F ||F_k^T||_F ||\widehat{S}^{2n,2k}||_F \\
& + ||\widehat{W}^{2n,2k}||_F ||E_k||_F ||\widehat{N_u}^{2k,2k}||_F \\
\leq \ & ||\widehat{S}^{2n,2k}||_F \left[||\widehat{K_u}^{2k,2k}||_F ||F_k||_F + ||E_k||_F ||\widehat{N_u}^{2k,2k}||_F \right] \\
\leq \ & \mathbf{u} \, ||\widehat{S}^{2n,2k}||_F^2 \left\{ 7 \, ||\widehat{K_u}^{2k,2k}||_F ||\widehat{N_u}^{2k,2k}||_F \right. \\
& \left. + (m + 5) \, ||\widehat{K_u}^{2k,2k}||_F ||\widehat{N_u}^{2k,2k}||_F ||M||_F^2 \right.
\end{aligned}
$$

$$+ 4 \, \|\widehat{K_u}^{2k,2k}\|_F$$
$$+ (m+2) \, \|\widehat{K_u}^{2k,2k}\|_F \|M\|_F^2$$
$$+ (m+8) \, \|\widehat{K_u}^{2k,2k}\|_F^2 \|M\|_F$$
$$+ 4 \, \|\widehat{N_u}^{2k,2k}\|_F^2 \|M\|_F$$
$$+ (m+6) \, \|\widehat{N_u}^{2k,2k}\|_F \|M\|_F \Big\} + \mathcal{O}(\mathbf{u}^2).$$

Summarizing, we obtain the following corollary, which gives an upper bound for $|\psi_1|$ and $|\psi_2|$.

COROLLARY 5.17 *Assume that* $\Delta_k = 0$ *in Theorem 5.15. Then*

$$
\begin{aligned}
|\psi| \;\leq\; & \mathbf{u}\,\mathrm{cond}(\lambda_j) \Big\{ (m+5) \, \|\widehat{K_u}^{2k,2k}\|_F \|\widehat{N_u}^{2k,2k}\|_F \|M\|_F^2 \\
& + 7 \, \|\widehat{K_u}^{2k,2k}\|_F \|\widehat{N_u}^{2k,2k}\|_F + 4 \|\widehat{K_u}^{2k,2k}\|_F \\
& + (m+2) \, \|\widehat{K_u}^{2k,2k}\|_F \|M\|_F^2 \\
& + \Big((m+8) \, \|\widehat{K_u}^{2k,2k}\|_F^2 + 4 \, \|\widehat{N_u}^{2k,2k}\|_F^2 \Big) \|M\|_F \\
& + (m+6) \, \|\widehat{N_u}^{2k,2k}\|_F \|M\|_F \Big\} + \mathcal{O}(\mathbf{u}^2),
\end{aligned}
$$

where $\psi \in \{\psi_1, \psi_2\}$ *and*

$$\mathrm{cond}(\lambda_j) = \mathrm{cond}(\lambda_j^{-1}) = \|\widehat{S}^{2n,2k}\|_F^2 \|y_{2j}\|_2 \|y_{2j-1}\|_2$$

is the condition number of the Ritz values λ_j *and* λ_j^{-1}.

Note that this bound is too pessimistic. In order to derive an upper bound for $\|U_k^{(4)}\|_F$, an upper bound for the matrix (5.1.58) is used, as $U_k^{(4)}$ is the strictly upper block triangular part of that matrix. This is a very generous upper bound for $\|U_k^{(4)}\|_F$. Moreover, the term

$$\|\widehat{K_u}^{2k,2k}\|_F \|\widehat{N_u}^{2k,2k}\|_F$$

is an upper bound for the norm of $\widehat{B}_P^{2k,2k}$. The squared terms $\|\widehat{K_u}^{2k,2k}\|_F^2$ and $\|\widehat{N_u}^{2k,2k}\|_F^2$ are introduced as the original equations derived in (5.1.27) and (5.1.34) are given in terms of $\widehat{K_u}^{2k,2k}$ and $\widehat{N_u}^{2k,2k}$, but not in terms of $\widehat{B}_P^{2k,2k}$.

Unfortunately, for the symplectic Lanczos process (as for any unsymmetric Lanczos-like process), near-breakdown may cause the norms of the vectors $\|\tilde{v}_j\|_2$ and $\|w_j\|_2$ to grow unboundedly. Accumulating the quantity

$\sum_{j=1}^{k}(\|\widehat{\widetilde{v}}_j\|_2^2 + \|\widehat{w}_j\|_2^2)$, which costs about $4nk$ flops, we can obtain a computable bound for $\mathrm{cond}(\lambda_j)$ and $\mathrm{cond}(\lambda_j^{-1})$ in practise. Theorem 5.15 and Corollary 5.17 indicate that if the J–orthogonality between $\widehat{r}_{k+1}$ and x_j (and z_j) is lost, then the value $\widehat{d}_{k+1}(e_{2k}^T y_{2j-1})$ is proportional to $|\psi_1|$ (and the value $\widehat{d}_{k+1}(e_{2k}^T y_{2j})$ is proportional to $|\psi_2|$). Given the upper bound from Corollary 5.17, and supposing that $\mathrm{cond}(\lambda_j)$ is reasonably bounded, the loss of J–orthogonality implies that $\widehat{d}_{k+1}(e_{2k}^T y_{2j-1})$ (and $\widehat{d}_{k+1}(e_{2k}^T y_{2j})$) are small. Therefore, in the best case we can state that if the effects of finite-precision arithmetic, E_k and F_k in (5.1.47) and (5.1.48), are small, then small residuals tell us that the computed eigenvalues are eigenvalues of matrices close to the given matrix.

EXAMPLE 5.18 *Example 5.14 and 5.16 are continued. Table 5.6 reports the value for ψ_2 and its upper bound from Corollary 5.17 using a random starting vector. The upper bound $|\psi|$ is too pessimistic, as already discussed above.*

Lanczos step	ψ_2	bound for $|\psi|$
1	$5.8679 * 10^{-17}$	$4.0869 * 10^{-08}$
2	$6.0217 * 10^{-18}$	$3.6108 * 10^{-07}$
3	$6.8293 * 10^{-19}$	$8.4543 * 10^{-07}$
4	$-4.8526 * 10^{-19}$	$1.5293 * 10^{-06}$
5	$-3.9411 * 10^{-19}$	$2.3792 * 10^{-06}$
6	$-3.9140 * 10^{-19}$	$3.5814 * 10^{-06}$
7	$-3.9135 * 10^{-19}$	$4.7835 * 10^{-06}$
8	$3.9133 * 10^{-19}$	$6.0405 * 10^{-06}$
9	$3.9133 * 10^{-19}$	$8.5821 * 10^{-06}$
10	$-3.9133 * 10^{-19}$	$1.4113 * 10^{-05}$
11	$-3.9133 * 10^{-19}$	$1.0338 * 10^{-04}$
12	$-3.9123 * 10^{-19}$	$4.9373 * 10^{-04}$
13	$-3.8875 * 10^{-19}$	$8.9149 * 10^{-04}$
14	$3.7786 * 10^{-19}$	$9.3192 * 10^{-04}$
15	$-2.7270 * 10^{-19}$	$9.5668 * 10^{-04}$
16	$-1.6512 * 10^{-18}$	$9.7898 * 10^{-04}$

Table 5.6. $|\psi_2|$ and its upper bound from Corollary 5.17, random starting vector

When using the special starting vector $v_1 = [1, 1, 10^{-11}, \ldots, 10^{-11}]^T$ the results are similar.

5.2   THE IMPLICITLY RESTARTED SYMPLECTIC LANCZOS ALGORITHM

In the previous sections we have discussed two algorithms for computing approximations to the eigenvalues of a symplectic matrix M. The symplectic Lanczos algorithm is appropriate when the matrix M is large and sparse. If only a small subset of the eigenvalues is desired, the length $2k$ symplectic Lanczos factorization may suffice. The analysis in the last chapter suggests that a strategy for finding $2k$ eigenvalues in a length $2k$ factorization is to find an appropriate starting vector that forces the residual r_{k+1} to vanish. The SR algorithm, on the other hand, computes approximations to all eigenvalues and eigenvectors of M. From Theorem 5.1 (an Implicit-Q-Theorem for the SR case) we know that in exact arithmetic, when using the same starting vector, the SR algorithm and the length $2n$ Lanczos factorization generate the same symplectic butterfly matrices (up to multiplication by a trivial matrix). Forcing the residual for the symplectic Lanczos algorithm to zero has the effect of deflating a sub-diagonal element during the SR algorithm: by Remark 5.2 $r_{k+1} = M_P v_{k+1}$ and from the symplectic Lanczos process we have $d_{k+1} = \|v_{k+1}\|_2$. Hence a zero residual implies a zero d_{k+1} such that deflation occurs for the corresponding butterfly matrix.

Our goal in this section will be to construct a starting vector for the symplectic Lanczos process that is a member of the invariant subspace of interest. Our approach is to implicitly restart the symplectic Lanczos factorization. This was first introduced by Sorensen [132] in the context of unsymmetric matrices and the Arnoldi process. The scheme is called implicit because the updating of the starting vector is accomplished with an implicit shifted SR mechanism on $B^{2j,2j}, j \leq n$. This allows to update the starting vector by working with a symplectic matrix in $\mathbf{R}^{2j \times 2j}$ rather than in $\mathbf{R}^{2n \times 2n}$ which is significantly cheaper.

5.2.1   THE IMPLICIT RESTART

The iteration starts by extending a length $2k$ symplectic Lanczos factorization by p steps. Next, $2p$ shifts are applied to $B^{2(k+p),2(k+p)}$ using double or quadruple implicit SR steps. The last $2p$ columns of the factorization are discarded resulting in a length $2k$ factorization. The iteration is defined by repeating this process until convergence.

For simplicity let us first assume that $p = 1$ and that a $2n \times 2(k+1)$ matrix $S_P^{2n,2k+2}$ is known such that

$$M_P S_P^{2n,2k+2} = S_P^{2n,2k+2} B_P^{2k+2,2k+2} + d_{k+2} r_{k+2} e_{2k+2}^T \qquad (5.2.59)$$

as in (5.1.5). Let μ be a real shift and

$$q_2(B^{2k+2,2k+2}) = SR,$$

where

$$q_2(B) = (B - \mu I)(B - \mu^{-1}I)B^{-1}.$$

Then (using Theorem 4.2) $S_P^{-1} B_P^{2k+2,2k+2} S_P$ will be a permuted symplectic butterfly matrix and S_P is an upper triangular matrix with two additional subdiagonals.

Now postmultiplying (5.2.59) by S_P yields

$$
\begin{aligned}
M_P(S_P^{2n,2k+2} S_P) &= (S_P^{2n,2k+2} S_P)(S_P^{-1} B_P^{2k+2,2k+2} S_P) \\
&\quad + d_{k+2} r_{k+2} e_{2k+2}^T S_P.
\end{aligned}
$$

Defining $\breve{S}_P^{2n,2k+2} = S_P^{2n,2k+2} S_P$ and $\breve{B}_P^{2k+2,2k+2} = S_P^{-1} B_P^{2k+2,2k+2} S_P$ this yields

$$
M_P \breve{S}_P^{2n,2k+2} = \breve{S}_P^{2n,2k+2} \breve{B}_P^{2k+2,2k+2} + d_{k+2} r_{k+2} e_{2k+2}^T S_P. \tag{5.2.60}
$$

The above equation fails to be a symplectic Lanczos factorization since the columns $2k$, $2k + 1$ and $2k + 2$ of the matrix

$$
d_{k+2}(b_{k+2} v_{k+2} + a_{k+2} w_{k+2}) e_{2k+2}^T S_P
$$

are nonzero. Let s_{ij} be the (i, j)th entry of S_P. The residual term in (5.2.60) is

$$
\begin{aligned}
&d_{k+2}(b_{k+2} v_{k+2} + a_{k+2} w_{k+2}) \\
&\quad \cdot (s_{2k+2,2k} e_{2k}^T + s_{2k+2,2k+1} e_{2k+1}^T + s_{2k+2,2k+2} e_{2k+2}^T).
\end{aligned}
$$

Rewriting (5.2.60) as

$$
M_P \breve{S}_P^{2n,2k+2} = [\breve{S}_P^{2n,2k}, \ \breve{v}_{k+1}, \ \breve{w}_{k+1}, \ v_{k+2}, \ w_{k+2}] Z
$$

where Z is blocked as

$$
\left[
\begin{array}{c|c|c}
\breve{B}_P^{2k,2k} & 0 & d_{k+1}(\breve{b}_k e_{2k-1} + \breve{a}_k e_{2k}) \\
\hline
b_{k+1} \breve{d}_{k+1} e_{2k}^T & b_{k+1} & b_{k+1} \breve{c}_{k+1} - \breve{a}_{k+1}^{-1} \\
\breve{a}_{k+1} \breve{d}_{k+1} e_{2k}^T & \breve{a}_{k+1} & \breve{a}_{k+1} \breve{c}_{k+1} \\
\hline
d_{k+2} b_{k+2} s_{2k+2,2k} e_{2k}^T & d_{k+2} b_{k+2} s_{2k+2,2k+1} & d_{k+2} b_{k+2} s_{2k+2,2k+2} \\
d_{k+2} a_{k+2} s_{2k+2,2k} e_{2k}^T & d_{k+2} a_{k+2} s_{2k+2,2k+1} & d_{k+2} a_{k+2} s_{2k+2,2k+2}
\end{array}
\right]
$$

we obtain as a new Lanczos identity

$$
M_P \breve{S}_P^{2n,2k} = \breve{S}_P^{2n,2k} \breve{B}_P^{2k,2k} + \breve{r}_{k+1} e_{2k}^T \tag{5.2.61}
$$

where

$$
\begin{aligned}
\breve{r}_{k+1} &= \breve{d}_{k+1}(\breve{b}_{k+1} \breve{v}_{k+1} + \breve{a}_{k+1} \breve{w}_{k+1}) \\
&\quad + d_{k+2} s_{2k+2,2k}(b_{k+2} v_{k+2} + a_{k+2} w_{k+2}).
\end{aligned}
$$

Here, $\breve{a}_j$, $\breve{b}_j$, $\breve{d}_j$ denote parameters of $\breve{B}_P^{2j,2j}$, while a_j, b_j, d_j are parameters of $B_P^{2j,2j}$. In addition, $\breve{v}_j, \breve{w}_j$ are the last two column vectors of $\breve{S}_P^{2n,2j}$, while v_j, w_j are the two last column vectors of $S_P^{2n,2j}$.

As the space spanned by the columns of

$$S^{2n,2k+2} = (P^{2n,2n})^T S_P^{2n,2k+2} P^{2(k+1),2(k+1)}$$

is J–orthogonal, and S_P is a permuted symplectic matrix, the space spanned by the columns of

$$\breve{S}^{2n,2k} = (P^{2n,2n})^T \breve{S}_P^{2n,2k} P^{2k,2k}$$

is J–orthogonal. Thus (5.2.61) is a valid symplectic Lanczos factorization. The new starting vector is $\breve{v}_1 = \rho q_2(M_P)v_1$ for some scalar $\rho \in \mathbf{R}$. This can be seen as follows: First note that for unreduced butterfly matrices $B^{2k+2,2k+2}$ we have $q_2(B_P^{2k+2,2k+2})e_1 \neq 0$. Hence, from $q_2(B_P^{2k+2,2k+2}) = S_P R_P$ we obtain $q_2(B_P^{2k+2,2k+2})e_1 = \rho S_P e_1$ for $\rho = e_1^T R_P e_1$ as R_P is an upper triangular matrix. As $q_2(B_P^{2k+2,2k+2})e_1 \neq 0$, we have $\rho \neq 0$. Using (5.2.61) it follows that

$$
\begin{aligned}
\breve{S}_P^{2n,2k} e_1 &= S_P^{2n,2k+2} S_P e_1 \\
&= \frac{1}{\rho} S_P^{2n,2k+2} q_2(B_P^{2k+2,2k+2}) e_1 \\
&= \frac{1}{\rho} S_P^{2n,2k+2} (B_P^{2k+2,2k+2} - \mu I)(B_P^{2k+2,2k+2} - \mu^{-1} I) \\
&\quad \cdot (B_P^{2k+2,2k+2})^{-1} e_1 \\
&= \frac{1}{\rho} (M_P S_P^{2n,2k+2} - d_{k+2} r_{k+2} e_{2k+2}^T - \mu S_P^{2n,2k+2}) \\
&\quad \cdot (I - \mu^{-1}(B_P^{2k+2,2k+2})^{-1}) e_1 \\
&= \frac{1}{\rho} (M_P S_P^{2n,2k+2} - \mu S_P^{2n,2k+2})(I - \mu^{-1}(B_P^{2k+2,2k+2})^{-1}) e_1
\end{aligned}
$$

as $r_{k+2} e_{2k+2}^T (I - \mu^{-1}(B_P^{2k+2,2k+2})^{-1}) e_1 = 0$. Thus using again (5.2.61) we get

$$\breve{S}_P^{2n,2k} e_1 = \frac{1}{\rho} (M_P - \mu I)(S_P^{2n,2k+2} - \mu^{-1} S_P^{2n,2k+2}(B_P^{2k+2,2k+2})^{-1}) e_1$$

$$= \frac{1}{\rho}(M_P - \mu I)(S_P^{2n,2k+2} - \mu^{-1}M_P^{-1}S_P^{2n,2k+2})e_1$$

$$\quad - \frac{1}{\rho}\mu^{-1}M_P^{-1}d_{k+2}r_{k+2}e_{2k+2}^T(B_P^{2k+2,2k+2})^{-1}e_1$$

$$= \frac{1}{\rho}(M_P - \mu I)(I - \mu^{-1}M_P^{-1})S_P^{2n,2k+2}e_1$$

$$= \frac{1}{\rho}q_2(M_P)v_1$$

as $e_{2k+2}^T(B_P^{2k+2,2k+2})^{-1}e_1 = 0$.

Note that in the symplectic Lanczos process the columns v_j of $S_P^{2n,2k}$ satisfy the condition $\|v_j\|_2 = 1$ and the parameters b_j are chosen to be one. This is no longer true for the odd numbered column vectors of S_P generated by the SR decomposition and the parameters $\breve{b}_j$ from $\breve{B}_P^{2k,2k}$ and thus for the new Lanczos factorization (5.2.61). Both properties could be forced using trivial factors. Numerical tests indicate that there is no obvious advantage in doing so.

Using standard polynomials as shift polynomials instead of Laurent polynomials as above results in the following situation: In

$$p_2(B_P^{2k+2,2k+2}) = (B_P^{2k+2,2k+2} - \mu I)(B_P^{2k+2,2k+2} - \mu^{-1}I) = S_P R_P$$

the permuted symplectic matrix S_P is upper triangular with four (!) additional subdiagonals. Therefore, the residual term in (5.2.60) has five nonzero entries. Hence not the last two, but the last four columns of (5.2.60) have to be discarded in order to obtain a new valid Lanczos factorization. That is, we would have to discard wanted information which is avoided when using Laurent polynomials.

This technique can be extended to the quadruple shift case using Laurent polynomials as the shift polynomials. The implicit restart can be summarized as given in Table 5.7. In the course of the iteration we have to choose p shifts $\Delta = \{\mu_1, \ldots, \mu_p\}$ in order to apply $2p$ shifts: choosing a real shift μ_k implies that μ_k^{-1} is also a shift due to the symplectic structure of the problem. Hence, μ_k^{-1} is not added to Δ as the use of the Laurent polynomial q_2 guarantees that μ_k^{-1} is used as a shift once $\mu_k \in \Delta$. In case of a complex shift μ_k, $|\mu_k| = 1$, this implies that $\overline{\mu_k}$ is also a shift not added to Δ. For complex shifts μ_k, $|\mu_k| \neq 1$, we include $\mu_k, \overline{\mu_k}$ in Δ.

Numerous choices are possible for the selection of the p shifts. One possibility is to choose p "exact" shifts with respect to $B_P^{2(k+p),2(k+p)}$. That is, first the eigenvalues of $B_P^{2(k+p),2(k+p)}$ are computed (by the SZ algorithm as the symplectic Lanczos algorithm computes the parameters that determine $B_P^{2(k+p),2(k+p)}$), then p unwanted eigenvalues are selected. One choice for this

---

Algorithm: k–step Restarted Symplectic Lanczos Method

Given an initial vector $\tilde{v}_1 \in \mathbf{R}^{2n}, \tilde{v}_1 \neq 0$, and a symplectic matrix $M \in \mathbf{R}^{2n \times 2n}$, this algorithm computes the parameters $a_1, \ldots, a_k$, $b_1, \ldots, b_k, c_1, \ldots, c_k, d_2, \ldots, d_k$ that determine a $2k \times 2k$ symplectic butterfly matrix $B^{2k,2k}$ with the property $\sigma(B^{2k,2k}) \subset \sigma(M)$ if no breakdown occurs.

    perform k steps of the symplectic Lanczos algorithm to compute
        $S_P^{2n,2k}$ and $B_P^{2k,2k}$
    obtain the residual vector r_{k+1}
    while $\|r_{k+1}\| > tol$
        perform p additional steps of the symplectic Lanczos method
           to compute $S_P^{2n,2(k+p)}$ and $B_P^{2(k+p),2(k+p)}$
        select p shifts μ_i
        compute $\breve{B}_P^{2k,2k}$ and $\breve{S}_P^{2n,2k}$ via implicitly shifted SR steps
        set $S_P^{2n,2k} = \breve{S}_P^{2n,2k}$ and $B_P^{2k,2k} = \breve{B}_P^{2k,2k}$
        obtain the new residual vector r_{k+1}
    end while

Table 5.7.   *k*–step Restarted Symplectic Lanczos Method

selection might be: sort the eigenvalues by decreasing magnitude. There will be $k + p$ eigenvalues with modulus greater than or equal to 1 which can be ordered as

$$|\lambda_1| \geq \ldots \geq |\lambda_k| \geq |\lambda_{k+1}| \geq \ldots |\lambda_{k+p}| \geq 1$$
$$\geq |\lambda_{k+p}^{-1}| \geq \ldots \geq |\lambda_{k+1}^{-1}| \geq |\lambda_k^{-1}| \geq \ldots \geq |\lambda_1^{-1}|.$$

Select the $2p$ eigenvalues with modulus closest to 1 as shifts. If λ_{k+1} is complex with $|\lambda_k| = |\lambda_{k+1}| \neq 1$, then we either have to choose $2p+2$ shifts or just $2p-2$ shifts, as λ_{k+1} belongs to a quadruple pair of eigenvalues of $B_P^{2(k+p),2(k+p)}$ and in order to preserve the symplectic structure either both λ_k and λ_{k+1} have to be chosen or none of them.

    A different possibility of choosing the shifts is to keep those eigenvalues that are good approximations to eigenvalues of M. That is eigenvalues for which (5.1.15) is small. Again we have to make sure that our set of shifts is complete in the sense described above.

    Choosing eigenvalues of $B_P^{2(k+p),2(k+p)}$ as shifts has an interesting consequence on the next iterate. Assume for simplicity that $B_P^{2(k+p),2(k+p)}$ is diagonalizable. Let

$$\sigma(B_P^{2(k+p),2(k+p)}) = \{\theta_1, \ldots, \theta_{2k}\} \cup \{\mu_1, \ldots, \mu_{2p}\}$$

be a disjoint partition of the spectrum of $B_P^{2(k+p),2(k+p)}$. Selecting the exact shifts $\mu_1, \ldots, \mu_{2p}$ in the implicit restart, following the rules mentioned above yields a matrix

$$\breve{B}_P^{2(k+p),2(k+p)} = \begin{bmatrix} \breve{B}_P^{2k,2k} & X \\ 0 & Y \end{bmatrix}$$

where $\sigma(\breve{B}_P^{2k,2k}) = \{\theta_1, \ldots, \theta_{2k}\}$ and $\sigma(Y) = \{\mu_1, \ldots, \mu_{2p}\}$. This follows from Theorem 4.2. Moreover, the new starting vector has been implicitly replaced by the sum of $2k$ approximate eigenvectors:

$$\begin{aligned} \breve{v}_1 &= S_P^{2n,2(k+p)} S_P e_1 \\ &= \frac{1}{\rho} S_P^{2n,2(k+p)} q(B_P^{2(k+p),2(k+p)}) e_1 \qquad \text{for } \rho = e_1^T R_P e_1 \\ &= \frac{1}{\rho} S_P^{2n,2(k+p)} \sum_{j=1}^{2k} \zeta_j y_j \end{aligned}$$

where $B_P^{2(k+p),2(k+p)} y_j = \theta_j y_j$ and ζ_j is properly chosen. The last equation follows since $q(B_P^{2(k+p),2(k+p)}) e_1$ has no components along an eigenvector of $B_P^{2(k+p),2(k+p)}$ associated with $\mu_j, 1 \leq j \leq 2p$. Hence $\breve{v}_1$ is a linear combination of the $2k$ Ritz vectors associated with the Ritz value that are kept

$$\breve{v}_1 = \rho \sum_{j=1}^{2k} \zeta_j x_j \qquad \text{where } S_P^{2n,2(k+p)} y_j = x_j.$$

It should be mentioned that the k–step restarted symplectic Lanczos method as in Table 5.7 with exact shifts builds a J–orthogonal basis for a number of generalized Krylov subspaces simultaneously. The subspace of length $2(k+p)$ generated during a restart using exact shifts contains all the Krylov subspaces of dimension $2k$ generated from each of the desired Ritz vectors. We have already seen that after the implicit restart the new starting vector of the Lanczos recursion is a combination of Ritz vectors. Assuming as above that $2p$ exact shifts are used, an induction argument using the same idea as above shows that the first $2k$ columns of $S_P^{2n,2(k+p)}$ are combinations of the desired $2k$ Ritz vectors. (The only difference to the proof above is in showing that for $2 \leq j \leq 2k$, $S_P e_j$ can be written as $q(B_P^{2(k+p),2(k+p)}) w$ for some vector w. Then $S_P^{2n,2(k+p)} S_P e_j$ is, like $\breve{v}_1$, a combination of the desired Ritz vectors.) Hence, during the next Lanczos run, the subspace of degree $2k$ is $\text{span}\{x_1, \ldots, x_{2k}\}$. Let the subspace generated during that run be given by

$$\mathcal{N} = \text{span}\{x_1, \ldots, x_{2k}, v_{k+1}, w_{k+1}, \ldots, v_{k+p}, w_{k+p}\}$$

or equivalently,

$$\mathcal{N} = \text{span}\{x_1, \ldots, x_{2k}, v_{k+1}, M_P^{-1}v_{k+1}, \ldots, M_P^{-(p-1)}v_{k+1},$$
$$M_P v_{k+1}, \ldots, M_P^p v_{k+1}\}.$$

We will now show that this subspace is equivalent to the subspaces

$$\mathcal{M}_j = \text{span}\{x_1, \ldots, x_{2k}, M_P^{-1}x_j, \ldots, M_P^{-p}x_j, M_P x_j, \ldots, M_P^p x_j\}$$

for all j. The Lanczos run under consideration starts from the equation

$$M_P S_P^{2n,2k} = S_P^{2n,2k} B_P^{2k,2k} + d_{k+1} r_{k+1} e_{2k}^T.$$

For a Ritz vector $x = S_P^{2n,2k} y$ and the corresponding Ritz value λ (that is, $B_P^{2k,2k} y = \lambda y$) we have

$$M_P x = \lambda x + d_{k+1} r_{k+1} e_{2k}^T y.$$

Hence, with $\alpha = e_{2k}^T y \in \mathbf{R}$ and Remark 5.2 we can rewrite

$$M_P x = \lambda x + \alpha d_{k+1} M_P v_{k+1}. \tag{5.2.62}$$

Therefore, for all Ritz vectors $x_j, j = 1, \ldots, 2k$, $M_P x_j \in \mathcal{N}$. Then

$$M_P^2 x = \lambda M_P x + \alpha d_{k+1} M_P^2 v_{k+1}.$$

Hence, $M_P^2 x$ is a combination of $M_P x$, and $M_P^2 v_{k+1}$, and therefore for all Ritz vectors $x_j, j = 1, \ldots, 2k$, $M_P^2 x_j \in \mathcal{N}$. Similar for other i, $M_P^i x$ is contained in the subspace $\mathcal{N}$. For example, $M_P^p x$ is a linear combination of $M_P^{p-1} x$ and $M_P^p v_{k+1}$, and therefore for all Ritz vectors $x_j, j = 1, \ldots, 2k$, $M_P^p x_j \in \mathcal{N}$. Moreover, from (5.2.62)

$$\lambda x = M_P x - \alpha d_{k+1} M_P v_{k+1}.$$

Hence

$$\lambda M_P^{-1} x = x - \alpha d_{k+1} v_{k+1},$$
$$\lambda M_P^{-2} x = M_P^{-1} x - \alpha d_{k+1} M_P^{-1} v_{k+1},$$
$$\vdots$$
$$\lambda M_P^{-p} x = M_P^{-(p-1)} x - \alpha d_{k+1} M_P^{-(p-1)} v_{k+1}.$$

Thus $M_P^{-1}x_j, M_P^{-2}x_j, \ldots, M_P^{-p}x_j \in \mathcal{N}$. As $\dim(\mathcal{N}) = \dim(\mathcal{M}_j)$, $\mathcal{N}$ and $\mathcal{M}_j$ span the same space.

A similar observation for Sorensen's restarted Arnoldi method with exact shifts was made by Morgan in [107]. For a discussion of this observation see [107] or [93]. Morgan infers 'the method works on approximations to all of the desired eigenpairs at the same time, without favoring one over the other' [107, p.1220, l. 7–8 from the bottom]. This remark can also be applied to the method presented here.

Moreover, the implicitly restarted symplectic Lanczos method can be interpreted as a non-stationary subspace iteration. An analogous statement for the implicitly restarted Arnoldi method is given in [91]. Assume that we have computed

$$M_P S_P^{2n,2m} = S_P^{2n,2m} B_P^{2m,2m} + d_{m+1}r_{m+1}e_{2m}^T, \tag{5.2.63}$$

a length $2m = 2(k + p)$ symplectic Lanczos reduction. As p shifts for the implicit restart we have chosen $\{\mu_1, \ldots, \mu_p\}$ where the shifts are sorted such that first all the complex shifts are given so that for a shift $\mu_{2j} \in \mathbb{C}, |\mu_{2j}| \neq 1$ we have $\overline{\mu_{2j-1}} = \mu_{2j}$, then all real and purely imaginary shifts are given. Hence during the implicit restart we want to apply the Laurent polynomial

$$q_{2p}(\nu) = \widehat{q}_p(\nu)\widehat{q}_{p-1}(\nu)\cdots\widehat{q}_1(\nu) = \widehat{q}_p(\nu)q_{2p-2}(\nu), \tag{5.2.64}$$

where

$$\widehat{q}_j(\nu) := (\nu - \mu_j I)(\nu - \mu_j^{-1}I)\nu^{-1}$$

and $\widehat{q}_0(\nu) = 1$.

LEMMA 5.19 *Assume that (5.2.63) and (5.2.64) are given. Then*

$$q_{2p}(M_P)S_P^{2n,2m}$$

$$= S_P^{2n,2m}q_{2p}(B_P^{2m,2m}) + d_{m+1}\sum_{\ell=2}^{p+1}\psi_\ell^p(M_P)r_{m+1}e_{2m}^T q_{2(\ell-2)}(B_P^{2m,2m})$$

$$- d_{m+1}\sum_{\ell=2}^{p+1}\psi_\ell^p(M_P)M_P^{-1}r_{m+1}e_{2m}^T(B_P^{2m,2m})^{-1}q_{2(\ell-2)}(B_P^{2m,2m}),$$

where

$$\psi_\ell^j(\nu) := \prod_{k=\ell}^{j}\widehat{q}_k(\nu)$$

with $\psi_{j+1}^j(\nu) = 1$. Moreover,

$$q_{2p}(M_P)S_P^{2n,2k} = S_P^{2n,2m}q_{2p}(B_P^{2m,2m})[e_1, e_2, \ldots, e_{2k}]. \tag{5.2.65}$$

PROOF: The proof is by induction. For $p = 1$ we have

$$
\begin{aligned}
q_2(M_P)&S_P^{2n,2m} \\
&= (M_P - \mu_1 I)(M_P - \mu_1^{-1} I)M_P^{-1}S_P^{2n,2m} \\
&= (M_P + M_P^{-1} - (\mu_1 + \mu_1^{-1})I)S_P^{2n,2m} \\
&= S_P^{2n,2m} B_P^{2m,2m} + d_{m+1}r_{m+1}e_{2m}^T + S_P^{2n,2m}(B_P^{2m,2m})^{-1} \\
&\quad - d_{m+1}M_P^{-1}r_{m+1}e_{2m}^T(B_P^{2m,2m})^{-1} - S_P^{2n,2m}(\mu_1 + \mu_1^{-1}) \\
&= S_P^{2n,2m}q_2(B_P^{2m,2m}) + d_{m+1}r_{m+1}e_{2m}^T \\
&\quad - d_{m+1}M_P^{-1}r_{m+1}e_{2m}^T(B_P^{2m,2m})^{-1}.
\end{aligned}
$$

As $\psi_2^1(M_P) = q_0(B_P^{2m,2m}) = I$ the case $p = 1$ is proven. For the induction step consider

$$
\begin{aligned}
q_{2j+2}&(M_P)S_P^{2n,2m} \\
&= \widehat{q}_{j+1}(M_P)q_{2j}(M_P)S_P^{2n,2m} \\
&= \widehat{q}_{j+1}(M_P) \left\{ S_P^{2n,2m}q_{2j}(B_P^{2m,2m}) \right. \\
&\qquad + d_{m+1}\sum_{\ell=2}^{j+1}\psi_\ell^j(M_P)r_{m+1}e_{2m}^T q_{2(\ell-2)}(B_P^{2m,2m}) \\
&\qquad \left. - d_{m+1}\sum_{\ell=2}^{j+1}\psi_\ell^j(M_P)M_P^{-1}r_{m+1}e_{2m}^T(B_P^{2m,2m})^{-1}q_{2(\ell-2)}(B_P^{2m,2m}) \right\}.
\end{aligned}
$$

Similar to the algebraic manipulations performed for $p = 1$ we obtain

$$
\widehat{q}_{j+1}(M_P)S_P^{2n,2m} = S_P^{2n,2m}\widehat{q}_{j+1}(B_P^{2m,2m}) + d_{m+1}r_{m+1}e_{2m}^T \\
- d_{m+1}M_P^{-1}r_{m+1}e_{2m}^T(B_P^{2m,2m})^{-1}.
$$

As

$$
\widehat{q}_{j+1}(B_P^{2m,2m})q_{2j}(B_P^{2m,2m}) = q_{2j+2}(B_P^{2m,2m})
$$

and

$$
\widehat{q}_{j+1}(M_P)\psi_\ell^j(M_P) = \psi_\ell^{j+1}(M_P)
$$

we have

$$q_{2j+2}(M_P)S_P^{2n,2m}$$

$$= S_P^{2n,2m}q_{2j+2}(B_P^{2m,2m}) + d_{m+1}r_{m+1}e_{2m}^T q_{2j}(B_P^{2m,2m})$$

$$- d_{m+1}M_P^{-1}r_{m+1}e_{2m}^T(B_P^{2m,2m})^{-1}q_{2j}(B_P^{2m,2m})$$

$$+ d_{m+1}\sum_{\ell=2}^{j+1}\psi_\ell^{j+1}(M_P)r_{m+1}e_{2m}^T q_{2(\ell-2)}(B_P^{2m,2m})$$

$$- d_{m+1}\sum_{\ell=2}^{j+1}\psi_\ell^{j+1}(M_P)M_P^{-1}r_{m+1}e_{2m}^T(B_P^{2m,2m})^{-1}q_{2(\ell-2)}(B_P^{2m,2m})$$

$$= S_P^{2n,2m}q_{2j+2}(B_P^{2m,2m})$$

$$+ d_{m+1}\sum_{\ell=2}^{j+2}\psi_\ell^{j+1}(M_P)r_{m+1}e_{2m}^T q_{2(\ell-2)}(B_P^{2m,2m})$$

$$- d_{m+1}\sum_{\ell=2}^{j+2}\psi_\ell^{j+1}(M_P)M_P^{-1}r_{m+1}e_{2m}^T(B_P^{2m,2m})^{-1}q_{2(\ell-2)}(B_P^{2m,2m}),$$

which establishes the first hypothesis of the lemma.

In order to prove the second hypothesis note that

$$q_{2j}(B_P^{2m,2m}) = p_{2j}(B_P^{2m,2m} + (B_P^{2m,2m})^{-1})$$

for

$$p_{2j}(\nu) = \prod_{i=1}^{j}(\nu - (\mu_i + \mu_i^{-1}))^i.$$

Since $B_P^{2m,2m}$ is an permuted symplectic butterfly matrix $B_P^{2m,2m} + (B_P^{2m,2m})^{-1}$ is an upper triangular matrix with two additional subdiagonals. Therefore $q_{2j}(B_P^{2m,2m})$ is represented by an upper triangular matrix with $2j$ additional subdiagonals. Hence the matrix

$$\sum_{\ell=2}^{p}\psi_\ell^{p+1}(M_P)r_{m+1}e_{2m}^T q_{2(\ell-2)}(B_P^{2m,2m})$$

is zero through its first $2k + 2$ columns, while the matrix

$$\sum_{\ell=2}^{p}\psi_\ell^{p+1}(M_P)M_P^{-1}r_{m+1}e_{2m}^T(B_P^{2m,2m})^{-1}q_{2(\ell-2)}(B_P^{2m,2m})$$

is zero through its first $2k$ columns. This yields the second hypothesis. $\checkmark$

Hence, $q_{2p}(M_P)$ applied to the first $2k$ columns of $S_P^{2n,2m}$ is equivalent to the basis representation given by the first $2k$ columns of $S_P^{2n,2m}q_{2p}(B_P^{2m,2m})$. Applying an implicit restart to (5.2.63) using the spectral transformation function q_{2p}, we essentially apply the SR algorithm with shifts $\mu_1, \mu_1^{-1}, \ldots, \mu_p, \mu_p^{-1}$ to $B_P^{2m,2m}$

$$B_P^{2m,2m} S_P = S_P \breve{B}_P^{2m,2m}.$$

$S_P \in \mathbf{R}^{2m \times 2m}$ is a symplectic, upper triangular matrix with $m - k$ additional subdiagonals. Write S_P as $S_P = [S_P^{[1]}\ S_P^{[2]}\ S_P^{[3]}]$ with $S_P^{[1]} \in \mathbf{R}^{2m \times 2k}, S_P^{[2]} \in \mathbf{R}^{2m \times 2}, S_P^{[3]} \in \mathbf{R}^{2m \times (2m-2k-2)}$. Then

$$B_P^{2m,2m} S_P^{[1]} = [S_P^{[1]}\ S_P^{[2]}\ S_P^{[3]}] \left[\begin{array}{c} \breve{B}_P^{2k,2k} \\ \hline \breve{b}_{k+1}\breve{d}_{k+1}e_{2k}^T \\ \breve{a}_{k+1}\breve{d}_{k+1}e_{2k}^T \\ \hline 0 \end{array}\right].$$

Postmultiplying (5.2.63) with $S_P^{[1]}$ and using $e_{2m}^T S_P^{[1]} = 0$ which is due to the special form of S_P (upper triangular with $m - k$ additional subdiagonals) we obtain

$$\begin{aligned} M_P S_P^{2n,2m} S_P^{[1]} &= S_P^{2n,2m} B_P^{2m,2m} S_P^{[1]} + r_{m+1}e_{2m}^T S_P^{[1]} \\ &= \breve{S}_P^{2n,2k} \breve{B}_P^{2k,2k} + \breve{r}_{k+1}e_{2k}^T. \end{aligned}$$

where $\breve{S}_P^{2n,2k} = S_P^{2n,2m} S_P^{[1]}$. This is just the implicitly restarted symplectic Lanczos recursion obtained by applying one implicit restart with the Laurent polynomial q_{2p}. Applying the SR algorithm with shifts $\mu_1, \mu_1^{-1}, \ldots, \mu_p, \mu_p^{-1}$ to $B_P^{2m,2m}$ is equivalent to computing the permuted SR decomposition

$$q_{2p}(B_P^{2m,2m}) = S_P R_P.$$

Substituting this into (5.2.65) we obtain

$$q_{2p}(M_P)S_P^{2n,2k} = S_P^{2n,2m}S_P R_P[e_1, e_2, \ldots, e_{2k}] = \breve{S}_P^{2n,2k} \breve{R}_P,$$

where $\breve{R}_P$ is a $2k \times 2k$ upper triangular matrix. This equation describes a nonstationary subspace iteration. As an step of the implicitly restarted symplectic Lanczos process computes the new subspace spanned by the columns of $\breve{S}_P^{2n,2k}$ from $S_P^{2n,2k}$ the implicitly restarted symplectic Lanczos algorithm can be interpreted as a nonstationary subspace iteration.

In the above discussion we have assumed that the permuted SR decomposition

$$q(B_P^{2(k+p),2(k+p)}) = S_P R_P$$

exists. Unfortunately, this is not always true. During the bulge-chase in the implicit SR step, it may happen that a diagonal element a_j of K_u^{-1} (3.2.8) is zero (or almost zero). In that case no reduction to symplectic butterfly form with the corresponding first column $\breve{v}_1$ does exist. In the next section we will prove that a serious breakdown in the symplectic Lanczos algorithm is equivalent to such a breakdown of the SR decomposition. Moreover, it may happen that a subdiagonal element d_j of the $(2,2)$ block of N_u (3.2.9) is zero (or almost zero) such that

$$\breve{B}_P^{2(k+p),2(k+p)} = \left[\begin{array}{cc} \breve{B}_P^{2j,2j} & \\ & \widehat{B}_P \end{array} \right].$$

The matrix $\breve{B}_P^{2(k+p),2(k+p)}$ is split, an invariant subspace of dimension j is found. If $j \geq k$ and all shifts have been applied, then the iteration is halted. Otherwise we can continue as in the procedure described by Sorensen in [132, Remark 3].

One important property for a stable implicitly restarted Lanczos method is that the Lanczos vectors stay bounded after possibly many implicit restarts. Neither for the symplectic Lanczos method nor for the symplectic SR algorithm it can be proved that the symplectic transformation matrix stays bounded. Hence the symplectic Lanczos vectors $S_P^{2n,2k}$ computed via an implicitly restarted symplectic Lanczos method may not stay bounded; this has to be monitored during the iteration. During the SR step on the $2k \times 2k$ symplectic butterfly matrix, all but $k-1$ transformations are orthogonal. These are known to be numerically stable. For the $k-1$ nonorthogonal symplectic transformations that have to be used, we choose among all possible transformations the ones with optimal (smallest possible) condition number (see [38]).

As the iteration progresses, some of the Ritz values may converge to eigenvalues of M long before the entire set of wanted eigenvalues have. These converged Ritz values may be part of the wanted or unwanted portion of the spectrum. In either case it is desirable to deflate the converged Ritz values and corresponding Ritz vectors from the unconverged portion of the factorization. If the converged Ritz value is wanted then it is necessary to keep it in the subsequent factorizations; if it is unwanted then it must be removed from the current and the subsequent factorizations. Lehoucq and Sorensen develop in [93, 133] locking and purging techniques to accomplish this in the context of unsymmetric matrices and the restarted Arnoldi method. These ideas can be carried over to the situation here.

5.2.2   NUMERICAL PROPERTIES OF THE IMPLICITLY RESTARTED SYMPLECTIC LANCZOS METHOD

It is well known that for general Lanczos-like methods the stability of the overall process is improved when the norm of the Lanczos vectors is chosen to be equal to 1 [116, 134]. Thus, following the approach of Freund and Mehrmann [55] in the context of a symplectic look-ahead Lanczos algorithm for the Hamiltonian eigenproblem, Banse proposes in [13] to modify the pre-requisite $S_P^T J_P S_P = J_P$ of the symplectic Lanczos method to

$$
S_P^T J_P S_P = \begin{bmatrix} 0 & \sigma_1 & & & & & \\ -\sigma_1 & 0 & & & & & \\ & & 0 & \sigma_2 & & & \\ & & -\sigma_2 & 0 & & & \\ & & & & \ddots & & \\ & & & & & 0 & \sigma_n \\ & & & & & -\sigma_n & 0 \end{bmatrix} =: \Sigma
$$

and

$$
\|v_j\|_2 = \|w_j\|_2 = 1, \qquad j = 1, \ldots, n.
$$

For the resulting algorithm and a discussion of it we refer to [13]. It is easy to see that $S_P^{-1} B_P S_P$ is no longer a permuted symplectic matrix, but it still has the desired form of a butterfly matrix. Unfortunately, an SR step does not preserve the structure of $S_P^{-1} B_P S_P$ and thus, this modified version of the symplectic Lanczos method can not be used in connection with our restart approaches.

Without some form of reorthogonalization any Lanczos algorithm is numerically unstable. We re–J_P–orthogonalize each Lanczos vector as soon as it is computed against the previous ones as discussed in Section 5.1. The cost for the re–J–orthogonalization will be reasonable as k, the number of eigenvalues sought, is in general of modest size.

Another important issue is the numerical stability of the SR step employed in the restart. During the SR step on the $2k \times 2k$ symplectic butterfly matrix, all but $k - 1$ transformations are orthogonal. These are known to be numerically stable. For the $k - 1$ nonorthogonal symplectic transformations that have to be used, we choose among all possible transformations the ones with optimal (smallest possible) condition number. But there is the possibility that one of the Gauss transformations might not exist, i.e., that the SR algorithm breaks down.

If there is a starting vector $\breve{v}_1 = \rho q(M) v_1$ for which the explicitly restarted symplectic Lanczos method breaks down, then it is impossible to reduce the symplectic matrix M to symplectic butterfly form with a transformation matrix whose first column is $\breve{v}_1$. Thus, in this situation the SR decomposition of $q(B)$ can not exist.

As will be shown in the rest of this section, this is the only way that break-downs in the SR decomposition can occur. In the SR step, most of the transformations used are orthogonal symplectic transformations; their computation can not break down. The only source of breakdown can be one of the symplectic Gauss eliminations L_j. For simplicity, we will discuss the double shift case. Only the following elementary elimination matrices are used in the implicit SR step: elementary symplectic Givens matrices G_k, elementary symplectic Householder transformations H_k, and elementary symplectic Gauss elimination matrices L_k as introduced in Section 2.1.2.

Assume that k steps of the symplectic Lanczos algorithm are performed, then from (5.1.5)

$$M_P S_P^{2n,2k} = S_P^{2n,2k} B_P^{2k,2k} + d_{k+1} r_{k+1} e_{2k}^T. \qquad (5.2.66)$$

Now an implicit restart is to be performed using an implicit double shift SR step. In the first step of the implicit SR step, a symplectic Householder matrix H_1 is computed such that

$$H_1^T q(B^{2k,2k}) e_1 = \alpha e_1.$$

H_1 is applied to $B^{2k,2k}$

$$H_1^T B^{2k,2k} H_1$$

introducing a small bulge in the butterfly form: additional elements are found in the positions $(2,1)$, $(1,2)$, $(n+2,n+1)$, $(n+1,n+2)$, $(1,n+3)$, $(3,n+1)$, $(n+1,n+3)$, and $(n+3,n+1)$. The remaining implicit transformations perform a bulge-chasing sweep down the subdiagonal to restore the butterfly form. An algorithm for this is given in Section 3; it can be summarized for the situation here

> for $\ell = 1 : n-1$
>   compute $G_{\ell+1}$ such that $(G_{\ell+1} B^{2k,2k})_{n+\ell+1,\ell} = 0$
> $B^{2k,2k} = G_{\ell+1} B^{2k,2k} G_{\ell+1}^T$
>   compute $L_{\ell+1}$ such that $(L_{\ell+1} B^{2k,2k})_{\ell+1,\ell} = 0$
> $B^{2k,2k} = L_{\ell+1} B^{2k,2k} L_{\ell+1}^{-1}$
>   compute $G_{\ell+1}$ such that $(B^{2k,2k} G_{\ell+1})_{\ell,\ell+1} = 0$
> $B^{2k,2k} = G_{\ell+1}^T B^{2k,2k} G_{\ell+1}$
>   compute $H_{\ell+1}$ such that $(B^{2k,2k} H_{\ell+1})_{\ell,n+\ell+2} = 0$
> $B^{2k,2k} = H_{\ell+1}^T B^{2k,2k} H_{\ell+1}$
> end

Suppose that the first $j-1$ Gauss transformations, $j < k$, exist and that we have computed

$$\widehat{S} = H_1 G_2^T L_2^{-1} G_2 \ldots H_{j-2} G_{j-1}^T L_{j-1}^{-1} G_{j-1} H_{j-1} G_j^T.$$

Note that, using the same convention as in the previous chapter, the two G_i that appear in the formulae denote matrices of the same form but with different entries for c_i and s_i. In order to simplify the notation, we switch to the permuted version and rewrite the permuted symplectic matrix $\widehat{S}_P$ as

$$\widehat{S}_P = \begin{bmatrix} S_P & 0 \\ 0 & I^{2n-2j-2,2n-2j-2} \end{bmatrix},$$

where $S_P \in \mathbf{R}^{(2j+2)\times(2j+2)}$ making use of the fact that the accumulated transformations affect only the rows 1 to j and $n+1$ to $n+j$. The leading $(2j+2)\times(2j+2)$ principal submatrix of

$$\widehat{S}_P^{-1} B_P^{2k,2k} \widehat{S}_P$$

is given by $\widetilde{B}_P^{2j+2,2j+2}$, that is

$$
\left[
\begin{array}{cccc|ccc|cc}
\multicolumn{2}{c}{\multirow{2}{*}{$\breve{B}_P^{2j-4,2j-4}$}} & 0 & \breve{b}_{j-2}\breve{d}_{j-1}e_{2j-5} & & & & & \\
 & & 0 & \breve{a}_{j-2}\breve{d}_{j-1}e_{2j-4} & & & & & \\
\hline
\breve{b}_{j-1}\breve{d}_{j-1}e_{2j-4}^T & \breve{b}_{j-1} & \breve{b}_{j-1}\breve{c}_{j-1}-\breve{a}_{j-1}^{-1} & 0 & \widehat{x} & & & \\
\breve{a}_{j-1}\breve{d}_{j-1}e_{2j-4}^T & \breve{a}_{j-1} & \breve{a}_{j-1}\breve{c}_{j-1} & 0 & \widehat{x} & & & \\
\hline
 & 0 & \widehat{y}_1 & \widehat{b}_j & \widehat{x} & \widehat{x} & \widehat{x} \\
 & 0 & \widehat{y}_2 & \widehat{a}_j & \widehat{x} & \widehat{x} & \widehat{x} \\
\hline
 & 0 & \widehat{x}_1 & \widehat{x}_2 & \widehat{x} & \widehat{x} & \widehat{x} \\
 & 0 & 0 & 0 & \widehat{x} & \widehat{x} & \widehat{x} \\
\end{array}
\right],
\quad (5.2.67)
$$

where the hatted quantities denote unspecified entries that would change if the SR update could be continued. Next, the $(2j+1, 2j-1)$ entry should be annihilated by a permuted symplectic Gauss elimination. This elimination will fail to exist if $\widehat{a}_j = 0$; the SR decomposition of $q(B^{2k,2k})$ does not exist.

For later use, let us note that $\widehat{a}_j = 0$ implies $\widehat{y}_2 = 0$. This follows as $\widetilde{B}_P^{2j+2,2j+2}$ is J_P–orthogonal: From

$$e_{2j-2}^T \widetilde{B}_P^{2j+2,2j+2} J_P (\widetilde{B}_P^{2j+2,2j+2})^T e_{2j} = e_{2j-2}^T J_P e_{2j} = 0$$

we obtain

$$
0 = [0 \cdots 0\, 0\, \breve{a}_{j-1}\breve{d}_{j-1}\ \breve{a}_{j-1}\ \breve{a}_{j-1}\breve{c}_{j-1}\ 0\ \widehat{x}\ 0\ 0] J_P
\begin{bmatrix} 0 \\ \vdots \\ 0 \\ \widehat{y}_2 \\ \widehat{a}_j \\ \widehat{x} \\ \widehat{x} \\ \widehat{x} \end{bmatrix}
$$

$$= -\breve{a}_{j-1}\widehat{y}_2 - \widehat{x}\widehat{a}_j.$$

If $\widehat{a}_j = 0$ we have $\widehat{y}_2 = 0$ as $\breve{a}_{j-1} \neq 0$ (otherwise the last Gauss transformation L_{j-1} did not exist).

Next we show that this breakdown in the SR decomposition implies a breakdown in the Lanczos process started with the starting vector $\breve{v}_1 = \rho q(M_P)v_1$.

For this we have to consider (5.2.66) postmultiplied by $\widehat{S}_P$. From the derivations in the last section we know that the starting vector of that recursion is given by $\breve{v}_1 = \rho q(M_P)v_1$. As the trailing $(2n - 2j - 2) \times (2n - 2j - 2)$ principal submatrix of $\widehat{S}_P$ is the identity, we can just as well consider

$$M_P S_P^{2n,2j+2} = S_P^{2n,2j+2} B_P^{2j+2,2j+2} + d_{j+2}r_{j+2}e_{2j+2}^T,$$

postmultiplied by S_P

$$M_P \breve{S}_P^{2n,2j+2} = \breve{S}_P^{2n,2j+2} \breve{B}_P^{2j+2,2j+2} + d_{j+2}r_{j+2}e_{2j+2}^T S_P, \qquad (5.2.68)$$

where

$$\breve{B}_P^{2j+2,2j+2} = S_P^{-1} B_P^{2j+2,2j+2} S_P$$

corresponds to the matrix in (5.2.67) (no butterfly form!) and

$$\breve{S}_P^{2n,2j+2} = S_P^{2n,2j+2} S_P = [\breve{v}_1, \breve{w}_1, \ldots, \breve{v}_{j-1}, \breve{w}_{j-1}, \widehat{v}_j, \widehat{w}_j, \widehat{v}_{j+1}, \widehat{w}_{j+1}].$$

The columns of $\breve{S}_P^{2n,2j+2}$ are J_P–orthogonal,

$$(\breve{S}_P^{2n,2j+2})^T J_P^{2n,2n} \breve{S}_P^{2n,2j+2} = J_P^{2(j+1),2(j+1)}. \qquad (5.2.69)$$

The starting vector of the recursion (5.2.68) is given by $\breve{v}_1 = \rho q(M_P)v_1$. Deleting the last four columns of $\breve{S}_P^{2n,2j+2}$ in the same way as in the implicit restart we obtain a valid symplectic Lanczos factorization of length $2j - 2$.

In order to show that a breakdown in the SR decomposition of $q(B)$ implies a breakdown in the above symplectic Lanczos recursion, we need to show

$$\widehat{a}_j = 0 \implies \breve{a}_j = \breve{v}_j^T J_P^{2n,2n} M_P \breve{v}_j = 0.$$

From (5.2.67) and (5.2.68) we obtain

$$\begin{aligned} M_P \breve{w}_{j-1} = {} & \breve{b}_{j-2}\breve{d}_{j-1}\breve{v}_{j-2} + \breve{a}_{j-2}\breve{d}_{j-1}\breve{w}_{j-2} \\ & + (\breve{b}_{j-1}\breve{c}_{j-1} - \breve{a}_{j-1}^{-1})\breve{v}_{j-1} \\ & + \breve{a}_{j-1}\breve{c}_{j-1}\breve{w}_{j-1} + \widehat{y}_1\widehat{v}_j + \widehat{y}_2\widehat{w}_j + \widehat{x}_1\widehat{v}_{j+1} \end{aligned} \qquad (5.2.70)$$

and

$$M_P \breve{v}_k = \breve{b}_k \breve{v}_k + \breve{a}_k \breve{w}_k, \qquad k \leq j - 1. \qquad (5.2.71)$$

Further we know from the symplectic Lanczos algorithm

$$\breve{v}_j = -\breve{d}_{j-1}\breve{v}_{j-2} - \breve{c}_{j-1}\breve{v}_{j-1} + \breve{w}_{j-1} + \breve{a}_{j-1}^{-1}M_P^{-1}\breve{v}_{j-1}, \qquad (5.2.72)$$

all of these quantities are already known. Now consider

$$
\begin{aligned}
\breve{a}_j &= \breve{v}_j^T J_P M_P \breve{v}_j \\
&= \underbrace{-\breve{d}_{j-1}\breve{v}_j^T J_P M_P \breve{v}_{j-2}}_{x_1} \underbrace{-\breve{c}_{j-1}\breve{v}_j^T J_P M_P \breve{v}_{j-1}}_{x_2} \\
&\quad + \underbrace{\breve{v}_j^T J_P M_P \breve{w}_{j-1}}_{x_3} + \underbrace{\breve{a}_{j-1}^{-1}\breve{v}_j^T J_P \breve{v}_{j-1}}_{x_4}.
\end{aligned}
$$

Due to J_P–orthogonality, $x_4 = 0$. Using (5.2.71) we obtain

$$
\breve{v}_j^T J_P M_P \breve{v}_k = \breve{b}_k \breve{v}_j^T J_P \breve{v}_k + \breve{a}_k \breve{v}_j^T J_P \breve{w}_k = 0
$$

for $k = j - 1, j - 2$. Hence $x_1 = x_2 = 0$. Using (5.2.70) end (5.2.69) we can show that $x_3 = 0$:

$$
\begin{aligned}
\breve{v}_j^T J_P M_P \breve{w}_{j-1} &= \breve{b}_{j-2}\breve{d}_{j-1}\breve{v}_j^T J_P \breve{v}_{j-2} + \breve{a}_{j-2}\breve{d}_{j-1}\breve{v}_j^T J_P \breve{w}_{j-2} \\
&\quad + (\breve{b}_{j-1}\breve{c}_{j-1} - \breve{a}_{j-1}^{-1})\breve{v}_j^T J_P \breve{v}_{j-1} \\
&\quad + \breve{a}_{j-1}\breve{c}_{j-1}\breve{v}_j^T J_P \breve{w}_{j-1} \\
&\quad + \widehat{y}_1 \breve{v}_j^T J_P \widehat{v}_j + \widehat{y}_2 \breve{v}_j^T J_P \widehat{w}_j + \widehat{x}_1 \breve{v}_j^T J_P \widehat{v}_{j+1} \\
&= \underbrace{\widehat{y}_1 \breve{v}_j^T J_P \widehat{v}_j}_{z_1} + \underbrace{\widehat{y}_2 \breve{v}_j^T J_P \widehat{w}_j}_{z_2} + \underbrace{\widehat{x}_1 \breve{v}_j^T J_P \widehat{v}_{j+1}}_{z_3}.
\end{aligned}
$$

As $\widehat{a}_j = 0$, $\widehat{y}_2 = 0$ and therefore $z_2 = 0$. With (5.2.72) we get

$$
\begin{aligned}
z_1 &= -\widehat{y}_1 \widehat{v}_j^T J_P \breve{v}_j \\
&= \widehat{y}_1 (\breve{d}_{j-1}\widehat{v}_j^T J_P \breve{v}_{j-2} - \breve{c}_{j-1}\widehat{v}_j^T J_P \breve{v}_{j-1} \\
&\quad - \widehat{v}_j^T J_P \breve{w}_{j-1} - \breve{a}_{j-1}^{-1}\widehat{v}_j^T J_P M_P^{-1} \breve{v}_{j-1}) \\
&= -\widehat{y}_1 \breve{a}_{j-1}^{-1}\widehat{v}_j^T M_P^T J_P \breve{v}_{j-1}.
\end{aligned}
$$

From (5.2.68) we obtain

$$
M_P \widehat{v}_j = \widehat{b}_j \widehat{v}_j + \widehat{a}_j \widehat{w}_j + \widehat{x}_2 \widehat{v}_{j+1}.
$$

Hence using (5.2.69) yields

$$
z_1 = -\widehat{y}_1 \breve{a}_{j-1}^{-1}(\widehat{b}_j \widehat{v}_j^T J_P \breve{v}_{j-1} + \widehat{a}_j \widehat{w}_j^T J_P \breve{v}_{j-1} + \widehat{x}_2 \widehat{v}_{j+1}^T J_P \breve{v}_{j-1}) = 0.
$$

Similar, it follows that $z_3 = 0$. Hence $x_3 = 0$, and therefore $\breve{v}_j^T J_P M_P \breve{v}_j = 0$.

This derivation has shown that an SR breakdown implies a serious Lanczos breakdown. The opposite implication follows from the uniqueness of the Lanczos factorization. The result is summarized in the following theorem.

THEOREM 5.20 *Suppose the symplectic butterfly matrix $B^{2k,2k}$ corresponding to (5.1.5) is unreduced and let $\mu \in \mathbf{R}$. Let L_j be the jth symplectic Gauss transformation required in the SR step on $(B^{2k,2k} - \mu I)(B^{2k,2k} - \mu^{-1}I)(B^{2k,2k})^{-1}$. If the first $j-1$ symplectic Gauss transformations of this SR step exist, then L_j fails to exist if and only if $\breve{v}_j^T J_P M_P \breve{v}_j = 0$ with $\breve{v}_j$ as in (5.2.61).*

5.3 NUMERICAL EXPERIMENTS

Some examples to demonstrate the abilities of the (implicitly restarted) symplectic Lanczos method are presented here. The computational results are quite promising but certainly preliminary. All computations were done using MATLAB Version 5.1 on a Sun Ultra 1 with IEEE double-precision arithmetic and machine precision $\epsilon = 2.2204 \times 10^{-16}$.

Our code implements exactly the algorithm as given in Table 5.7. In order to detect convergence in the restart process, the rather crude criterion

$$||r_{k+1}|| \le ||M|| * 10^{-6}$$

was used. This ad hoc stopping rule allowed the iteration to halt quite early. Usually, the eigenvalues largest in modulus (and their reciprocals) of the wanted part of the spectrum are much better approximated than the ones of smaller modulus. In a black-box implementation of the algorithm this stopping criterion has to be replaced with a more rigorous one to ensure that all eigenvalues are approximated to the desired accuracy (see the discussion in Section 5.1.2). Benign breakdown in the symplectic Lanczos process was detected by the criterion

$$||v_{m+1}|| \le \epsilon * ||M|| \quad \text{or} \quad ||w_{m+1}|| \le \epsilon * ||M||,$$

while a serious breakdown was detected by

$$v_{m+1} \ne 0, \quad w_{m+1} \ne 0, \quad |a_{m+1}| \le \epsilon * ||M||.$$

Our implementation intends to compute the k eigenvalues of M largest in modulus and their reciprocals. Hence, in the implicit restart the $2p$ eigenvalues of $B_P^{2(k+p),2(k+p)}$ with modulus closest to 1 are chosen as shifts. That is, the eigenvalues of $B_P^{2(k+p),2(k+p)}$ are computed (by MATLAB's eig function, a better choice here would be to use the butterfly SZ algorithm as discussed in Section 4.1), then the eigenvalues are sorted by decreasing magnitude. There will be $k+p$ eigenvalues with modulus greater than or equal to 1,

$$|\lambda_1| \ge \ldots \ge |\lambda_k| \ge |\lambda_{k+1}| \ge \ldots |\lambda_{k+p}| \ge 1 \tag{5.3.73}$$
$$\ge |\lambda_{k+p}^{-1}| \ge \ldots \ge |\lambda_{k+1}^{-1}| \ge |\lambda_k^{-1}| \ge \ldots \ge |\lambda_1^{-1}|.$$

We select the $2p$ eigenvalues with modulus closest to 1 as shifts.

Our observations have been the following.

- Re–J–orthogonalization is necessary; otherwise after a few steps J–orthogonality of the computed Lanczos vectors is lost, and ghost eigenvalues (see, e.g., [58]) appear. That is, multiple eigenvalues of $B^{2k,2k}$ correspond to simple eigenvalues of M. See also the discussion in Section 5.1.4.

- The implicit restart is more accurate than the explicit one.

- The leading end of the 'wanted' Ritz values (that is, the eigenvalues largest in modulus and their reciprocals) converge faster than the tail end (closest to cut off of the sort). The same behavior was observed by Sorensen for his implicitly restarted Arnoldi method [132]. In order to obtain faster convergence, it seems advisable (similar to the implementation of Sorensen's implicitly restarted Arnoldi method in MATLAB's eigs) to increase the dimension of the computed Lanczos factorization. That is, instead of computing $S_P^{2n,2k}$ and $B_P^{2k,2k}$ as a basis for the restart, one should compute a slightly larger factorization, e.g., of dimension $2(k + 3)$ instead of dimension $2k$. When 2ℓ eigenvalues have converged, a subspace of dimension $2(k + 3 + \ell)$ should be computed as a basis for the restart, followed by p additional Lanczos steps to obtain a factorization of length $2(k + 3 + \ell + p)$. Using implicit SR steps this factorization should be reduced to one of length $2(k + 3 + \ell)$. If the symplectic Lanczos method would be implemented following this approach, the convergence check could be done using only the k Ritz values of largest modulus (and their reciprocals) or those that yield the smallest Ritz estimate

$$|d_{k+1}|\, |e_{2k}^T y_j|\, \|b_{k+1}\hat{v}_{k+1} + a_{k+1}\hat{w}_{k+1}\|$$

where the y_j are the eigenvectors of $B^{2k,2k}$.

- It is fairly difficult to find a good choice for k and p. Not for every possible choice of k there exists an invariant subspace of dimension $2k$ associated to the k eigenvalues λ_i largest in modulus and their reciprocals. If λ_k is complex and $\overline{\lambda_{k+1}} = \lambda_k$ then we can not choose the $2p$ eigenvalues with modulus closest to 1 as shifts as this would tear a quadruple of eigenvalues apart resulting in a shift polynomial p such that $p(B_P^{2(k+p),2(k+p)})$ is not real. All we can do is to choose the $2p - 2$ eigenvalues with modulus closest to 1 as shifts. In order to get a full set of $2p$ shifts we add as the last shift a real eigenvalue with largest Ritz estimate. Depending on how good that real eigenvalue approximates an eigenvalue of M, this strategy works, but the resulting subspace is no longer the subspace corresponding to the k eigenvalues largest in modulus and their reciprocals. If the real eigenvalue has converged to an eigenvalue of M, it is unlikely to remove that eigenvalue just by restarting, it will keep coming back. Only a purging technique like the one discussed by Lehoucq and Sorensen [93, 133] will be able to remove this eigenvalue. Moreover, there is no guarantee that there is a real eigenvalue of $B_P^{2(k+p),2(k+p)}$ that can be used here. Hence, in a black box implementation one should either try to compute an invariant subspace of dimension $2(k-1)$ or of dimension $2(k+1)$. As this is not known a priori, the algorithm should adapt k during the iteration process appropriately. This is no problem, if as suggested above, one always computes a slightly larger Lanczos factorization than requested.

EXAMPLE 5.21 *Tests were done using a* 100×100 *symplectic matrix with the eigenvalues*

$$200, 100, 50, 47, \ldots, 4, 3, 2 \pm i,$$
$$\frac{1}{2 \pm i}, \frac{1}{3}, \frac{1}{4}, \ldots, \frac{1}{47}, \frac{1}{50}, \frac{1}{100}, \frac{1}{200}.$$

(See also the tests reported in Examples 5.14 – 5.18.) A symplectic block-diagonal matrix with these eigenvalues was constructed and a similarity transformation with a randomly generated orthogonal symplectic matrix was performed to obtain a symplectic matrix M.

The first test performed concerned the loss of J–orthogonality of the computed Lanczos vectors during the symplectic Lanczos method and the ghost eigenvalue problem (see, e.g., [58]). For a more detailed discussion see Section 5.1.4. As expected, when using a random starting vector M*'s eigenvalues largest in modulus (and the corresponding reciprocals) tend to emerge right from the start, e.g., the eigenvalues of* $B^{10,10}$ *are*

$$199.99997, \ 100.06771, \ 48.71752, \ 26.85083, \ 8.32399$$

and their reciprocals. Without any form of re–J–orthogonalization, the J–orthogonality of the Lanczos vectors is lost after a few iterations as indicated in Figure 5.3.

The loss of J–orthogonality in the Lanczos vectors results, as in the standard Lanczos algorithm, in ghost eigenvalues. That is, multiple eigenvalues of $B^{2k,2k}$ *correspond to simple eigenvalues of* M. *For example, using no re–J–orthogonalization, after 17 iterations the 6 eigenvalues largest in modulus of* $B^{34,34}$ *are*

$$207.63389, \ 200, \ 100, \ 49.99982, \ 47.04542, \ 45.85367.$$

Using complete re–J–orthogonalization, this effect is avoided:

$$200, \ 100, \ 49.99992, \ 47.02461, \ 45.93018, \ 42.31199.$$

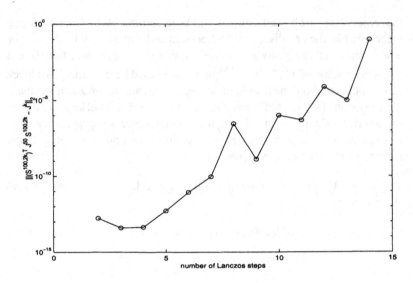

Figure 5.3.   loss of J–orthogonality after k symplectic Lanczos steps

The second test performed concerned the question whether an implicit restart is more accurate than an explicit one. After nine steps of the symplectic Lanczos method (with a random starting vector) the resulting 18×18 symplectic butterfly matrix $B^{18,18}$ had the eigenvalues (using the MATLAB function eig*)*

$$200.000000000000$$
$$99.999999841718$$
$$50.070648930465$$
$$41.873264094053$$
$$35.891491504806$$
$$23.654512559868$$
$$13.344815062428$$
$$3.679215125563 \pm 5.750883779240i$$

and their reciprocals. Removing the 4 complex eigenvalues from $B^{18,18}$ using an implicit restart as described in Section 5.2, we obtain a symplectic butterfly matrix $B_{impl}^{14,14}$ whose eigenvalues are

$$200.000000000000$$
$$99.99999984171\underline{9}$$
$$50.07064893046\underline{4}$$
$$41.873264094053$$
$$35.891491504806$$
$$23.654512559868$$
$$13.344815062428$$

and their reciprocals. From Theorem 4.2 it follows that these have to be the 14 real eigenvalues of $B^{18,18}$ which have not been removed. As can be seen, we lost one digit during the implicit restart (indicated by the underbar under the 'lost' digits in the above table). Performing an explicit restart with the explicitly computed new starting vector $\check{v}_1 = (M_P - \mu I)(M_P - \overline{\mu} I)(M_P - \mu^{-1} I)(M_P - \overline{\mu}^{-1} I) M_P^{-2} v_1$ yields a symplectic butterfly matrix $B_{expl}^{14,14}$ whose eigenvalues are

$$200.000000000000$$
$$99.9999998417\underline{93}$$
$$50.070648\underline{885030}$$
$$41.8732\underline{47045627}$$
$$35.891922701991$$
$$23.6545\underline{09163541}$$
$$13.3448\underline{10484061}$$

and their reciprocals. This time we lost up to nine digits.

The last set of tests performed on this matrix concerned the k–step restarted symplectic Lanczos method as given in Table 5.7. As M has only one quadruple of complex eigenvalues, and these eigenvalues are smallest in magnitude there is no problem in choosing $k \ll n$. For every such choice there exists an invariant symplectic subspace corresponding to the k eigenvalues largest in magnitude and their reciprocals. In the tests reported here, a random starting vector was used. Figure 5.4 shows a plot of $\|r_{k+1}\|$ versus the number of iterations performed. Iteration step 1 refers to the norm of the residual after the first k Lanczos steps, no restart is performed. The three lines in Figure 5.4 present three different choice for k and p: $k = p = 8, k = 8, p = 16$ and $k = 5, p = 10$. Convergence was achieved for all three choices (and many more, not shown here). Obviously, the choice $k = 8, p = 2k$ results in faster convergence than the choice $k = p = 8$. Convergence is by no means monotonic, during the major part of the iteration the norm of the residual is changing quite dramatically. But once a certain stage is achieved, the norm of the residual decreases monotonically. Although convergence for $k = 8, p = k$ or $p = 2k$ was quite fast, this does not imply that convergence is as fast for other choices of k and p. The third line in Figure 5.4 demonstrates that the convergence for $k = 5, p = 10$ does need twice as many iteration steps as for $k = 8, p = 16$.

EXAMPLE 5.22 *Symplectic generalized eigenvalue problems occur in many applications, e.g., in discrete linear-quadratic optimal control, discrete Kalman filtering, the solution of discrete algebraic Riccati equations, discrete stability radii and H_∞–norm computations (see, e.g., [83, 104] and the references therein) and discrete Sturm-Liouville equations (see, e.g., [27]).*

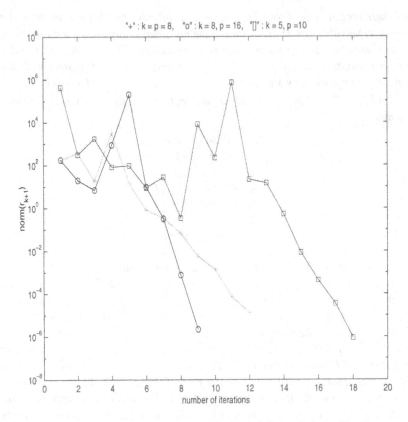

Figure 5.4. k–step restarted symplectic Lanczos method, different choices of k and p

Symplectic matrix pencils that appear in discrete-time linear-quadratic optimal control problems are typically of the form

$$K - \lambda N = \begin{bmatrix} F & 0 \\ C^T C & I \end{bmatrix} - \lambda \begin{bmatrix} I & -BB^T \\ 0 & F^T \end{bmatrix},$$

where $F \in \mathbf{R}^{n \times n}, C \in \mathbf{R}^{p \times n}$, and $B \in \mathbf{R}^{n \times m}$. (Note: For $F \neq I$, K and N
are not symplectic, but $K - \lambda N$ is a symplectic matrix pencil.) Assuming that
K and N are nonsingular (that is, F is nonsingular), solving this generalized
eigenproblem is equivalent to solving the eigenproblem for the symplectic
matrix

$$N^{-1} K = \begin{bmatrix} I & -BB^T \\ 0 & F^T \end{bmatrix}^{-1} \begin{bmatrix} F & 0 \\ C^T C & I \end{bmatrix}.$$

If one is interested in computing a few of the eigenvalues of $K - \lambda N$, one
can use the restarted symplectic Lanczos algorithm on $M = N^{-1} K$. In each

step of the symplectic Lanczos algorithm, one has to compute matrix-vector products of the form Mx and $M^T x$. Making use of the special form of K and N this can be done without explicitly inverting N: Let us consider the computation of $y = Mx$. First compute

$$Kx = \begin{bmatrix} F & 0 \\ C^T C & I \end{bmatrix} \begin{bmatrix} x_1 \\ x_2 \end{bmatrix} = \begin{bmatrix} Fx_1 \\ C^T Cx_1 + x_2 \end{bmatrix} =: \begin{bmatrix} z_1 \\ z_2 \end{bmatrix} = z$$

where $x \in \mathbf{R}^{2n}$ is written as $x = [x_1 \ x_2]^T, x_1, x_2 \in \mathbf{R}^n$. Next one has to solve the linear system $Ny = z$. Partition $y \in \mathbf{R}^{2n}$ analogous to x and z, then from $Ny = z$ we obtain

$$y_2 = F^{-T} z_2, \qquad y_1 = z_1 + BB^T y_2.$$

In order to solve $y_2 = F^{-T} z_2$ we compute the LU decomposition of F and solve the linear system $F^T y_2 = z_2$ using backward and forward substitution. Hence, the explicit inversion of N or F is avoided. In case F is a sparse matrix, sparse solvers can be employed. In particular, if the control system comes from some sort of discretization scheme, F is often banded which can be used here by computing an initial band LU factorization of F in order to minimize the cost for the computation of y_2. Note that in most applications, $p, m \ll n$ such that the computational cost for $C^T Cx_1$ and $BB^T y_2$ is significantly cheaper than a matrix-vector product with an $n \times n$ matrix. In case of single-input ($m = 1$) or single-output ($p = 1$) the corresponding operations come down to two dot products of length n each.

Using MATLAB's sparse matrix routine sprandn *sparse normally distributed random matrices F, B, C (here, $p = m = n$) of different dimensions and with different densities of the nonzero entries were generated. Here an example of dimension $2n = 1000$ is presented, where the density of the different matrices was chosen to be*

matrix	$\approx$ nonzero entries
F	$0.5n^2$
B	$0.2n^2$
C	$0.3n^2$

MATLAB *computed the norm of the corresponding matrix $M = N^{-1}K$ to be $\approx 5.3 \times 10^5$.*

In the first set of tests k was chosen to be 5, and we tested $p = k$ and $p = 2k$. As can be seen in Figure 5.5, for the first 3 iterations, the norm of the residual decreases for both choice of p, but then increases quite a bit. During the first step, the eigenvalues of $B^{10,10}$ are approximating the 5 eigenvalues of $K - \lambda N$ largest in modulus and their reciprocals. In step 4, a 'wrong' choice of the shifts is done in both cases. The extended matrices $B^{20,20}$ and $B^{30,30}$ both still approximate the 5 eigenvalues of $K - \lambda N$ largest in modulus, but there

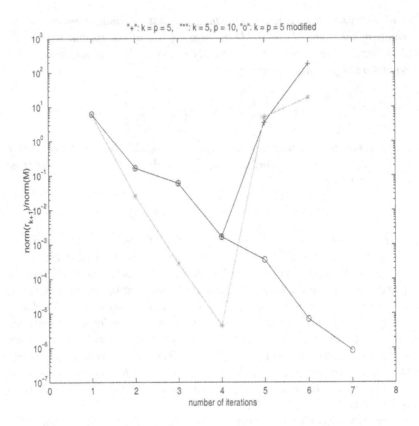

Figure 5.5. k–step restarted symplectic Lanczos method, different choices of the shifts

is a new real eigenvalue coming in, which is not a good approximation to an eigenvalue of $K - \lambda N$. But, due to the way the shifts are chosen here, this new eigenvalue is kept, while an already good approximated eigenvalue — a little smaller in magnitude — is shifted away, resulting in a dramatic increase of $\|r_{k+1}\|$. Modifying the choice of the shifts such that the good approximation is kept, while the new real eigenvalue is shifted away, the problem is resolved, the 'good' eigenvalues are kept and convergence occurs in a few steps (the 'o'-line in Figure 5.5).

Using a slightly larger Lanczos factorization as a basis for the restart, e.g., a factorization of length $2(k+3)$ instead of length $2k$ and using a locking technique to decouple converged approximate eigenvalues and associated invariant subspaces from the active part of the iteration, this problem is avoided.

Figure 5.6 displays the behavior of the k–step restarted symplectic Lanczos method for different choices of k and p, where k is quite small. Convergence is achieved in any case.

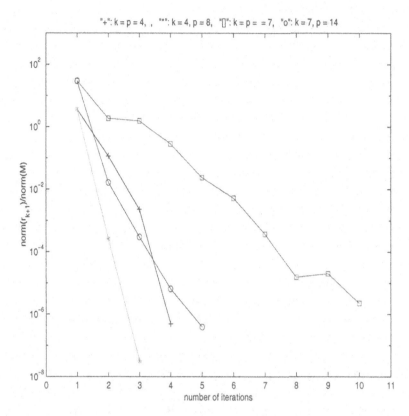

Figure 5.6. k–step restarted symplectic Lanczos method, different choices of k and p

So far, in the tests presented, k was always chosen such that there exists a deflating subspace of $K - \lambda N$ corresponding to the k eigenvalues largest in modulus and their reciprocals. For $k = 20$, there is no such deflating subspace (there is one for $k = 19$ and one for $k = 21$). See Figure 5.7 for a convergence plot. The eigenvalues of $B^{2(k+p),2(k+p)}$ in the first iteration steps approximate the $k+j$ eigenvalues of largest modulus and their reciprocals (where $5 \leq j \leq p$) quite well. Our choice of shifts is to select the 2p eigenvalues with modulus closest to 1, but as λ_{k+1} is complex with $|\lambda_{k+1}| = |\lambda_k| \neq 1$, we can only choose $2(p - 1)$ shifts that way. The last shift is chosen according to the strategy explained above. This eigenvalue keeps coming back before it is annihilated. A better idea to resolve the problem is to adapt k appropriately.

The examples presented demonstrate that the implicitly restarted symplectic Lanczos method is an efficient method to compute a few extremal eigenvalues of symplectic matrices and the associated eigenvectors or invariant subspaces.

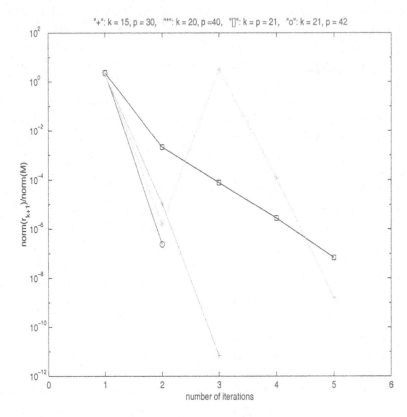

Figure 5.7. k–step restarted symplectic Lanczos method, different choices of k and p

The residual of the Lanczos recursion can be made to zero by choosing proper shifts. It is an open problem how these shifts should be chosen in an optimal way. The preliminary numerical tests reported here show that for exact shifts, good performance is already achieved.

Before implementing the symplectic Lanczos process in a black-box algorithm, some more details need consideration: in particular, techniques for locking of converged Ritz values as well as purging of converged, but unwanted Ritz values, needs to be derived in a similar way as it has been done for the implicitly restarted Arnoldi method.

Chapter 6

CONCLUDING REMARKS

Several structure-preserving algorithms for the symplectic eigenproblem have been discussed. All algorithms were based on the symplectic butterfly form.

First different aspects of the symplectic butterfly form have been considered in detail. The $2n \times 2n$ symplectic butterfly form contains $8n - 4$ nonzero entries and is determined by $4n - 1$ parameters. The reduction to butterfly form can serve as a preparatory step for the SR algorithm, as the SR algorithm preserves the symplectic butterfly form in its iterations. Hence, its role is similar to that of the reduction of an arbitrary unsymmetric matrix to upper Hessenberg form as a preparatory step for the QR algorithm. We have shown that an unreduced symplectic butterfly matrix in the context of the SR algorithm has properties similar to those of an unreduced upper Hessenberg matrix in the context of the QR algorithm. The SR algorithm not only preserves the symplectic butterfly form, but can be rewritten in terms of the $4n - 1$ parameters that determine the symplectic butterfly form. Therefore, the symplectic structure, which will be destroyed in the numerical computation due to roundoff errors, can be restored in each iteration step.

We have presented SR and SZ algorithms for the symplectic butterfly eigenproblem based on the symplectic butterfly form. The first algorithm presented, an SR algorithm, works with the $8n - 4$ nonzero entries of the butterfly matrix. Laurent polynomials are used to drive the SR step as this results in a smaller bulge and hence less arithmetic operations than using standard shift polynomials. Forcing the symplectic structure of the iterates whenever necessary, the algorithm works better than the SR algorithm for symplectic J-Hessenberg matrices proposed in [53]. The convergence rate of the butterfly SR algorithm is typically cubic; this can be observed nicely in numerical experiments. Making use of the factorization of the symplectic

247

butterfly matrix B into $K^{-1}N$ as in (3.2.7), we developed a parameterized SR algorithm for computing the eigeninformation of a parameterized symplectic butterfly matrix. The algorithm works only on the $4n - 1$ parameters that determine the symplectic butterfly matrix B. Finally a second algorithm that works only on the parameters was developed. Making once more use of the fact that the symplectic butterfly matrix B can be factored into $B = K^{-1}N$ as in (3.2.7), the idea was that instead of considering the eigenproblem for B, we can just as well consider the generalized eigenproblem $(K - \lambda N)x = 0$. An SZ algorithm was developed to solve these generalized eigenproblems. It works with the $4n - 1$ parameters that determine K and N. The algorithm reduces K and N to a form that decouples into a number of 2×2 and 4×4 symplectic eigenproblems. There is no need to force the symplectic structure of K and N. The algorithm ensures that the matrices K and N remain symplectic separately.

We have derived a connection between the HR iteration for sign-symmetric matrices and the SR algorithm for symplectic butterfly matrices. Transforming symplectic butterfly matrices into the canonical form introduced in Section 3.3, it was shown that the SR iterations for the so obtained matrices with a special choice of shifts are equivalent to an HR iteration on a sign-symmetric matrix of half the size. The result is mainly of theoretical interest, as the resulting method suffers from a possible loss of half the significant digits during the transformation to canonical form.

While the butterfly SR and SZ algorithms are suitable for small to medium sized dense matrices, they should not be the methods of choice when dealing with large and sparse symplectic matrices. Therefore we presented a symplectic Lanczos method for symplectic matrices based on the butterfly form. The symplectic Lanczos method can be used to compute a few eigenvalues and eigenvectors of a symplectic matrix. The symplectic matrix is reduced to a symplectic butterfly matrix pencil $K - \lambda N$ of lower dimension, whose eigenvalues, computed, e.g., via the butterfly SZ algorithm, can be used as approximations to the eigenvalues of the original matrix. We discussed conditions for the symplectic Lanczos method terminating prematurely such that an invariant subspace associated with certain desired eigenvalues is obtained. The important question of determining stopping criteria was also considered. An error analysis of the symplectic Lanczos algorithm in finite-precision arithmetic analogous to the analysis for the unsymmetric Lanczos algorithm presented by Bai [11] was given. As to be expected, it follows that (under certain assumptions) the computed Lanczos vectors loose J–orthogonality when some Ritz values begin to converge. In order to deal with the numerical difficulties inherent to any unsymmetric Lanczos process we proposed an implicitly restarted symplectic Lanczos method. Employing the technique of implicitly restarting for the symplectic Lanczos method using double or quadruple shifts as zeros of the driving Laurent polynomials results in an efficient method to compute a few extremal eigenvalues of symplectic matrices and the associated eigenvectors or invariant subspaces. The residual of the Lanczos recursion can be made to zero by choosing proper shifts. It is an open problem how these shifts should be chosen in an optimal way. The preliminary numerical tests reported here show that for exact shifts, good performance is already achieved.

Numerical tests for all methods presented here show very satisfactory results. Future tests will demonstrate the usefulness of the proposed methods for real life problems. In close cooperation with users black-box implementations will be developed to suit their needs.

REFERENCES

[1] B.C. Levy A. Ferrante. Hermitian solutions of the equation $X = Q + NX^{-1}N^*$. *Linear Algebra Appl.*, 247:359–373, 1996.

[2] G.S. Ammar and W.B. Gragg. Schur flows for orthogonal Hessenberg matrices. In *Hamiltonian and gradient flows, algorithms and control*, pages 27–34. A. Bloch, Providence, RI : American Mathematical Society, Fields Inst. Commun. 3, 1994.

[3] G.S. Ammar, W.B. Gragg, and L. Reichel. On the eigenproblem for orthogonal matrices. In *Proc. 25th IEEE Conference on Decision and Control*, pages 1963–1966, Athens, Greece, 1986.

[4] G.S. Ammar, W.B. Gragg, and L. Reichel. Constructing a unitary Hessenberg matrix from spectral data. In G.H. Golub and P. Van Dooren, editors, *Numerical Linear Algebra, Digital Signal Processing, and Parallel Algorithms*, pages 385–396. Springer-Verlag, Berlin, 1991.

[5] G.S. Ammar and V. Mehrmann. On Hamiltonian and symplectic Hessenberg forms. *Linear Algebra Appl.*, 149:55–72, 1991.

[6] G.S. Ammar, L. Reichel, and D.C. Sorensen. An implementation of a divide and conquer algorithm for the unitary eigenproblem. *ACM Trans. Math. Software*, pages 292–307, 1992.

[7] B.D.O. Anderson and J. B. Moore. *Linear Optimal Control*. Prentice-Hall, Englewood Cliffs, NJ, 1971.

[8] B.D.O. Anderson and J. B. Moore. *Optimal Filtering*. Prentice-Hall, Englewood Cliffs, NJ, 1979.

[9] B.D.O. Anderson and J. B. Moore. *Optimal Control – Linear Quadratic Methods*. Prentice-Hall, Englewood Cliffs, NJ, 1990.

[10] E. Anderson, Z. Bai, C. Bischof, J. Demmel, J. Dongarra, J. Du Croz, A. Greenbaum, S. Hammarling, A. McKenney, S. Ostrouchov, and D. Sorensen. *LAPACK Users' Guide.* SIAM Publications, Philadelphia, PA, 2nd edition, 1995.

[11] Z. Bai. Error analysis of the Lanczos algorithm for the nonsymmetric eigenvalue problem. *Mathematics of Computation*, 62:209–226, 1994.

[12] G. Banse. Eigenwertverfahren für symplektische Matrizen zur Lösung zeitdiskreter optimaler Steuerungsprobleme. *Z. Angew. Math. Mech.*, 75(Suppl. 2):615–616, 1995.

[13] G. Banse. *Symplektische Eigenwertverfahren zur Lösung zeitdiskreter optimaler Steuerungsprobleme.* PhD thesis, Fachbereich 3 - Mathematik und Informatik,Universität Bremen, Bremen, Germany, 1995.

[14] G. Banse. Condensed forms for symplectic matrices and symplectic pencils in optimal control. *Z. Angew. Math. Mech.*, 76(Suppl. 3):375–376, 1996.

[15] G. Banse and A. Bunse-Gerstner. A condensed form for the solution of the symplectic eigenvalue problem. In U. Helmke, R. Menniken, and J. Sauer, editors, *Systems and Networks: Mathematical Theory and Applications*, pages 613–616. Akademie Verlag, 1994.

[16] P. Benner. Accelerating Newton's method for discrete-time algebraic Riccati equations. In A. Beghi, L. Finesso, and G. Picci, editors, *Mathematical Theory of Networks and Systems*, pages 569–572, Il Poligrafo, Padova, Italy, 1998.

[17] P. Benner and H. Faßbender. An implicitly restarted Lanczos method for the symplectic eigenvalue problem. *to appear in SIAM J. Matrix Anal. Appl.* (See also *Berichte aus der Technomathematik, Report 98–01*, (1998), Zentrum für Technomathematik, FB 3 – Mathematik und Informatik, Universität Bremen, 28334 Bremen, FRG).

[18] P. Benner and H. Faßbender. An implicitly restarted symplectic Lanczos method for the Hamiltonian eigenvalue problem. *Linear Algebra Appl.*, 263:75–111, 1997.

[19] P. Benner and H. Faßbender. The symplectic eigenvalue problem, the butterfly form, the SR algorithm, and the Lanczos method. *Linear Algebra Appl.*, 275–276:19–47, 1998.

[20] P. Benner, H. Faßbender, and D.S. Watkins. Two connections between the SR and HR eigenvalue algorithms. *Linear Algebra Appl.*, 272:17–32, 1997.

[21] P. Benner, H. Faßbender, and D.S. Watkins. SR and SZ algorithms for the symplectic (butterfly) eigenproblem. *Linear Algebra Appl.*, 287:41–76, 1999.

[22] P. Benner, A. Laub, and V. Mehrmann. A collection of benchmark examples for the numerical solution of algebraic Riccati equations II: Discrete-time case. Technical Report SPC 95_23, Fak. f. Mathematik, TU Chemnitz–Zwickau, 09107 Chemnitz, FRG, 1995. Available from http://www.tu-chemnitz.de/sfb393/spc95pr.html.

[23] P. Benner, V. Mehrmann, and H. Xu. A new method for computing the stable invariant subspace of a real Hamiltonian matrix. *J. Comput. Appl. Math.*, 86:17–43, 1997.

[24] P. Benner, V. Mehrmann, and H. Xu. A numerically stable, structure preserving method for computing the eigenvalues of real Hamiltonian or symplectic pencils. *Numer. Math.*, 78(3):329–358, 1998.

[25] R. Bhatia. Matrix factorizations and their perturbations. *Linear Algebra Appl.*, 197–198:245–276, 1994.

[26] S. Bittanti, A. Laub, , and J. C. Willems, editors. *The Riccati Equation*. Springer-Verlag, Berlin, 1991.

[27] M. Bohner. Linear Hamiltonian difference systems: Disconjugacy and Jacobi–type conditions. *J. Math. Anal. Appl.*, 199:804–826, 1996.

[28] B. Bohnhorst. Beiträge zur numerischen Behandlung des unitären Eigenwertproblems. Phd thesis, Fakultät für Mathematik, Universität Bielefeld, Bielefeld, Germany, 1993.

[29] M.A. Brebner and J. Grad. Eigenvalues of $Ax = \lambda Bx$ for real symmetric matrices A and B computed by reduction to pseudosymmetric form and the HR process. *Linear Algebra Appl.*, 43:99–118, 1982.

[30] J.R. Bunch. The weak and strong stability of algorithms in numerical algebra. *Linear Algebra Appl.*, 88:49–66, 1987.

[31] W. Bunse and A. Bunse-Gerstner. *Numerische lineare Algebra*. Teubner, 1985.

[32] A. Bunse-Gerstner. Berechnung der Eigenwerte einer Matrix mit dem HR-Verfahren. In *Numerische Behandlung von Eigenwertaufgaben, Band 2*, pages 26–39, Birkhäuser Verlag Basel, 1979.

[33] A. Bunse-Gerstner. An analysis of the HR algorithm for computing the eigenvalues of a matrix. *Linear Algebra Appl.*, 35:155–173, 1981.

[34] A. Bunse-Gerstner. Matrix factorizations for symplectic QR-like methods. *Linear Algebra Appl.*, 83:49–77, 1986.

[35] A. Bunse-Gerstner and L. Elsner. Schur parameter pencils for the solution of the unitary eigenproblem. *Linear Algebra Appl.*, 154–156:741–778, 1991.

[36] A. Bunse-Gerstner and H. Faßbender. Error bounds in the isometric Arnoldi process. *J. Comput. Appl. Math.*, 86:53–72, 1997.

[37] A. Bunse-Gerstner and C. He. On a Sturm sequence of polynomials for unitary Hessenberg matrices. *SIAM J. Matrix Anal. Appl.*, 16:1043–1055, 1995.

[38] A. Bunse-Gerstner and V. Mehrmann. A symplectic QR-like algorithm for the solution of the real algebraic Riccati equation. *IEEE Trans. Automat. Control*, AC-31:1104–1113, 1986.

[39] A. Bunse-Gerstner, V. Mehrmann, and D.S. Watkins. An SR algorithm for Hamiltonian matrices based on Gaussian elimination. *Methods of Operations Research*, 58:339–356, 1989.

[40] R. Byers and V. Mehrmann. Symmetric updating of the solution of the algebraic Riccati equation. *Methods of Operations Research*, 54:117–125, 1985.

[41] D. Calvetti, L. Reichel, and D.C. Sorensen. An implicitly restarted Lanczos method for large symmetric eigenvalue problems. *Electr. Trans. Num. Anal.*, 2:1–21, 1994.

[42] X.-W. Chang. On the sensitivity of the SR decomposition. *Linear Algebra Appl.*, 282:297–310, 1998.

[43] J. Della-Dora. *Sur quelques Algorithmes de recherche de valeurs propres.* Thése, L'Université Scientifique et Medicale de Grenoble, 1973.

[44] J. Della-Dora. Numerical linear algorithms and group theory. *Linear Algebra Appl.*, 10:267–283, 1975.

[45] J. Demmel and B. Kågström. Computing stable eigendecompositions of matrix pencils. *Linear Algebra Appl.*, 88/89:139–186, 1987.

[46] P.J. Eberlein and C.P. Huang. Global convergence of the QR algorithm for unitary matrices with some results for normal matrices. *SIAM J. Numer. Anal.*, 12:97–104, 1975.

[47] L. Elsner. On some algebraic problems in connection with general eigenvalue algorithms. *Linear Algebra Appl.*, 26:123–138, 1979.

[48] L. Elsner and C. He. Perturbation and interlace theorems for the unitary eigenvalue problem. *Linear Algebra Appl.*, 188/189:207–229, 1993.

[49] H. Faßbender. A parameterized SR algorithm for symplectic (butterfly) matrices. *to appear in Mathematics of Computation*. (See also *Berichte aus der Technomathematik, Report 99–01*, (1999), Zentrum für Technomathematik, FB 3 – Mathematik und Informatik, Universität Bremen, 28334 Bremen, FRG).

[50] H. Faßbender. *Symplectic Methods for Symplectic Eigenproblems*. Habilitationsschrift, Universität Bremen, Fachbereich 3 – Mathematik und Informatik, 28334 Bremen, Germany, 1998.

[51] H. Faßbender. Error analysis of the symplectic Lanczos method for the symplectic eigenproblem. *BIT*, 40:471–496, 2000.

[52] H. Faßbender and P. Benner. A hybrid method for the numerical solution of discrete-time algebraic Riccati equation. Berichte aus der Technomathematik, Report 99–12, Zentrum für Technomathematik, FB 3 – Mathematik und Informatik, Universität Bremen, 28334 Bremen, FRG, 1999.

[53] U. Flaschka, V. Mehrmann, and D. Zywietz. An analysis of structure preserving methods for symplectic eigenvalue problems. *RAIRO Automatique Productique Informatique Industrielle*, 25:165–190, 1991.

[54] J.G.F. Francis. The QR transformation, Part I and Part II. *Comput. J.*, 4:265–271 and 332–345, 1961.

[55] R. Freund and V. Mehrmann. A symplectic look-ahead Lanczos algorithm for the Hamiltonian eigenvalue problem. manuscript.

[56] R.W. Freund. Transpose-free quasi-minimal residual methods for non-Hermitian linear systems. In G. Golub et al., editor, *Recent advances in iterative methods. Papers from the IMA workshop on iterative methods for sparse and structured problems, held in Minneapolis, MN, February 24-March 1, 1992.*, volume 60 of *IMA Vol. Math. Appl.*, pages 69–94, New York, NY, 1994. Springer–Verlag.

[57] R.W. Freund, M.H. Gutknecht, and N.M. Nachtigal. An implementation of the look-ahead Lanczos algorithm for non-Hermitian matrices. *SIAM J. Sci. Comput.*, 14(1):137–158, 1993.

[58] G.H. Golub and C.F. Van Loan. *Matrix Computations*. Johns Hopkins University Press, Baltimore, 3rd edition, 1996.

[59] G.H. Golub and J.H. Wilkinson. Ill-conditioned eigensystems and the computation of the Jordan canonical form. *SIAM Review*, 18:578–619, 1976.

[60] W.B. Gragg. Positive definite Toeplitz matrices, the Arnoldi process for isometric operators, and Gaussian quadrature on the unit circle (in russian). In E.S. Nikolaev, editor, *Numerical Methods in Linear Algebra*, pages 16–32. Moscow University Press, Moscow, 1982. See also J. Comput. Appl. Math. 46, No. 1-2, 183-198(1993).

[61] W.B. Gragg. The QR algorithm for unitary Hessenberg matrices. *J. Comp. Appl. Math.*, 16:1–8, 1986.

[62] W.B. Gragg and L. Reichel. A divide and conquer algorithm for the unitary eigenproblem. In M.T. Heath, editor, *Hypercube Multiprocessors*, pages 639–647. SIAM Publications, Philadelphia, PA, 1987.

[63] W.B. Gragg and L. Reichel. A divide and conquer method for the unitary and orthogonal eigenproblem. *Numer. Math.*, 57:695–718, 1990.

[64] W.B. Gragg and T.L. Wang. Convergence of the shifted QR algorithm for unitary Hessenberg matrices. Report NPS-53-90-007, Naval Postgraduate School, Monterey, CA, 1990.

[65] W.B. Gragg and T.L. Wang. Convergence of the unitary Hessenberg QR algorithm with unimodular shifts. Report NPS-53-90-008, Naval Postgraduate School, Monterey, CA, 1990.

[66] M. Green and D.J.N Limebeer. *Linear Robust Control*. Prentice-Hall, Englewood Cliffs, NJ, 1995.

[67] E.J. Grimme, D.C. Sorensen, and P. Van Dooren. Model reduction of state space systems via an implicitly restarted Lanczos method. *Numer. Algorithms.*, 12:1–31, 1996.

[68] M. Gutknecht. A completed theory of the unsymmetric Lanczos process and related algorithms, Part I. *SIAM J. Matrix Anal. Appl.*, 13:594–639, 1992.

[69] M. Gutknecht. A completed theory of the unsymmetric Lanczos process and related algorithms, Part II. *SIAM J. Matrix Anal. Appl.*, 15:15–58, 1994.

[70] M. Heath. Whole-system simulation of solid rockets is goal of ASCI center at Illinois. *SIAM News*, 1998.

[71] G.A. Hewer. An iterative technique for the computation of steady state gains for the discrete optimal regulator. *IEEE Trans. Automat. Control*, AC-16:382–384, 1971.

[72] N.J. Higham. *Accuracy and Stability of Numerical Algorithms*. SIAM Publications, Philadelphia, PA, 1996.

[73] D. Hinrichsen and N.K. Son. Stability radii of linear discrete-time systems and symplectic pencils. *Int. J. Robust Nonlinear Control*, 1:79–97, 1991.

[74] A.S. Householder. *The Theory of Matrices in Numerical Analysis*. Blaisdell, New York, 1964.

[75] C. Jagels and L. Reichel. The isometric Arnoldi process and an application to iterative solution of large linear systems. In *Iterative methods in linear algebra. Proceedings of the IMACS international symposium, Brussels, Belgium, 2-4 April, 1991*, pages 361–369, 1992.

[76] W. Kahan, B.N. Parlett, and E. Jiang. Residual bounds on approximate eigensystems of nonnormal matrices. *SIAM J. Numer. Anal.*, 19:470–484, 1982.

[77] H.W. Knobloch and H. Kwakernaak. *Lineare Kontrolltheorie*. Springer-Verlag, Berlin, 1985. In German.

[78] V.A. Kozlov, V.G. Maz'ya, and J. Roßmann. Spectral properties of operator pencils generated by elliptic boundary value problems for the Lamé system. *Rostocker Math. Kolloq.*, 51:5–24, 1997.

[79] V.N. Kublanoskaja. On some algorithms for the solution of the complete eigenvalue problem. *U.S.S.R. Comput. Math. and Math. Phys.*, 3:637–657, 1961.

[80] V. Kučera. *Analysis and Design of Discrete Linear Control Systems*. Academia, Prague, Czech Republic, 1991.

[81] H. Kwakernaak and R. Sivan. *Linear Optimal Control Systems*. Wiley-Interscience, New York, 1972.

[82] P. Lancaster. Strongly stable gyroscopic systems. *Electr. J. Linear Algebra*, 5:53–66, 1999.

[83] P. Lancaster and L. Rodman. *The Algebraic Riccati Equation*. Oxford University Press, Oxford, 1995.

[84] C. Lanczos. An iteration method for the solution of the eigenvalue problem of linear differential and integral operators. *J. Res. Nat. Bur. Standards*, 45:255–282, 1950.

[85] A.J. Laub. A Schur method for solving algebraic Riccati equations. LIDS Rept. LIDS-R-859, MIT, Lab. for Info. and Decis. Syst., Cambridge, MA, 1978. (including software).

[86] A.J. Laub. A Schur method for solving algebraic Riccati equations. *IEEE Trans. Automat. Control*, AC-24:913–921, 1979. (See also *Proc. 1978 CDC (Jan. 1979)*, pp. 60-65).

[87] A.J. Laub. Algebraic aspects of generalized eigenvalue problems for solving Riccati equations. In C.I. Byrnes and A. Lindquist, editors, *Computational and Combinatorial Methods in Systems Theory*, pages 213–227. Elsevier (North-Holland), 1986.

[88] A.J. Laub. Invariant subspace methods for the numerical solution of Riccati equations. In S. Bittanti, A.J. Laub, and J.C. Willems, editors, *The Riccati Equation*, pages 163–196. Springer-Verlag, Berlin, 1991.

[89] A.J. Laub and K.R. Meyer. Canonical forms for symplectic and Hamiltonian matrices. *Celestial Mechanics*, 9:213–238, 1974.

[90] D. Leguillon. Computation of 3d-singularities in elasticity. In M. Costabel et al., editor, *Boundary value problems and integral equations in nonsmooth domains*, pages 161–170, volume 167 of Lect. Notes Pure Appl. Math., Marcel Dekker, New York, 1995.

[91] R. B. Lehoucq. On the convergence of an implicitly restarted Arnoldi method. preprint, Sandia National Laboratory, P.O. Box 5800, MS 1110, Albuquerque, NM 87185-1110, 1999.

[92] R.B. Lehoucq. *Analysis and Implementation of an implicitly restarted Arnoldi Iteration*. PhD thesis, Rice University, Dep. Computational and Applied Mathematics, Houston, Texas, 1995.

[93] R.B. Lehoucq and D.C. Sorensen. Deflation techniques for an implicitly restarted Arnoldi iteration. *SIAM J. Matrix Anal. Appl.*, 17:789–821, 1996.

[94] R.B. Lehoucq, D.C. Sorensen, and C. Yang. *ARPACK user's guide. Solution of large-scale eigenvalue problems with implicitly restarted Arnoldi methods*. SIAM Publications, Philadelphia,PA, 1998.

[95] B.C. Levy, R. Frezza, and A.J. Krener. Modeling and estimation of discrete-time gaussian reciprocal processes. *IEEE Trans. Automat. Control*, 35:1013–1023, 1990.

[96] W.-W. Lin. A new method for computing the closed loop eigenvalues of a discrete-time algebraic Riccati equation. *Linear Algebra Appl.*, 6:157–180, 1987.

[97] W.-W. Lin and T.-C. Ho. On Schur type decompositions for Hamiltonian and symplectic pencils. Technical report, Institute of Applied Mathematics, National Tsing Hua University, 1990.

[98] W.-W. Lin, V. Mehrmann, and H. Xu. Canonical forms for Hamiltonian and symplectic matrices and pencils. *Linear Algebra Appl.*, 301–303:469–533, 1999.

[99] W.-W. Lin and C.-S. Wang. On computing stable Lagrangian subspaces of Hamiltonian matrices and symplectic pencils. *SIAM J. Matrix Anal. Appl.*, 18:590–614, 1997.

[100] Z.-S. Liu. On the extended HR algorithm. Technical report, Center for Pure and Applied Mathematics, University of California, Berkeley, 1992.

[101] K. Meerbergen. Locking and restarting quadratic eigenvalue solvers. Technical Report Report RAL-TR-1999-011, CLRC, Rutherford Appleton Laboratory, Dept. of Comp. and Inf., Atlas Centre, Oxon 0QX, GB, 1999.

[102] V. Mehrmann. Der SR-Algorithmus zur Berechnung der Eigenwerte einer Matrix. Diplomarbeit, Universität Bielefeld, Bielefeld, FRG, 1979.

[103] V. Mehrmann. A symplectic orthogonal method for single input or single output discrete time optimal linear quadratic control problems. *SIAM J. Matrix Anal. Appl.*, 9:221–248, 1988.

[104] V. Mehrmann. *The Autonomous Linear Quadratic Control Problem, Theory and Numerical Solution*. Number 163 in Lecture Notes in Control and Information Sciences. Springer-Verlag, Heidelberg, 1991.

[105] V. Mehrmann and E. Tan. Defect correction methods for the solution of algebraic Riccati equations. *IEEE Trans. Automat. Control*, 33:695–698, 1988.

[106] V. Mehrmann and D. Watkins. Structure-preserving methods for computing eigenpairs of large sparse skew-hamiltonian/hamiltonian pencils. Technical Report SFB393/00-02, Fak. f. Mathematik, TU

Chemnitz–Zwickau, 09107 Chemnitz, FRG, 2000. Available from http://www.tu-chemnitz.de/sfb393/.

[107] R.B. Morgan. On restarting the Arnoldi method for large nonsymmetric eigenvalue problems. *Mathematics of Computation*, 65:1213–1230, 1996.

[108] C.C. Paige. *The Computation of Eigenvalues and Eigenvectors of Very Large Sparse Matrices*. PhD thesis, University of London (UK), 1971.

[109] C.C. Paige. Error analysis of the Lanczos algorithm for tridiagonalizing a symmetric matrix. *J. Inst. Math. Applics.*, 18:341–349, 1976.

[110] C.C. Paige and C.F. Van Loan. A Schur decomposition for Hamiltonian matrices. *Linear Algebra Appl.*, 14:11–32, 1981.

[111] T. Pappas, A.J. Laub, and N.R. Sandell. On the numerical solution of the discrete-time algebraic Riccati equation. *IEEE Trans. Automat. Control*, AC-25:631–641, 1980.

[112] B.N. Parlett. Canonical decomposition of Hessenberg matrices. *Mathematics of Computation*, 21:223–227, 1967.

[113] B.N. Parlett. A new look at the Lanczos algorithm for solving symmetric systems of linear equations. *Linear Algebra Appl.*, 29:323–346, 1980.

[114] B.N. Parlett. *The Symmetric Eigenvalue Problem*. Prentice-Hall, Englewood Cliffs, New Jersey, 1980.

[115] B.N. Parlett and W.G. Poole. A geometric theory for the QR, LU, and power iterations. *SIAM Journal on Numerical Analysis*, 10:389–412, 1973.

[116] B.N. Parlett, D.R. Taylor, and Z.A. Liu. A look-ahead Lanczos algorithm for unsymmetric matrices. *Mathematics of Computation*, 44(169):105–124, 1985.

[117] R.V. Patel. Computation of the stable deflating subspace of a symplectic pencil using structure preserving orthogonal transformations. In *Proceedings of the 31st Annual Allerton Conference on Communication, Control and Computing*, University of Illinois, 1993.

[118] R.V. Patel. On computing the eigenvalues of a symplectic pencil. *Linear Algebra Appl.*, 188/189:591–611, 1993. See also: Proc. CDC-31, Tuscon, AZ, 1992, pp. 1921–1926.

[119] P.H. Petkov, N.D. Christov, and M.M. Konstantinov. *Computational Methods for Linear Control Systems*. Prentice-Hall, Hertfordshire, UK, 1991.

[120] H. Rutishauser. Der Quotienten-Differenzen-Algorithmus. *Zeitschrift für angewandte Mathematik und Physik*, 5:233–251, 1954.

[121] H. Rutishauser. Solution of eigenvalue problems with the LR-transformation. *National Bureau of Standards Applied Mathematics Series*, 49:47–81, 1958.

[122] H. Rutishauser. Bestimmung der Eigenwerte orthogonaler Matrizen. *Numer. Math.*, 9:104–108, 1966.

[123] Y. Saad. *Numerical methods for large eigenvalue problems: theory and applications*. John Wiley and Sons, New York, 1992.

[124] A. Saberi, P. Sannuti, and B.M. Chen. H_2 *Optimal Control*. Prentice-Hall, Hertfordshire, UK, 1995.

[125] R. Sandhu and K. Pister. A variational principle for linear coupled field problems in continuum mechanics. *Internat. J. Eng. Sci.*, 8:989–999, 1970.

[126] H. Schmitz, K. Volk, and W.L. Wendland. On the three-dimensional singularties of elastic fields near vertices. *Numer. Methods Partial Differ. Equations*, 9:323–337, 1993.

[127] H.R. Schwarz. *Methode der finiten Elemente*. Teubner, Stuttgart, Germany, 1984.

[128] V. Sima. Algorithm GRICSR solving continuous-time algebraic Riccati equations using Gaussian symplectic transformations. *Stud. Res. Comp. Inf.*, 1:237–254, 1992.

[129] V. Sima. *Algorithms for Linear-Quadratic Optimization*, volume 200 of *Pure and Applied Mathematics*. Marcel Dekker, Inc., New York, NY, 1996.

[130] B. Simon, J. Wu, O. Zienkiewicz, and D. Paul. Evaluation of u-w and u-p finite element methods for the dynamic response of saturated porous media using one-dimensional models. *Internat. J. Numer. Anal. Methods Geomech.*, 10:461–482, 1986.

[131] H. Simon. Analysis of the symmetric Lanczos algorithm with reorthogonalization methods. *Linear Algebra Appl.*, 61:101–132, 1984.

[132] D.C. Sorensen. Implicit application of polynomial filters in a k-step Arnoldi method. *SIAM J. Matrix Anal. Appl.*, 13(1):357–385, 1992.

[133] D.C. Sorensen. Deflation for implicitly restarted Arnoldi methods. Technical report, Department of Computational and Applied Mathematics, Rice University, Houston, Texas, 1998.

[134] D.R. Taylor. *Analysis of the look ahead Lanczos algorithm*. PhD thesis, Center for Pure and Applied Mathematics, University of California, Berkley, CA, 1982.

[135] P. Van Dooren. The computation of Kronecker's canonical form of a singular pencil. *Linear Algebra Appl.*, 27:103–140, 1979.

[136] P. Van Dooren. A generalized eigenvalue approach for solving Riccati equations. *SIAM J. Sci. Statist. Comput.*, 2:121–135, 1981.

[137] C.F. Van Loan. A symplectic method for approximating all the eigenvalues of a Hamiltonian matrix. *Linear Algebra Appl.*, 16:233–251, 1984.

[138] D.S. Watkins. Understanding the QR algorithm. *SIAM Review*, 24:427–440, 1982.

[139] D.S. Watkins. *Fundamentals of matrix computations*. John Wiley & Sons, Inc., New York, 1991.

[140] D.S. Watkins. HZ and SZ. Manuscript, 1998.

[141] D.S. Watkins and L. Elsner. Chasing algorithms for the eigenvalue problem. *SIAM J. Matrix Anal. Appl.*, 12:374–384, 1991.

[142] D.S. Watkins and L. Elsner. Convergence of algorithms of decomposition type for the eigenvalue problem. *Linear Algebra Appl.*, 143:19–47, 1991.

[143] D.S. Watkins and L. Elsner. Theory of decomposition and bulge chasing algorithms for the generalized eigenvalue problem. *SIAM J. Matrix Anal. Appl.*, 15:943–967, 1994.

[144] J.H. Wilkinson. *The Algebraic Eigenvalue Problem*. Clarendon Press, Oxford, England, 1965.

[145] H.K. Wimmer. Normal forms of symplectic pencils and discrete-time algebraic Riccati equations. *Linear Algebra Appl.*, 147:411–440, 1991.

[146] W.M. Wonham. *Linear Multivariable Control: A Geometric Approach*. Springer-Verlag, New York, second edition, 1979.

[147] H. Xu. The relation between the QR and LR algorithms. *SIAM J. Matrix Anal. Appl.*, 19:551–555, 1998.

[148] K. Zhou, J.C. Doyle, and K. Glover. *Robust and Optimal Control.* Prentice-Hall, Upper Saddle River, NJ, 1996.

ABOUT THE AUTHOR

Heike Faßbender received the Diplom in mathematics from the University of Bielefeld, Germany in 1989 and the M.S. degree in Computer Science from the State University of New York at Buffalo in 1991. In 1994, she received her Ph.D. and in 1999 her Habilitation, both in mathematics, from the University of Bremen, Germany.

From 1995 - 1999 she held a position as 'Wissenschaftliche Assistentin' (comparable to assistant professor) at the Department of Mathematics and Computer Science of the University of Bremen, Germany. In 2000 she joined the Department of Mathematics of the University of Technology in Munich (TUM), Germany, where she is a professor of numerical analysis.

Her research interests are in numerical analysis and scientific computing, with specialties in numerical linear algebra and parallel computing. Currently she is mainly working on problems arising in control, signal processing and systems theory.

INDEX